Gas Machinery

Gas Machinery

Library of Congress
Catalog Card Number 70-149760
ISBN—0-87201-309-X

Gas Machinery

Lyman F. Scheel

Gulf Publishing Company
Houston, Texas

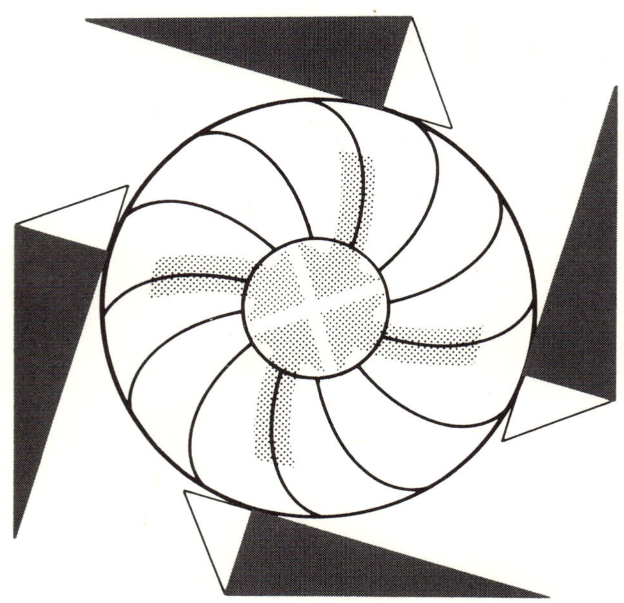

Dedication

This textbook is dedicated to the engineer who wants to make an independent and comprehensive evaluation of the performance capabilities of gas machinery.

The author searched the first 30 years of his career for a more concrete procedure and a common method of evaluating gas compressor capabilities. He has spent the past 15 years analyzing and describing a new unilateral technique which produces realistic compressor performance data.

Contents

Preface

I was motivated to write this text because comprehensive technical information concerning the performance of gas machinery is lacking. The contents include the following innovations and features:

1. A series of computer programs to resolve most every piston and centrifugal compressor application.
2. A comprehensive analysis for all types of turbomachinery.
3. The design limitations and metallurgy for rotors.
4. Specific diameters and specific speeds projected to optimum efficiency for fans, blowers and all forms of turbomachinery.
5. Comprehensive methods for projecting the performance data of axial and radial fans, lobe blowers, sliding-vane, spiral-axial, helical-screw and liquid-liner applications.
6. A series of calculation sheets with supporting equations for quick sizing and power appraisal of piston, rotary and turbomachinery.
7. An evaluation of the pulse intensity experienced with piston machinery and a method of sizing the attenuation chambers.
8. Shaft seal leakage and balanced piston evaluations.
9. Simplified charts for evaluating piston compression efficiency.
10. An evaluation of the minimum capabilities of centrifugal compressors in refrigeration service versus piston compressors.

The objective of this text is to advance a technology that produces substantial performance data for all types of gas machinery. These methods are rooted in fundamental aerodynamics and thermodynamics, using empirical data as secondary support material.

The method of rating piston compressors has been successfully applied for over 40 years. The intrinsic valve loss concept has been infallible for the past 11 years. The Baljé system of *specific* performance has proven to be an invaluable tool for over 10 years.

The hypothetical equations concerning the performance of rotary machines needs to be sustained by experience. These data should prove useful in advancing this technology and the reference literature. The analysis which relates the charge and exhaust head losses to the rotor tip speed poses a novel solution, similar to the NPSH requirement of a process type centrifugal pump.

The top of the acknowledgment and gratitude list must go to Dr. O. E. Baljé. His classical treatise on *Design Criteria of Turbomachines*, ASME Papers, numbers 60-WA-230 and 231, formed the matrix of the turbomachinery chapters.

The next parties to whom I am indebted are P.M. Huemmer and C.W. Pace, the ranking officers of Ehrhart & Associates, a division of Procon, affiliate of Universal Oil Products. Without their indulgence and appreciation of my objective, this book would not have been written.

Thanks to my fellow workers and contemporaries, R.F. Neerken, H.M. Rubenstein, J.P. Buchwold, E.S. Perkins and D.R. Jacobson who proofread and appraised the contents. R.F. Cline was invaluable in organizing the computer programs.

Thanks also to the manufacturers who have contributed substance to the text, mainly D. E. Steel of Allis Chalmers and many of his staff; W.F. Hartwick, W.G. McCachen and others from Cooper Bessemer; D.W. Schmitt and others from Joy Mfg.; H.H. Boettcher and T.A. Ammer of Chicago Pneumatic Tool Company.

A last word of appreciation for the feminine touch of the typewriter for Valerie Jelinek and my good wife.

San Gabriel, Calif. 91775 — Lyman F. Scheel, June, 1972

Gas Machinery

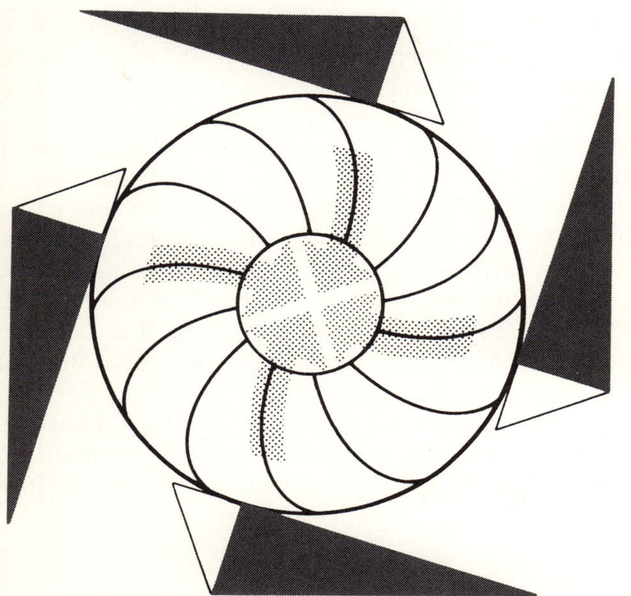

1 A Briefing in Compression

Gas is aeriform; it is without form. It must be confined in a vessel or pipe system in order to have identity. Its mass must be evaluated by a definition of conditions. The American Gas Association, American Standard Association, American Society of Mechanical Engineers and the most credible references in the United States accept the definition that one standard cubic foot (scf) of 60°F dry air at sea level, where the mean barometer is 14.696 psia, must weigh 0.0763 pounds. The European counterpart is the *normal cubic meter* which weighs 2.846 pounds at 14.696 or 760 Torr (*mm*) and zero Centigrade. The British *standard cubic meter* weighs 2.70 pounds at 14.73 psia or 30 inches of mercury barometric and 60°F.

The term *free air* is used to describe the capacity of compressors. It is the net volumetric efficiency corrected displacement at the local ambient conditions. For example, a specific volume of one pound of 60°F dry air at the San Diego Airport contains 13.1 cf; in mile high Denver, it contains 15.4 cf at a 12.5 psia (25.4 inches Hg) barometer. (See Chart 14 for elevation density derating.) The equivalent term for *free air* when concerned with a process gas compressor is actual cubic feet per minute, acfm. The specific volume is determined from the equation:

$$v_s = 10.73\ T(Z/m)P. \qquad (1.1)$$

Equation 1.1 is developed from Avogadro's law. It states that the volume of all gases are proportional to their respective molecular weight, providing the pressure and temperature are held constant. It follows that the molecular weight of air, 28.97, when multiplied by the specific volume, 13.1, produces a volume of 379.5 cf. The volume is common to all gases when the compressibility correction is unity. A vessel containing 379.5 cf of hydrogen gas would contain a net weight of 2 pounds. If filled with propylene, the net weight would be 42 pounds. The Universal Gas Constant is derived from the same mol volume:

$$R = 379.5(14.70)/520(m) = 10.73/m. \qquad (1.2)$$

When the pressure is referred to the pounds per square foot (psf), the gas constant is 1,545/*m*.

Gas Fluidity

Gas possesses instant fluidity. Any release in pressure from a gas vessel will create a velocity in accordance with the Torricelli (1643) equation:

$$V = F(2gL)^{0.5}, \text{fps}. \qquad (1.3)$$

Table 1.1
Common Nozzle and Orifice Coefficients

Nozzle Description	F-factor
Conoidal mouthpiece	0.98
Short cylindrical nozzle	
Rounded edge	0.92
Sharp edge	0.82
Pitot tube	0.86
Thin orifice plate, sharp edge	
d/D ratio, 0.10 to 0.60	0.65
d/D ratio, 0.61 to 0.74	0.68
d/D ratio limit, 0.75	0.72

The symbol "g" represents the gravitational acceleration, 32.2 ft/sec/sec. The letter "L" represents the head in feet of gas. It is the product of the differential pressure supporting the flow, the specific volume and 144 square inches per square foot. The term gas head has the same significance as a centrifugal pump head; i.e., (144/62.4) = 2.31 feet of water head is equivalent to one psi and so is (144/0.0763) = 1,885 feet of standard air. For example, a nozzle velocity from a 20 psig air tank exhausting into standard ambient conditions is determined in this manner. The average specific volume is

$$v_{sa} = 10.73(520)/29(14.7 + 20/2) = 7.8 \text{ cf/lb.}$$

The head of air is 144(7.8)20 = 22,500 feet. The ideal nozzle spouting velocity is $(64.4 \times 22{,}500)^{0.5}$ = 1,200 fps (by Equation 1.3). The actual velocity would be reduced by the amount of the flow coefficient, F. Several common nozzles and orifice coefficients are given in Table 1.1.

Viscosity is another factor to consider in reference to fluidity. Viscosity is the resistance of fluid particles to shear and the flow of these molecules against the molecular boundary layer. The tenacity of this attraction is measured by the viscosity and manifest in the Reynolds (R_e) number:

$$R_e = 0.105\ (\text{ppm})/d\mu \qquad (1.4)$$

Where ppm is the pounds per minute of flow, d is the diameter in inches and μ is the absolute viscosity in centipoise. The effects of viscosity cause two different conditions of flow. The smooth streamline flow which occurs at low velocities of approximately one fps for water and 4 fps for air is known as *laminar flow* in conventional sized pipe 2 to 10 inches. The upper limit of laminar flow is between a R_e of 2,700 and 4,000. Turbulent flow occurs at greater velocities and R_e values. The fluid particles are retained in symmetric layers by viscous action when in laminar flow. The particles move in heterogeneous fashion when in turbulent flow.

Velocity Head (Velad)

Not only does the pressure drop provide the necessary impetus to support the high velocity through a nozzle, but it also provides necessary motive power to propel fluids through long pipelines. The design of new installations requires a valuation of realistic resistances that may constitute the flow line circuits serving the compressor. Such nondescript pipelines can be assumed to have a resistance of 500(L/D). This parameter would include the equivalent length of 2,000 feet of 3-inch pipeline to 390 feet of 16-inch pipeline. This (L/D) ratio is multiplied by a frictional factor of 0.015 to give the number of velocity heads consumed in supporting the required velocity experienced in most plants where the Reynolds number varies from 300,000 to 3,000,000 using 3 to 16-inch pipelines (1). The term *velocity head* has been coined, *velad*, and is identified as K. It has become a convenient resistance reference for fittings, pipeline facilities, etc. A list of common resistances are listed in Table 1.2.

A fitting having a resistance of one K consumes one velad at that respective velocity. The velocity of gas flowing in pipeline is readily determined from the equation.

$$V = 3.06(\text{acfm})/d^2\text{, fps.} \qquad (1.5)$$

The last six items in Table 1.2 are the common accessories serving compressor installations. Presume that several compressor cylinders are operating in series. They are required to handle an airflow of 1,000 lb/min at 100 psia and 300°F to the intercooler. The air is cooled to 100°F and returned for the second stage of compression. The facilities also include a gas separator and a pulse damper with a choke on each end of the circuit. The line selection and pressure drop is developed in this same manner.

The hot gas density is 10.73(760)/29(100) = 2.81 cf/lb. The gas head is 144(2.81) = 405 ft/psi. The volume flow is 2.81 (1,000) = 2,810 acfm, and the velocity is 2,810(3.06)/(12 x 12) = 60 fps in a 12-inch pipeline. The unit of frictional resistance in terms of velads is $V^2/2g$ or (60 x 60)/64.4 = 56 ft/K or per velad. The resistance of the circuit from the cylinder through the intercooler is 500(L/D)0.015 = 7.5K; plus 12K for one surge bottle with a choke tube; plus 17K for an intercooler, which makes a total resistance of 36K. The friction is 56(36)/405 = 5.0 psi using a 12-inch pipeline. The cool gas has a density of 2.07 cf/lb. The gas head is 298 ft/psi and volume flow of 2,070 acfm. The velocity is 63 fps in a 10-inch pipeline, and the frictional resistance is 62 ft/K. The resistance of the 10-inch return connection is 500(12/10)0.015 = 9K plus the separator, 7K; plus another PD = 12K, making the total 28K. The pressure drop is 62(28)/298 = 5.8 psi. This frictional loss is too great. A 12-inch pipeline can be substituted and reduce the friction to 2.55 psi, making the total friction of 7.5 psi, or 7.5 percent of the system pressure (100 psia) is consumed in transmitting the gas from the first stage to the second stage cylinder.

Critical Velocity

The ideal spouting velocity example given earlier is only valid within certain limitations. There is a limiting pressure which establishes the maximum velocity flow. Where P_a is the reduced pressure, presumably 14.7 psia, and P_v is the vessel presure, the maximum velocity is attained at

$$P_v = P_a(2/k + 1)^{k/(k-1)}. \tag{1.6}$$

The limiting ratio for several gases are

Helium, P_v = 14.7(2.04) = 30.0 psia;
Air, P_v = 14.7(1.88) = 27.7 psia;
Natural gas, P_v = 14.7(1.82) = 26.8 psia;
Butane, P_v = 14.7(1.67) = 24.6 psia.

The velocities attained from these pressures are the *sonic* velocities for each gas. The sonic velocity is determined

$$\tau = 224(kT_o/m)^{0.5}. \tag{1.7}$$

Consistent flow characteristics are attained where critical or pertinent flow velocities are related to the decimal fraction of sonic velocity for that respective gas. The critical flow equation is

$$\text{acfm} = 73.3d^2F(kT/m)^{0.5}. \tag{1.8}$$

Specific Heat

The specific heat of a gas is the amount of heat required to raise one pound of gas one degree F.

Table 1.2
Common Resistances

Resistance Description	K-factor
Reducer contraction	
0.75	0.2
0.50	0.3
Reducer enlargement	
0.75	0.5
0.50	0.6
0.25	0.9
Gate valve	
Fully open	0.15
0.25 open	25.0
Elbow	
Long radius	0.15
Short radius	0.25
Miter	1.10
Close return bend	0.5
Swing check or ball valve	2.2
Tee flow through bull-head	1.8
Angle valve, open	3.0
Globe valve, open	5.0
Filters	
Clean	4.0
Foul	20.0
Intercoolers	17.0
Gas separators	7.0
Surge bottles	
No choke tube	4.0
With choke tube	12.0

There are two forms of specific heats for gases, defined as (a) specific heat at constant volume and (b) specific heat at constant pressure. The heating of a gas under constant pressure requires more energy to satisfy the increased volume change. The specific heat at constant pressure, C_p, is always greater than C_v for this reason. Several simple Perfect Gas Law (PGL) relations between C_p and C_v are as follows:

$$k = C_p/C_v = C_{pm}/(C_{pm} - 1.986)$$
$$= C_{pm}/C_{pm} - 0.993\ (Z_s + Z_d) \tag{1.9}$$
$$C_p - C_v = 1{,}545/m788 = (Z_s + Z_d)0.993/m \tag{1.10}$$
$$H = C_p\ (T_2^* - T_1) \tag{1.11}$$
$$\text{Isehp} = WC_p\ (T_2 - T_1)/42.5 = W\Delta H/42.5 \tag{1.12}$$
$$P_1V_1{}^k = P_2V_2{}^k = \text{constant}. \tag{1.13}$$

*Where T_2 is isentropic or adiabatic temperature rise.

Table 1.3
Gas Identification by Molecular Structure

Molecular Structure	"k" value	Typical Gas	Mean Temp. Degree
Monatomic	1.67	Argon, helium	220
Diatomic	1.40	Air, hydrogen	180
Triatomic	1.30	Carbon dioxide, steam	170
Polyatomic	1.20	Acetylene, ethane	135
Heavy Organic	1.10	Butane, benzene	105

The last equation is the fundamental adiabatic equation which indicates the realistic change of state. A further extension of the PV behavior follows:

$$T_2/T_1 = (P_2/P_1)^{(k-1)k} = (V_1/V_2)^{(k-1)} \quad (1.14)$$
$$(V_1/V_2) = (P_2/P_1)^{1/k}. \quad (1.15).$$

The molecular structure is the simplest category to identify a gas with an approximate exponential value "k." Table 1.3 lists these common categories. The mean temperature of a 3 R_c compression from an intake of 80°F is shown in the right column (2).

Discharge Temperature

For all practical design and operating purposes, the temperature rise in a piston machine follows the adiabatic Equation 1.14. The Rankine discharge temperature is the product of the Rankine suction temperature and the ratio of compression raised to the $\sigma = (k - 1)/k$ exponent; i.e.,

$$t_2 = T_1[(Z_1/Z_2)R_c)]^{\sigma} - 460.$$

Several well-documented references support this premise (2, 6 and 10). If the enthalpy differential and the discharge temperature can be abstracted from a Mollier chart (Isentropic Method), it is recommended over the adiabatic method using Equations 1.16 and 1.17.

During the CNGA Compressor Tests in 1936, it was noted that the actual temperature rise was less than the adiabatic function at R_c greater than 4.0. This same deviation was observed in a high pressure air test. A high vacuum air test showed a similar deviation at R_c values > 2.5. This indicates that the gas density has a significant effect on the temperature rise. The trends are shown in Figure 1.1 and on Chart 13. The deviation from the abiabatic temperature rise is a result of the reduced mass flow because of the smaller volumetric efficiencies at the higher R_c values. The influence of the heat conduction from the attached pipe header system and from the cylinder jackets is more pronounced with the low mass flow than it is at the normal 3 R_c design point. In all three cases the cylinders were cooled with an abundance of relatively cool water. The compression line AB in Figure 1.2 is relatively longer for a plus 4 R_c than a R_c value < 2. This provides a greater time and opportunity to dissipate more heat from the >4 R_c operation.

The vacuum cylinders used in the test were 42 inches in diameter and had 2.4 percent clearance. They showed an ultimate temperature rise of 1.50 at 7.5 R_c. Beyond this asymptotic crest, the heat leak dominated the greater exponential effect. The mole weight of the gas as well as the gas density in the cylinders apparently affect this deviation. In retrospect, these asymptotic crests are consistent with maximum discharge temperatures experienced under high R_c conditions. It is a rare set of conditions where discharge temperature for an air compressor exceeds 450°F, or 380°F for a natural gas operation. These temperatures are consistent with the 1.5 and 1.9 exponential values for the asymptotic crests shown on Figure 1.1.

An examination of the compression curve AB on Figure 1.2 suggests that the discharge temperature factor could be (P_4/P_1) in lieu of the visual (P_2/P_1) value. The reference material supports the latter, visual R_c basis. An appreciation of the valve action

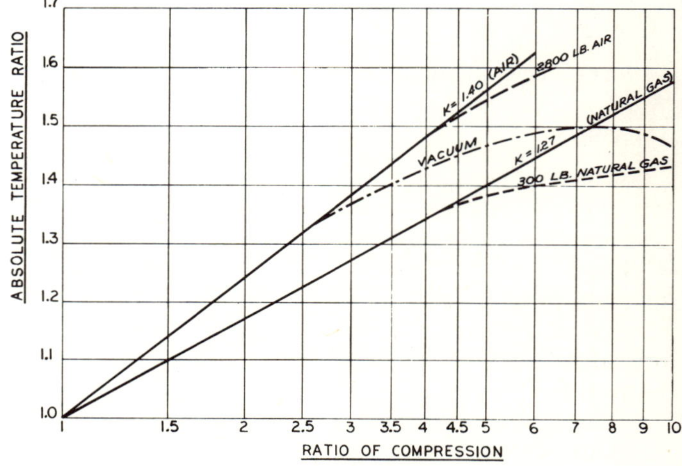

Figure 1.1. Temperature of compression factors vs. ratio of compression. (From Gas and Air Compression Machinery, *Courtesy of McGraw-Hill Book Company.)*

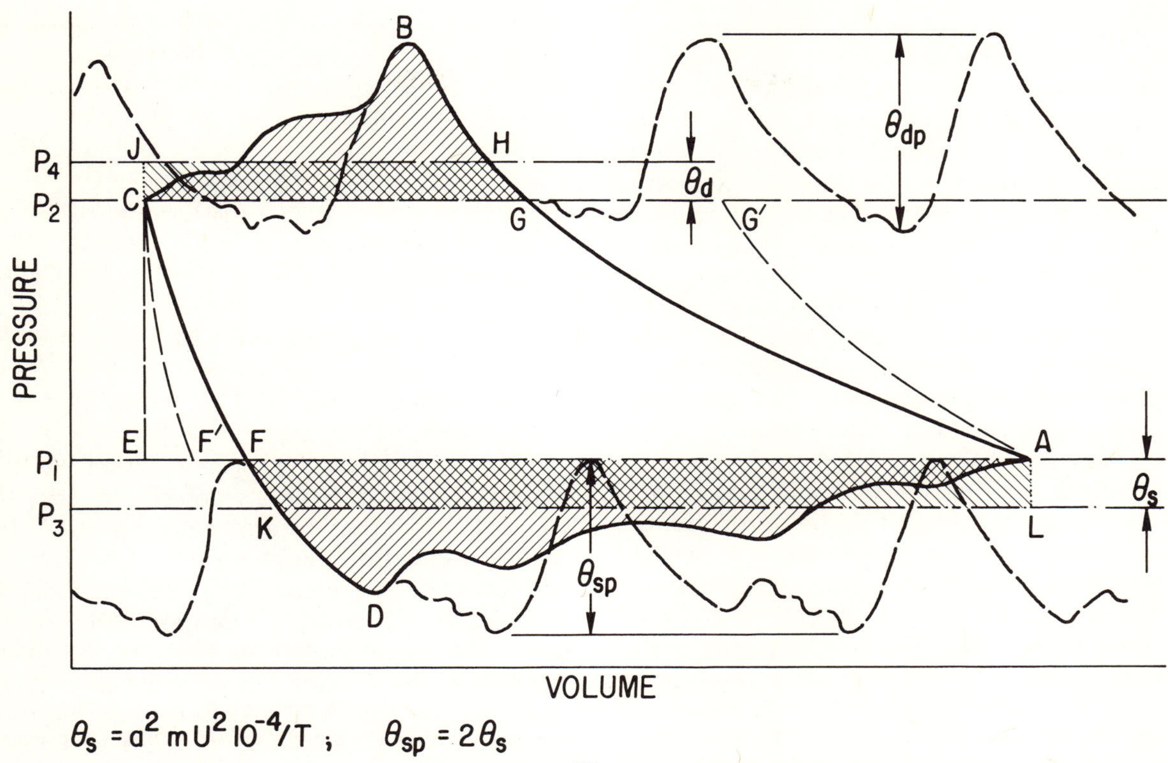

$$\theta_s = a^2 m U^2 10^{-4}/T; \quad \theta_{sp} = 2\theta_s$$

$$\theta_d = \theta_s/R_c^{\sigma}; \quad \theta_{dp} = 4\theta_s/R_c^{1/K}; \quad B = (1+\theta_d)/(1-\theta_s)$$

COMPREHENSIVE $R_c = P_4/P_3 = B(P_2/P_1) = BR_c$

THE DASH CURVES ILLUSTRATE THE OSCILLOSCOPE PRESSURE WAVES TAKEN AT THE CYLINDER FLANGE

Figure 1.2. A pressure-volume diagram illustrating the gas behavior inside a piston compressor.

offers an explanation for the discharge temperature rise conforming to the visual R_c. The valve action was once thought to be a matter of pneumatic leverage. The discharge valve would open when the product of the area under the discharge valve seat passage and the internal pressure in the cylinder exceeds the product of the full disk or strip area and the pressure in the cylinder chamber. To unseat the element under these conditions would require internal pressures 30 to 90 percent greater than the discharge pressure. Such pressure differentials are unrealistic. An evaluation of realistic pulse pressures is given in Chapter 3.

The lapped finish of the valve seat and the valve element is seldom equal or better than 10 RMS. It is reasonable to assume that a mean clearance of 16 micro-inches exists between these two components. This clearance provides an area of 0.02 percent of a normal valve area having a lift of 0.080 inch. The minimal blow-by leakage volume is 0.02 percent of the normal flow. The actual head against the closed valves is the full R_c which would make leakage velocity approach the sonic velocity or about 10 times the nominal open valve velocity. The valve leakage would be about 0.2 percent of the cylinder capacity under these conditions. With a minute flow always passing under the valve, the slightest favorable unbalance in pressures would cause the valve to lift. It is presumed that the adiabatic compression function "per se" terminates at point G, Figure 1.2. Greater pressures above line CG are in effect providing the necessary energy to expell the gas (isothermally) into the cylinder discharge channel. The reduced pressures below AF, Figure 1.2, represent the velocity heads that are expended to charge the cylinder.

There is a new arbitrary pressure drop correction that is being applied to the suction pressure of a piston compressor. These corrections may range from one to 10 psi. This presumedly represents the "spring-load" applied to the valve disks. The purpose of the springs was to damp the valve flutter. The force to move the spring 0.080 inch seldom exceeds 2 pounds which would limit the resistance to a fractional psi. The writer has only known of one instance where the compressor diagram toe (point A, Figure 1.2) was less than the system pressure. The latter was 24 inches of mercury, and the toe was 27 inches. These measurements were admittedly on the nebulous side. The valve spring tension was designed for dense gases. When the stiff springs were replaced with standard light-tension springs, the normal capacity was recuperated.

This arbitrary "spring loss" should not be misconstrued to be the equivalent of pulse differential ($\theta_s P_1$), the valve loss factor, expressed as a decimal fraction of the system pressure, times that pressure, P_1. (See Equation 2.10 and Equation 3.1.) It would be sheer coincidence that the arbitrary spring loss was equal to $\theta_s P_1$. The latter is a comprehensive evaluation of the contributing factors. It is used to appraise the extraneous power required in excess of the minimal adiabatic. The comprehensive valve loss is not a constant value, nor does it affect the cylinder charge pressure or the position of Point A, Figure 1.2. When such a pressure is applied, the correction is charging density and the capacity is reduced. Such correction is not consistent with the past 45 years of experience in rating gas compressors.

There is one further condition which can increase the temperature rise above the normal adiabatic function in a piston machine. A broken piston ring and/or a scarred valve disk or seat will permit a back-flow of 2 percent or more of the cylinder capacity. The hot discharge gas leaks isothermally from the compression cycle to the suction cycle. This dilution warms the suction gas which extends the discharge temperature.

An illustrative example follows. Assume that R_c = 3.0, k = 1.28 and σ = 0.2185. The incoming gas is 60°F (520 R); all cylinders are discharging at 202°F (662 R), except one which is 232°F (692 R). Determine the X percent of blow-by:

$$R_c^\sigma = 3.0^{0.2185} = 1.272$$

Normal discharge should be 1.272(520) = 662 R. The state of conditions follows:

$$[(1.00 - X)520 + X(692)]\ 1.272 = 1.00(692)$$
$$[520 + (692 - 520)X]\ 1.272 = 692$$
$$219\,X = 30$$
$$X = 14 \text{ percent leakage.}$$

This method of evaluating valve leakage is also valid for larger percentages. Using the same conditions, except that the discharge temperature is 363° F or $823R = T_x$

$$[T_o + (T_x - T_o)X]\ R_c^\sigma = T_x \qquad (1.16)$$
$$[520 + (823 - 520)X]\ 1.272 = 823$$
$$385\,X = 823 - 662$$
$$X = 42 \text{ percent leakage.}$$

Compression Efficiency

The area of "cap" BCGB on Figure 1.2 and the "sole" area AFDA represents the power added to the basic adiabatic power diagram AGCFA. It appears to be completely logical to add the area GHJCG, which has an area equal to BCGB to represent the added discharge burden. Likewise, the suction pressure is reduced to P_3, and the cross-hatched area AFKLA is equal to the area AFDA. The alternate and conventional method is to steepen the slope of curve AG and CF until the area AG′CF′A is equal to the full ABCDA. Reference 6 gives a case where the real "*k*" value of 1.28 was supplanted with a quasi "*n*" value of 1.375. A complete energy balance revealed that the Joule's mechanical equivalent heat unit must be 642 ft-lb per Btu in lieu of the universally accepted Joule's mechanical heat equivalent of 778 ft-lb per Btu (7). A comprehensive method of solving for the compresssion efficiency is given in the next chapter.

It should be borne in mind that the exponent and its corrections only effect the slope of AG and CD. The exponent σ applied to R_c in the denominator is partially cancelled by the σ factor in the numerator of the adiabatic head equation:

$$L_{ad} = (R_c^\sigma - 1)1{,}545TZ/\sigma m. \qquad (1.17)$$

The significance of the exponent correction has been abused. The valve losses are of much greater consequence, and their resistance affects are ignored. Equation 1.9 and 1.10 show the manner in which the compressibility corrections are applied as a pseudo polytropic function. In lieu of using the average compressibility as proposed by R. S. Ridgway, many application engineers prefer to use a correction factor of $(Z_s + Z_d)2Z_s$. This is subterfuge to inflate the power requirement. Ridgway's averaging method has been entirely satisfactory for a quarter of a century in solving the gamut of applications (8). The calculated (PGL) volumes are nebulous along the saturated vapor line. Mollier charts offer more reliable data. This also applies to heavy hydrocarbons operating at ultra-high pressures.

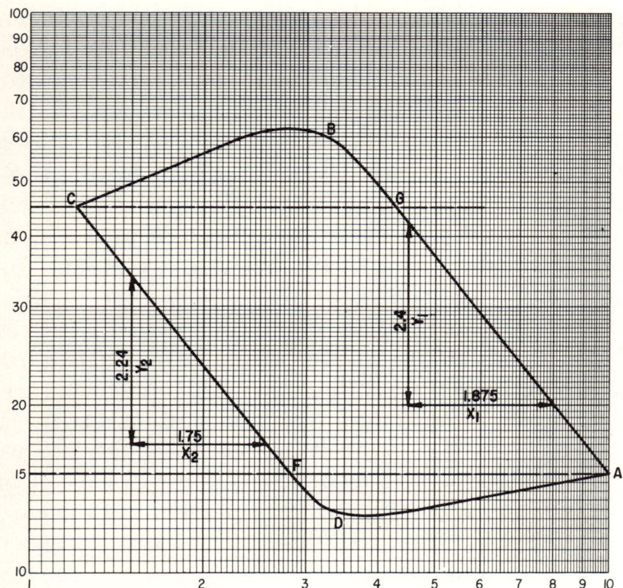

Figure 1.3. Compressor P/V diagram plotted on log/log chart.

These few cases would constitute the exception rather than form the rule. It is regretable that issues of this order which have little significance on the general procedures can create such a flurry of interest, while the gas technology in general is so lax in other comprehensive fundamentals.

Thermodynamic Behavior

Considerable dissertation has been given in the previous section concerning the four thermodynamic modes of *P—V* change. The first premise presumes that the temperature remains constant during a change of conditions where $P_1 V_1 = P_2 V_2 =$ Constant, for the isothermal mode. An instance where isothermal flow can be found is in the ideal throttling from an elevated pressure to a lower pressure through a control valve. The expansion of the gas after the compressor valves are open, points ≫ to *B* to *C*, and from points *F* to *D* to *A* in Figure 1.2, are other examples.

Adiabatic Behavior

The adiabatic (or isentropic) mode of compression requires the least power, except the isothermal, which is not a realistic mode. The adiabatic mode constitutes the best reference for all efficiency statements. Equations 1.13, 1.14 and 1.16 govern the realistic *adiabatic* change of state. These ideal changes presume that no heat is added or extracted. Piston friction is presumed to be negligible. When the added concession of reversibility is included in a cycle, it is known as an *isentropic* process. The piston compressor is the closest approach to an ideal adiabatic process that can be found in nature or in fact. When the power is calculated from the gas head using the Universal Gas Constant, the head is referred to as the *adiabatic* head and the power as the *adiabatic* horsepower. This is the second mode.

When the gas head is determined from enthalpy table or Mollier charts, complete reversibility is presumed. Power calculated from the enthalpy is identified as *isentropic* horsepower and constitutes the third mode. This terminology is used to provide each term with a more specific meaning (3, 4 and 5).

Polytropic Behavior

Figure 1.2 can be transposed to a log/log graph and have the general appearance of Figure 1.3. The slope of Y_1/X_1 is the compression line AG. The slope of Y_2/X_2 is the expansion line CF. Both of these slopes on an ideal diagram are equal to the ratio of specific heats, "*k*." The *cap* BCG and the *sole* AFD represent the isothermal flow through the valves. The area of the rectangle represents the power of compression. The slopes of AG and CF are the unadulterated true behavior of an ideal compressor. Leakages up to 2 percent are not apparent and consequently are negligible. Where the dominate leakage is past the piston rings, the compression and the expansion slopes are decreased. This should reduce the "*k*" value, but actually the "warm-up" dilution of the suction charge has the effect of a larger "*k*" valve. Where the dominate leakage is through the suction valve, the affects are the same. Where the discharge valve leakage is dominate, both slopes are increased as well as apparently increasing the "*k*" value.

Consider the preceding thermal-leakage example and apply that data to Figure 1.3. Whereas the slope "*n*," Y_1/X_1, would be 1.350 and $\sigma' = 0.259$ to match the abnormal discharge temperature of 232° F.

The pseudo polytropic efficiency for the greater slope is $\sigma/\sigma' = 0.2185/0.259 = 84.5$ percent. This ratio develops the spot polytropic efficiency, but it does not represent a consistent margin required between the basic adiabatic power and the actual power requirement. It more accurately represents the affect of a 14 percent internal leakage. The effect of further exponential corrections can only increase the slope of the compression and expansion curves. The slopes can be raised toward the perpendicular to include an area equivalent to the "cap" BCG and the "sole" ADF to create a new rectangle AG′ CFA′. This would only serve this one particular special case, and the results could not be projected to other instances with any degree of confidence.

Comprehensive methods of evaluating the magnitude of the valve (cap and sole area) losses are developed in subsequent chapters. It is important to know that all piston and rotary displacement compressions are essentially adiabatic processes. The abnormal discharge temperature rises ($<4R_c$) are most likely the pseudo polytropic effect of internal by-passing. The only true polytropic compression is experienced in an axial or centrifugal compressor. The super-adiabatic heating effects are the result of intrinsic friction at high velocities, >0.4 MACH.

The polytropic efficiency is an important device to project the discharge temperature of a centrifugal compressor. In fact, the abnormal amount of heating experienced over the adiabatic is the method used to evaluate the efficiency of centrifugal compressors. This data is largely reproducible (10). Again referring to the leakage example, assume that it is a centrifugal compressor and has a discharge temperature of 253°F. The polytropic discharge multiplier is 520/(460 + 258) = 1.38. The exponent is $l_n 1.38/l_n 3.0 = 0.292 = \sigma'$. The polytropic efficiency is σ/σ' or 0.2185/0.292 = 75 percent.

The rotary displacement compressors are also adiabatic machines that have a pseudo polytropic effect on the discharge temperature. Again, repeating the advise that it is better to evaluate the machine as an adiabatic function and examine the discharge temperature as an index of the quantity of internal by-passing.

Capacity

The volume sweep of the piston is termed the cylinder displacement. It is measured in cubic feet per minute (cfm). It is determined from the equations:

$$\text{cfm} = 0.00091LN(D^2 - 0.5d^2) \quad (1.18)$$
$$\text{cfm} = 0.327U(D^2 - 0.5d^2) \quad (1.19)$$
$$\text{cfm} = 4.2\,D^2\,(U/13.3\text{ fps}). \quad (1.20)$$

Where L = length of stroke D is the diameter of the cylinder and d is piston rod diameter, all in inches. U is the average piston speed in feet per second (fps.). Equation 1.20 offers a quick displacement with acceptable accuracy for 10-inch and larger cylinders having 2 to 4-inch piston rods. The capacity of cylinders having bores smaller than 8 inches should be calculated with Equation 1.19. The piston speed is the product of the stroke (in inches) and the rpm divided by 360. The average piston speed is 13.3 fps. The maximum rated speed is 18.3 fps. Speeds of 15 fps are common. Speeds quoted below 10 fps are rare. Cylinders range from 4 to 42 inches in diameter. The small cylinders are generally used for operating pressures in excess of 2,500 psi and are made from steel forgings. Cylinders that range from 12 to 6 inches in diameter are used for 800 to 3,000 psi and are made of cast steel. The rest of the cylinders are made of cast iron, ductile iron or Meehanite. Most cylinders have replaceable cast iron liners. The large cylinders—36 through 42-inch—are usually applied to vacuum service and limited to about 40 psig or less. They will have clearance volumes as low as 2.5 percent and have a blank-off at 30 R_c or 0.5 psia.

The volume retained in the cylinder at the end of the stroke is called the clearance gas. It is expressed as a percentage of the volume sweep. The minimum clearance for a typical cylinder is about 10 percent. The clearance can be estimated by totaling the volume of void space within the cylinder. The clearance between the piston and the head is about 0.100 inch at the head and crank end dead centers. Each valve occupies a volume equal to its port area and one inch deep. The sum of all valve volumes in each end plus the piston clearance divided by the piston sweep (piston area times stroke) equals the percent of clearance. For example, a 20 x 15-inch cylinder has a cross-sectional area of 314 square inches and a sweep of 4,710 cubic inches. The cylinder has four valves in each end that are 8 inches in diameter. The 8-inch valve area is 50 square inches or a total of 200 cubic inches. The estimated clearance is (200 + 31.4)/4,710 = 4.9 percent. The clearance is increased by the manufacturer when the cylinder displacement is too large for the design conditions. This is accomplished by either turning down the piston thickness or adding to the cylinder head spacer. The clearance gas trapped at the end of the stroke expands in reversible mode as illustrated by the curve CD on Figure 1.2. The amount of gas pumped with each stroke is represented by the ratio of lines AF/AE. This is known as the volumetric efficiency. The equation that best depicts this function is

$$E_v = 100 + C\,(1 - \Lambda R_c 1/k) \text{ or}$$
$$E_v = 100 + C - C\Lambda R_c{}^{1/k}\,(Z_s/Z_d). \quad (1.21)$$

The decimal fraction of Lambda (Λ) represents the percentage of clearance gas leakage that escaped past rings and suction valve seats. Well-seated valves and smooth piston rings may have a clearance volume leakage of 5 percent or a lambda factor of 1.05. Nonlapped or worn rings and disks may have a leakage of 10 percent or a lambda factor of 1.10. These tolerances hold up to 7 R_c, unless the piston rings or valve elements are poorly seated or defaced.

The cylinder capacity (acfm) is the product of the displacement and the volumetric efficiency. The unit weight flow (lb/min) is cfm/v_{so}. The "P" in Equation 1.1 for determining the specific volume, v_{so}, is point A on Figure 1.2. The entire capacity concept is based upon these simple statements. Point A is as close to being the same as the suction line or source pressure as present day instrumentation can provide. Likewise, point C represents the discharge line pressure. The pressures above the discharge line CG represent the required head necessary to overcome the valve resistance. The same explanation is offered for the pressure mode below the suction line pressure AF.

Capacity Control

There are six methods of varying the pumping capacity of a reciprocating compressor. 1. The suction pressure can be throttled to reduce the capacity by diminishing the gas density. This control is usually energized from a fixed discharge pressure controller. The power load is diminished by this device. 2. The same controller can be energized by the suction pressure, maintaining it constant with a by-pass control valve returning cooled discharge gas to the suction. The former method is more economical than the discharge by-pass system, which maintains a constant power load.

There are two methods of partially unloading one end of a cylinder. 3. One method is to provide a port in the side wall of the cylinder at a mid-stroke position. This port is connected to the suction chamber through an angle valve flanged on the cylinder. Half of the cylinder displacement is thereby returned to the suction with minimum head loss. The diameter of the by-pass line must be at least 5 percent of the cylinder diameter. A 20-inch cylinder should have a 4.5-inch diameter passage. 4. An extraneous pocket is frequently provided in the cylinder head as illustrated in Figure 2.1. The manual operated valve can be replaced with an automatic pneumatic valve. Such a pocket usually contains a volume of 1 to 2 gallons and adds about 20 percent clearance. This is ample to half-load the head-end. The percent of unloading can be readily developed from Equation 1.21. A variable volume clearance pocket is illustrated in Figure 1.4. This device can be remotely controlled by a pneumatic motor. It provides finite control.

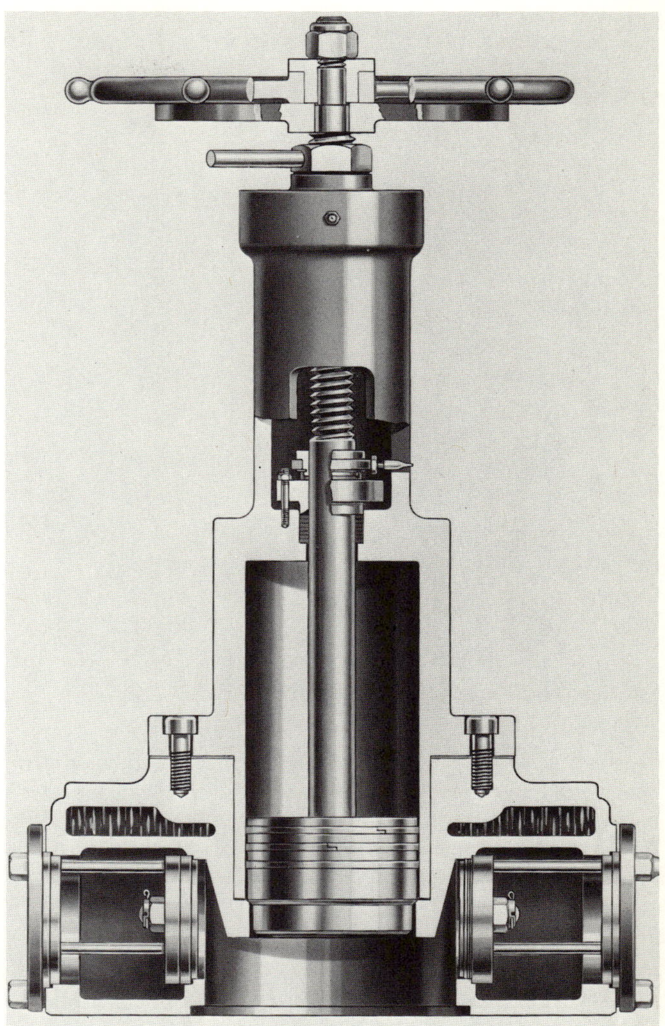

Figure 1.4. A variable volume clearance pocket permits infinite capacity control. The control can be made to operate automatically. (Courtesy of Chicago Pneumatic Tool Co.)

There are two methods of completely unloading either or both ends of a cylinder. 5. The most common method of finger lift unloaders is illustrated in Figure 1.5. The extended fingers in the right hand section depress the suction valve elements off the seat, permitting the gas to surge back and forth, in and out of the cylinder. Continued idling for periods of 30 minutes or longer can cause serious overheating and coking of the lube oil and unsaturated gases. For this reason, it is not a popular control for hydrocarbon processing. Gases containing unsaturates tend to polymerize at temperature in

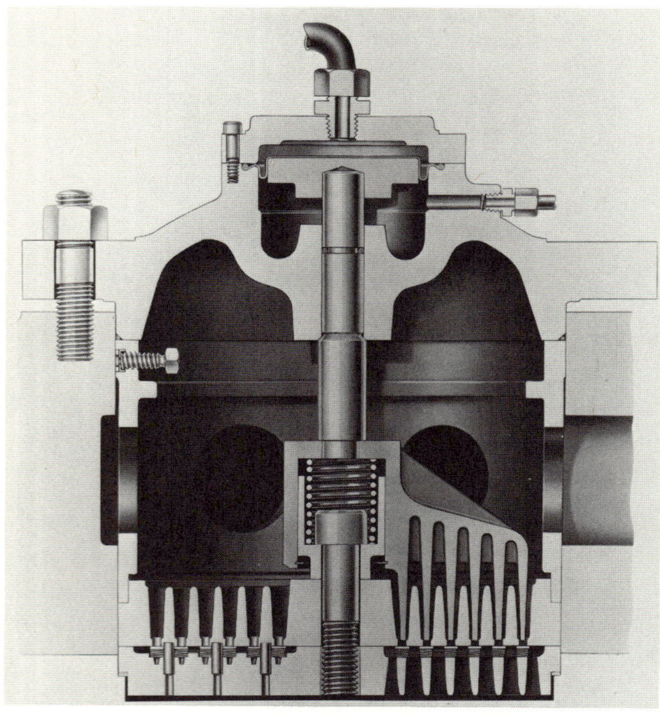

Figure 1.5 (above). The finger projections shown on the right-hand prong section are extended through the valve ports, contacting and holding the valve elements off seat. This permits the gas to return to suction on the return discharge stroke. (Courtesy of Chicago Pneumatic Tool Co.)

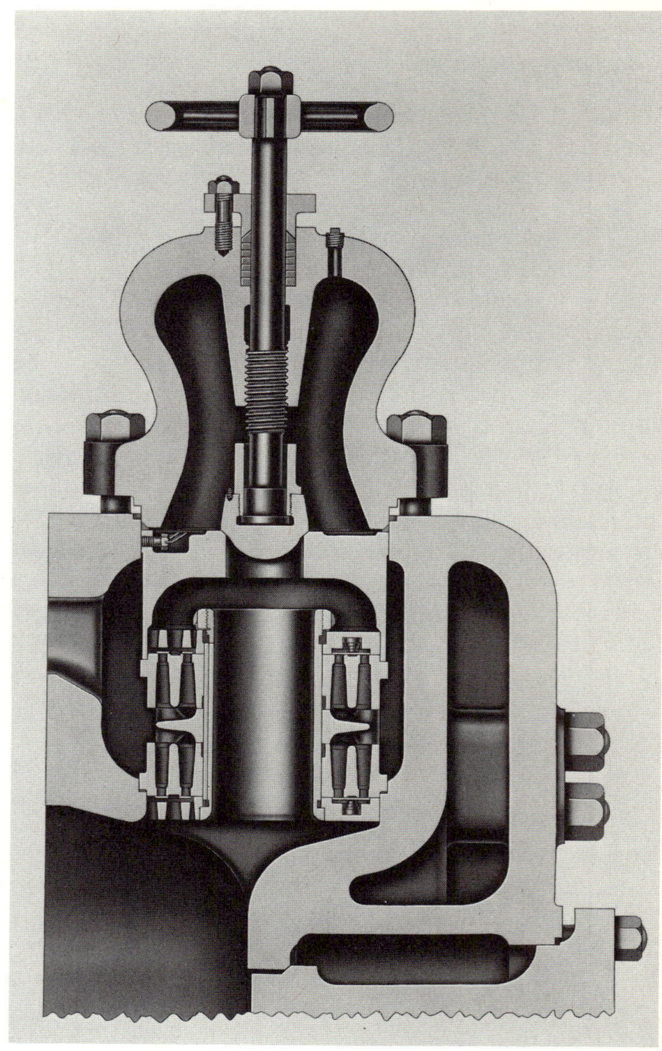

Figure 1.6 (right). Sectional view of a dual decked suction valves with a manual by-pass plug valve for unloading one end of a compressor cylinder. (Courtesy of Cooper-Bessemer Co.)

excess of 200° F. It is often necessary to flush such cylinders with a mist of compatible solvent or steam. 6. The alternate method provides a retrievable chair attached to a suction valve. A swivel screw stem working through a gland on the valve cover lifts the entire valve off the port seat. This method of by-passing avoids the possibilities of overheating and the resulting maintenance. The area of the passage channels serving these pockets should be greater than 5 percent of the piston area.

Another version of a cylinder-end unloader is shown in Figure 1.6. In addition to illustrating the by-pass plug valve, this figure shows the application of dual parallel suction valves. Figure 1.7 illustrates the nylon poppet valve, believed to be the most efficient valve in piston compressor service. One of the latest innovations in compressor valve design is shown in Figure 1.8. A sectional view of a ring channel valve is given in Figure 1.9. This valve is designed to reduce impact stress and improve gas flow.

Power of Compression

The brake horsepower required to pump gas is the product of weight flow and the gas head, divided by 33,000 ft-lb per minute, the dynamic and the mechanical efficiency. In addition to the several previous examples of isothermal change of state, the Fan horsepower equation presumes that the fluid is incompressible. Most Europeans and some U.S. manufacturers use the isothermal mode for calculating the compressor horsepower. The fan and the isothermal horsepower equations are

$$\text{Fan hp} = \text{cfm (inches of water)}/6{,}356\ \eta_s; \qquad (1.22)$$

$$\text{Iso hp} = 1{,}545\ W(ln\ R_c)T_o/m. \qquad (1.23)$$

Figure 1.7. The nylon poppet valve is believed to be the most efficient valve in piston compressor service. It has a resistance coefficient of half that of a concentric steel disk valve. The springs seated on the nylon poppets fit in the drilled holes of the guard which is turned 180° over the seat. (Courtesy of Clark Bros. Co.)

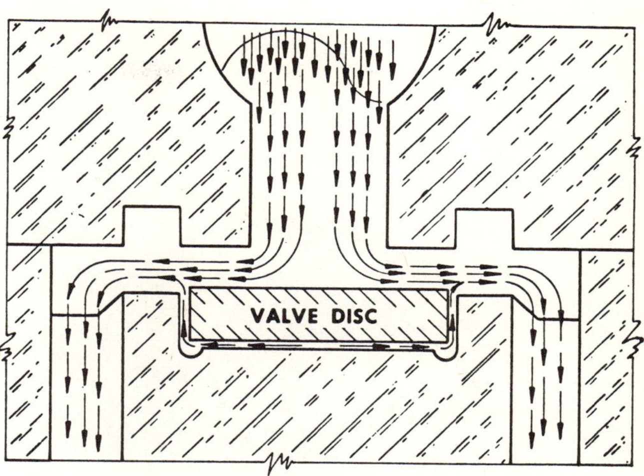

Figure 1.8 (above). One of the latest innovations in compressor valve design. The valve disk lift action is pneumatically cushioned in the recess groove. (Courtesy of Cooper-Bessemer Corp.)

Figure 1.9 (below). Section view of a ring channel valve, designed to reduce impact stress and improve gas flow. (Courtesy of Cooper-Bessemer Corp.)

W is the weight flow in pounds per minute, and ($ln\ R_c$) is the natural logarithm of the ratio of compression. The adiabatic and polytropic horsepower are the products of the respective gas head and the weight flow, pounds per minute divided by 33,000 ft-lb/min/hp. The polytropic head is obtained by using an exponent σ/η in lieu of σ, in both the radical and denominator application in Equation 1.17. All four modes of change produce a common brake horsepower when the respective thermal efficiencies are applied. The isothermal efficiency is the furthest removed from actual performance. Projecting a likely efficiency for a new machine would depend upon the similitude of data obtained from a prior built machine operating under similar conditions and gas densities. There is no reference data available concerning the isothermal or the polytropic behavior of gases. The premise of an isothermal change of state precludes a temperature change. This would preclude an entropy and enthalpy relationship and deny modern concepts of thermodynamics.

The polytropic properties of a gas would require a separate table or chart for each efficiency and would obviously be impractical. Most manufacturers chart their centrifugal compressor performance in units of polytropic feet of head. This data can be converted to the adiabatic status as explained in Chapter 4. The simplest solution is to stay with the adiabatic evaluation and reserve the use of the polytropic efficiency for projecting the discharge temperature of centrifugal compressors and for measuring the amount of internal by-passing in rotary displacement compressors.

Figure 1.10. Exploded view of double plate valve. (Courtesy of Cooper-Bessemer Corp.)

Conventional Power Evaluation

The first major application for gas compressors was pumping the low pressure natural gas from the casing heads and gas treating plants in the oil fields into the public utility pipelines. The conventional unit of measuring the flow is millions of standard cubic feet per day. (MMscfd). This unit is still used as the equivalent weight flow in the basic conventional horsepower equation. The head requirement was related to the ratio of compression. Empirical data is compiled to show the relationship between the R_c and the brake horsepower per MMscfd/d(9). The most recent conventional power calculation procedure uses a basic adiabatic mode in lieu of the previous technique of applying empirical factors. The brake horsepower is derived from seven new empirical overall efficiency charts (9). Because the specific gravities used in the displacement equation and in the head equation are cancelled, the conventional practice disregards the effect of the gas density. An aerodynamic energy balance of the gas flow in and out of the cylinder makes the mol weight a significant factor. Such an analysis is given in Chapter 2. The (Σ) adiabatic horsepower per million scf/day for air having a "k" value of 1.40 and m = 29; natural gas having a "k" value of 1.28 and m = 18; wet gas having a "k" value of 1.22 and m = 24; LPG having a "k" value of 1.13 and m = 42 are given below:

$$\Sigma air = 155(R_c{}^{0.286} - 1) \quad (1.24)$$
$$\Sigma ng = 203(R_c{}^{0.22} - 1) \quad (1.25)$$
$$\Sigma wg = 246(R_c{}^{0.18} - 1) \quad (1.26)$$
$$\Sigma lpg = 386(R_c{}^{0.115} - 1). \quad (1.27)$$

The dynamic efficiency is given by Equation 2.4 and illustrated in Figures 2.2, 2.3, 2.4 and 2.5. These efficiencies are equally applicable to the previous power factors. The mechanical efficiency is appraised by the following equations for single and multiple cylinder units:

$$\eta_m = 1 - 1/(\text{HP/CYL})^{0.5} \quad (1.28)$$
$$\eta_m = (\text{HP/UNIT}) - (\text{HP/CYL})^{0.5} (\text{CYL/UNIT})/(\text{HP/UNIT}). \quad (1.29)$$

Figure 1.10 shows an exploded view of a double plate valve. Most of the process operations have a low tolerance for lube oil carry-over from piston compressors. This is especially true for chillers, refrigeration, instrumentation and where catalytic converters are concerned. The nonlube equipment has been developed for two to four years of service. Figure 1.11 gives an exploded view of a typical Teflon rider type piston which is an important part of the nonlube system. Figure 1.12 illustrates the type of multistage balanced opposed compressors that are concerned in this text.

Closure

The objective of Operation Head-Start was to present a comprehensive, yet simple procedure of preparing a piston compressor application. A background of events, fluid dynamics, thermodynamics and editorial comment has been presented to this point. The closure for section on Operation Head--Start should include an application of the material presented and introduce some of the features presented in the subsequent chapters.

The design conditions for an exemplary application requires the following information.

Design Conditions

Flow rate: W, moles/hr; lb/min, 100; MMscfd
Commodity Air: Mol Wt 29; "k" 1.40
Suction: 100 psia; 1.00 Z_s; 60°F; 520 R Discharge:
Discharge: 300 psia; 1.00 Z_d; Amb 14.5 psia; σ = 0.286.
Compressor Features:

- Posture, Balanced opposed, 6 throws;
- Power/throw, 200 Bhp; Stroke, 6 inches, 720 fpm;
- Piston rod diameter, inches 1.75; Thrust capacity, pounds 24,000.

Power Requirement

$$R_c = 300/100 = 3.00;\ X_\sigma = 1{,}545(520)/29(0.286) = 97{,}000 \qquad (1.17)$$

$$V_s = 10.73(520)/29(100) = 1.925 \text{ cf/lb} \qquad (1.1)$$

$$R_c = 3.00^{0.286} = 1.37;\ R_c{}^{1/k} = 3.00^{0.715} = 2.19$$

$$L_{ad} = X_\sigma(R_c - 1) = 97{,}000(0.37) = 35{,}900 \qquad (1.17)$$

$$\text{Adhp} = (\text{lb/min})L_{ad}/33{,}000 \qquad (1.30)$$

Alternate equations are

$$\text{Adhp} = 3.03(\text{lb/min})L_{ad}(10)^{-5}$$

$$= 1.92m(\text{moles/hr})L_{ad}(10)^{-4} \qquad (1.31)$$

$$= 1.61m(\text{MMscfd})L_{ad}(10)^{-3} \qquad (1.32)$$

$$\text{Adhp} = 3.03(5{,}300)35{,}900(10)^{-5} = 5{,}760.$$

The conversion of moles per hour to pounds per minute is a simple correction of the constants: 3.03(29/60) = 1.47. The correction from MMscfd to lb/min is 3.03 $(10)^6$/380(1,440) = 5.55. Another handy conversion from MMscfd to pounds per second (pps) is

$$\text{Flow rate, pps} = \text{MMscfd}(m/32.8). \qquad (1.33)$$

The reader may have observed an inconsistency in using the terms MMscf/d and MMscfd interspersly. The former is the more accurate expression. Both terms have the same meaning; i.e., millions of standard cubic feet per day. Please credit the expressions as a colloquialism of the trade.

The (Adhp) adiabatic horsepower is the minimum energy requirement for a visual line pressure change. Its value should constitute the basis for all reference efficiencies. The sum of the adiabatic horsepower and the dynamic losses gives the Gas hp. The ratio of Adhp/Gas hp is the dynamic efficiency, η_d. This efficiency is developed from the piston/valve area ratio (a), the piston speed (U) and molecular weight (m), in subsequent chapters. The appendix contains a series of charts which quickly resolve the dynamic efficiency for typical operations. The a ratio for a typical cylinder should be about 10. This value is determined by measuring the minimum flowing area. Suppose a 17.75-inch diameter cylinder has four valves per quadrant, or a total of sixteen 8-inch valves for the cylinder. Presume that each valve contains three concentric disks: 6.25 x 5.25, 4.5 x 3.5, 2.75 x 1.75 inches in diameter, *OD* x *ID*. The total flowing diametrical inches are 24. It is reasonable to use a valve disk lift of 0.080 inch. This makes the minimum valve area 24 (π) 0.080 = 6 square inches or 24 square inches per quadrant. The piston area is 247 square inches. The piston/valve/quadrant "a" ratio is 10.3. A narrower 3/8-inch disk can accommodate four disks having an area of 33 square inches and an a ratio of 7.5. Valve lifts of 0.100 and 0.120 are common in slow speed machines <350 rpm. Such lifts would reduce the a ratio to 6 and 5, respectively. The latter appears to be an asymptotic minimum.

A short 6-inch stroke machine can be operated at 900, 700, 600 or 515 rpm with an 8, 10, 12 or 14-pole synchronous or induction direct drive motor. The higher speed machines are better suited for air and lighter gases. The lower speed units render more efficient service in handling propane, butanes and heavier refrigerants. Having selected a speed of 720 rpm, 720 fpm, a = 10, then by referring to Chart 2 for air, the dynamic or compression efficiency is observed to be 87 percent. The gas hp for the example is 109/0.87 = 125. The

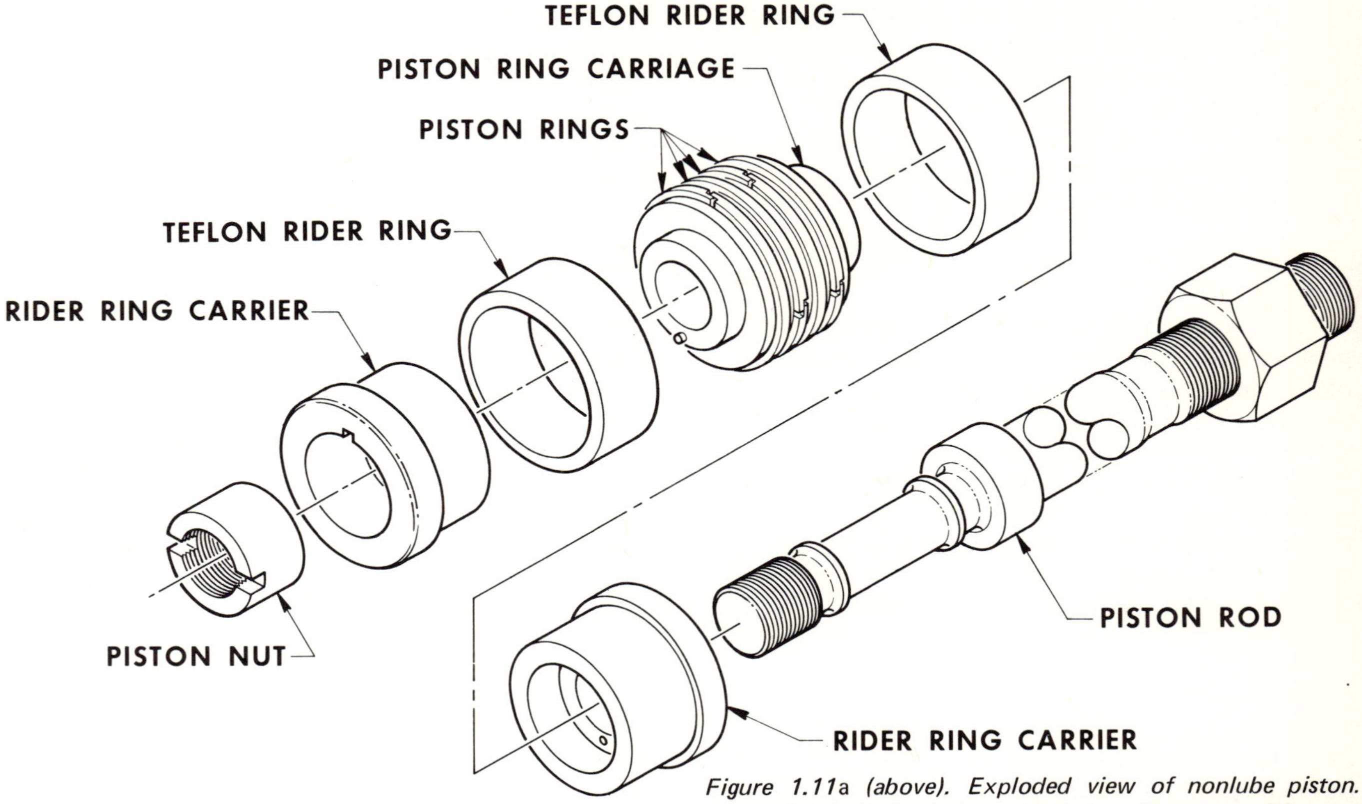

*Figure 1.11*a *(above). Exploded view of nonlube piston. Figure 1.11*b *(below). Assembled view. (Courtesy of Cooper-Bessemer Corp.)*

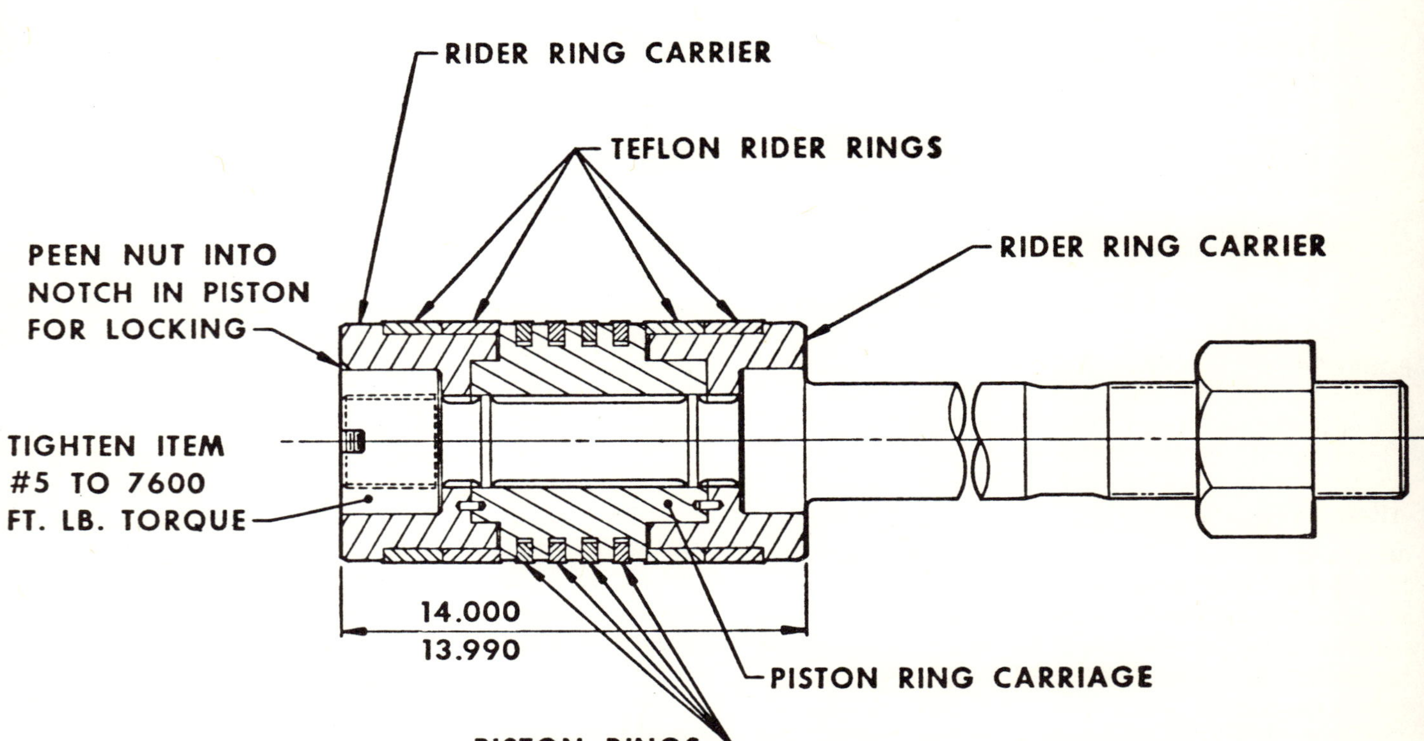

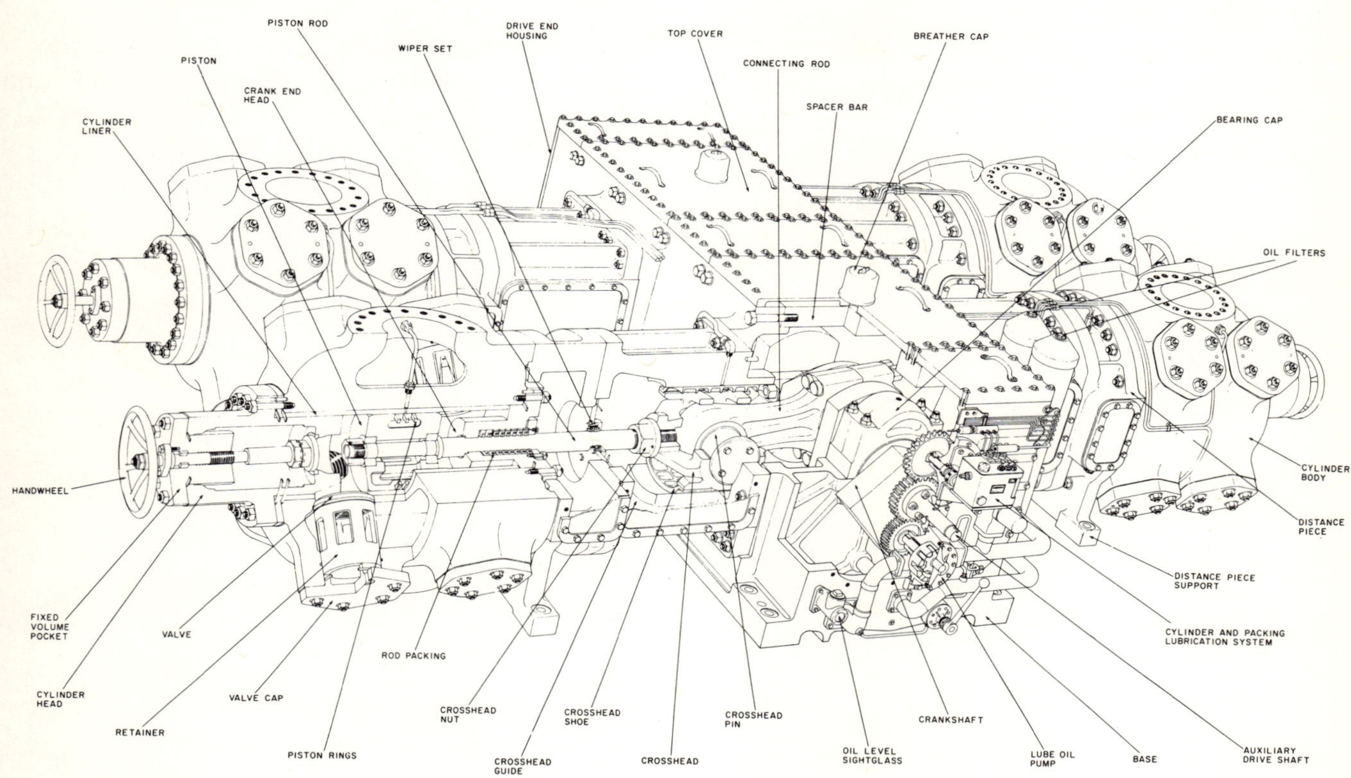

Figure 1.12. Line drawing of balanced-opposed compressor. (Courtesy of De Laval Turbine, Inc.)

brake horsepower is the result of applying the mechanical efficiency Equation 1-28. If the operation was handled with one cylinder, the mechanical efficiency would be $1.0 - 1/(125)^{0.5}$ or $1.0 - 0.089 = 0.911$, and the Bhp is $125/0.911 = 139$ Bhp. If six 200 hp cylinders of equal 200 hp loads were involved, the mechanical efficiency is $1{,}200 - 6(200)^{0.5}/1{,}200 = 93$ percent.

Cylinder Sizing

The cylinder capacity is the product of W(lb/min) and the specific suction volume, $100(1.93) = 193$ acfm. The next problem is to establish an accreditable volumetric efficiency which requires a realistic clearance volume. It has been observed that the minimum clearance of a compressor cylinder can be reasonably estimated by the expression, $42/(\text{stroke})^{0.5}$. The minimum clearance for a 6-inch stroke cylinder is $42/2.44 = 17.2$ percent. The volumetric efficiency is determined from Equation 1.21: $100 + 17.2 - 17.2(1.10)(2.19) = 76$ percent. The cylinder displacement is $193/0.76 = 254$ cfm. The d^2 value from Equation 1.20 is $254/4.32(12/13.3) = 66$. The cylinder diameter is the square root of 66 or 8.10 inches. Compressor cylinders are generally bored to the next nearest quarter inch for standard sizing. This cylinder would be sized at 8.25 inches. There is no maximum clearance limitation other than physical structure. Most low R_c cylinders as used in pipeline service have clearances in excess of 100 percent. The cylinder bore enlargement is corrected by adding to the given 17 percent clearance. The volumetric efficiency is first corrected: $76/(8.25/8.1)^2 = 73.3$ percent. The necessary clearance is $73.3 = 100 + C - C(1.1)(2.19)$, $1.41C = 26.7$, $C = 19$ percent.

The piston rod thrust capacity for the example machine is rated at 24,000 pounds for a piston rod,

1.75 inches in diameter. This represents a nominal tensile load of 10,000 psi. This is a conservative load for the alloy steel (SAE 4140) generally used for this service. The piston rod is the weakest link in the reciprocating system.

The piston rod thrust is determined:

$$\text{Thrust} = A_p(P_2 - P_1) - A_r(P_2 - 15). \tag{1.34}$$

For this example, the thrust is 53.8(300 — 100) — 2.4(285) = 10,060 pounds. The thrust load is only 42 percent of the rating. No difficulty would be anticipated from such nominal loading.

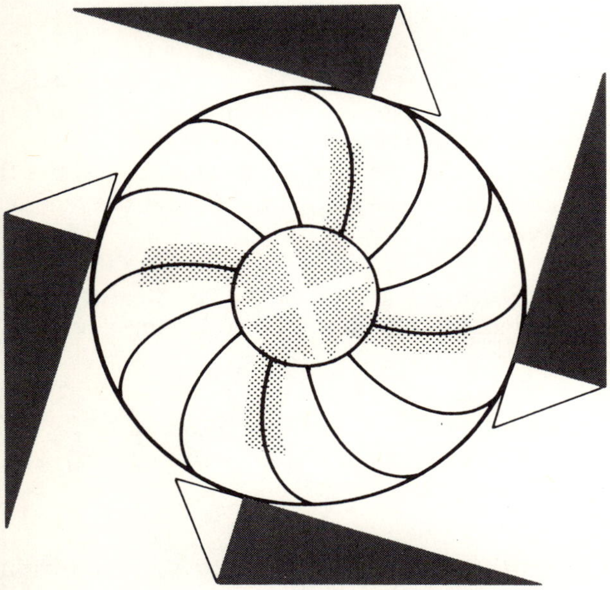

2 Piston Compressors

Introduction

This chapter introduces a comprehensive method of evaluating piston compressor performance. The principal feature is an evaluation of the compressor valve losses as influenced by the molecular weight and temperature of the gas, the effective valve area, the valve design, the piston speed and the effect of the piston diameter.

The new system is based on aerodynamics and thermodynamics rather than on custom and empiricism. The system is unique in presenting a mathematical evaluation of the *compression* and the *mechanical efficiency*. It has been successfully applied over the past 10 years in producing rational solutions to the most difficult compressor problems.

The 1966 NGPSA Data Book contains five charts for visual selection of an *overall efficiency* which includes a 93 percent mechanical efficiency as well as the *compression efficiency*. Earlier editions contained similar charts which included a 95 percent mechanical efficiency. This 2 percent difference in efficiency maneuvers a subtle $10,000 price increase for a 3,500-hp integral gas engine compressor. The previous 95 percent mechanical efficiency was adequate for a 400-hp cylinder and perhaps overgenerous for greater powered cylinders. A frictional allowance equal to the square root of the cylinder hp has been found to be quite realistic. For example, a 1,000-hp cylinder would have a 97 percent mechanical efficiency, $1 - (1{,}000)^{0.5}/1{,}000 = 0.9683$. A 100-hp cylinder would have a 90 percent mechanical efficiency.

When the API Standard 618 committee introduced the *no negative* capacity tolerance in the guarantee clause 55, the price increased another $15,000 for the same engine. Likewise, another 5 percent power penalty or $25,000 is added for nonlube features. There is serious doubt that 5 percent frictional heat from a 1,000-hp cylinder can be transmitted through a set of Teflon or carbon rings without being consumed. A sliding coefficient of 0.03 could account for about one percent nonlube frictional loss.

How to Enhance Compressor Ratings

The *adiabatic* horsepower (AD hp) constitutes the basic power reference for all forms of compression. There is no other media that is more efficient. An isothermal operation would require less power, but it does not exist in nature or in fact. The losses involved in charging and exhausting the cylinder must be evaluated. This is best accomplished by

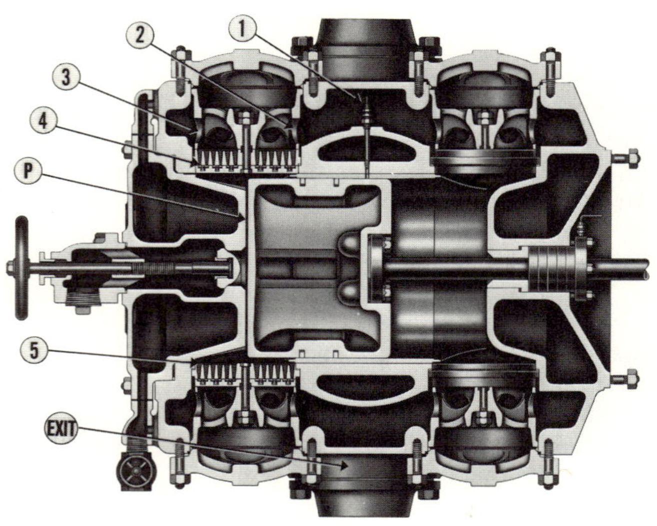

Figure 2.1. The five velocity changes involved in charging a compressor cylinder. (Courtesy of Cooper-Bessemer Co.)

relating such losses to the valve velocity. This concerns the maximum velocity through the minimum restriction. In order to make this evaluation, the specification must require (a) the type of valve, (b) the minimal valve area, (c) the piston area, (d) the average piston speed and (e) the mole weight of the gas.

The simple power pump equation is the easiest method of determining the hp for a gas compressor:

$$bhp = ppm(\text{adiabatic head})/33{,}000(\eta_d)\eta_m. \quad (2.1)$$

The flow is usually expressed in process parlance as pounds per minute (ppm). The adiabatic head (L_{ad}) is equivalent to the head lift for a pump and determined from Equations 2.2 and 2.3.

$$L_{ad} = X_\sigma(R_c^\sigma - 1) \quad (2.2)$$
$$X_\sigma = 1{,}545T_1 Z_a/m\sigma \quad (2.3)$$
$$\eta_d = (R_c^\sigma - 1)/(BR_c^\sigma - 1) \quad (2.4)$$
$$\eta_m = (bhp - bhp^{0.5})/bhp \quad (2.5)$$
$$\sigma = (k-1)/k \quad (2.6)$$
$$B = (1 + \theta_d)/(1 - \theta_s) \quad (2.7)$$
$$R_c = P_d/P_s. \quad (2.8)$$

The various abbreviations are explained in the glossary. The intrinsic correction factor B extends the normal R_c to represent the actual effective R_c within the cylinder. The symbol θ_d represents the mean psi required to exhaust the cylinder displacement into the header. The symbol θ_s represents the suction valve loss experienced in filling the cylinder. Both are expressed as a decimal fraction of the respective system pressure. The derivation of an equation to evaluate θ_s is given next.

Aerodynamic Valve Analysis

The vector flow path through a typical cylinder is shown in Figure 2.1. The arrows depict abrupt changes in velocity that are experienced in charging the cylinder.

1. V_1 is the line velocity of approximately three times the average piston velocity (U) and the velocity coefficient K_1 is assumed to be 0.3 velads. (Velocity head = $V^2/64.4$.)
2. V_2 is the cylinder channel flow to each valve port at an approximate velocity of $4U$ and the K_2 value is taken as 0.4 velads.
3. V_3 is the valve guard velocity of $5U$, and the frictional resistance factor K_3 is evaluated as 0.6 velads.
4. V_4 is the lifted valve element average velocity of $12U$, and the resistance K_4 value is 4.0 velads.
5. V_5 is the valve seat velocity of $5U$, and the resistance K_5 value is 0.7 velads.
6. Inside of the cylinder where the gas follows the piston action, the velocity is U, and the f value is 1.0 velad.

Substituting the equivalent velads for the respective piston velocities, the following equation is evolved:

$$\theta_S = [(0.3 \times 9)U^2 + (0.4 \times 16)U^2\ (0.6 \times 25)U^2 + (4 \times 144)U^2 + (0.7 \times 25)U^2 - U^2]\ /\ 288V_{S1}\ (g)$$
$$\theta_S = 616U^2\ /288V_{S1}\ (g).$$

Substituting $10.73\ T_1/mP_1$ for V_{S1} and dividing the sum of the suction valve resistances by P_1, so as to express the pressure drop in terms of a decimal percentage of the suction pressure, we have $\theta_S = 616mU^2/10^5\,T$. The largest and controlling resistance is the valve element loss of $576U^2$ velads, referred to the average piston speed. It can be equated to Ka^2 where K is the valve resistance factor of four and a piston/valve area ratio a factor of 12. The remaining $40U^2$ only represents 6 percent of the total resistance. Any deviation or correction in this evaluation is unlikely to effect the general premise. The equation can be modified to carry all secondary losses as equivalent to $40U^2$. It has been shown that the piston speed during the central 60 percent of the piston travel is 1.5 times the average full stroke speed (2). Since most of the valve flow occurs during the central travel, the arithmetical average

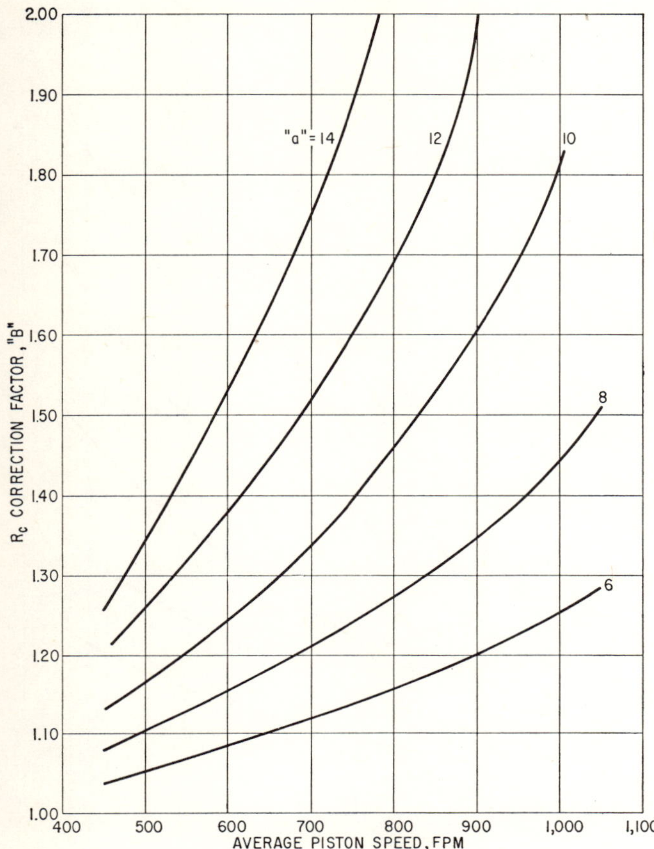

Figure 2.2. Correction B *factor to convert line* R_c *to intrinsic* R_c *for butane,* m = *58,* t_1 = *60° F,* R_c = *3 (a is the piston/valve area ratio).*

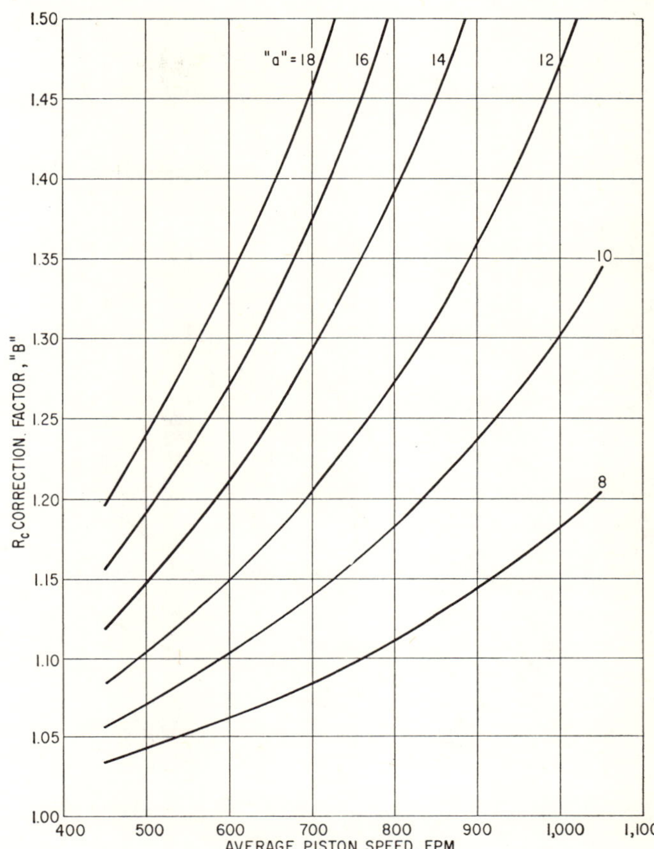

Figure 2.3. Correction factor B *to convert line* R_c *to intrinsic* R_c *for air,* m = *29,* R_c = *3,* t_1 = *60°F (a is the piston/valve area ratio).*

speed U can be increased by the square of 1.5 or 2.25. When the above equation is regrouped and the controlling valve resistance identified as $K_a{}^2$, we have

$$\theta_s = 2.25(40 + K_a{}^2)U^2 m/10^5 (T_1). \tag{2.9}$$

It is convenient to consider the typical disk type valve resistance factor as 4.44 velads. The most efficient nylon poppet valves have about half this resistance. The most inefficient valve design has a K value of about 50 percent greater. The congested flow through double-deck, parallel valves increases the K resistance by one-third. Using K = 4.44 for ring disk valves, the suction valve loss is

$\theta_s = (aU)^2 m/(T_1\ 10{,}000)$ or where K is other than 4.44, then

$$\theta_s = 2.4K(aU)^2 m/(T_1\ 100{,}000). \tag{2.10}$$

The only condition that is changed in equation 2.10 to make it applicable to the discharge valve is the temperature. This is corrected by

$$\theta_d = \theta_s / R_c^{\sigma} \tag{2.11}$$

The common lift for disk valves operating in 1,000 psig service is 0.080 inch. This is reduced to 0.050 for pressures in excess of 2,000 psig. When high pressure gas has a molecular weight less than 10, a lift of 0.030 inch may minimize the valve maintenance. Lifts of 0.100 inch are common for 100 psig and lower pressures. Nylon poppet type valves with 0.250-inch lift have rendered excellent service at speeds of 600 rpm in 1,000 psig service.

Modern compressor cylinders are usually provided with a piston/valve a ratio of 8 to 12. Early 1930 model cylinders equipped with strip valves had a ratios as high as 20. An a ratio less than 5 generally requires a high-lift poppet type valve. A set of reference charts, Figures 2.2 through 2.5, and Charts 1 through 6 are included, whereby B factors can be readily extracted. All charts are based on $3R_c$. The decimal fraction of a similar series of B factors at $1.5R_c$ are only 8 percent greater than the $3R_c$ decimal fractions. The decimal portion of the B

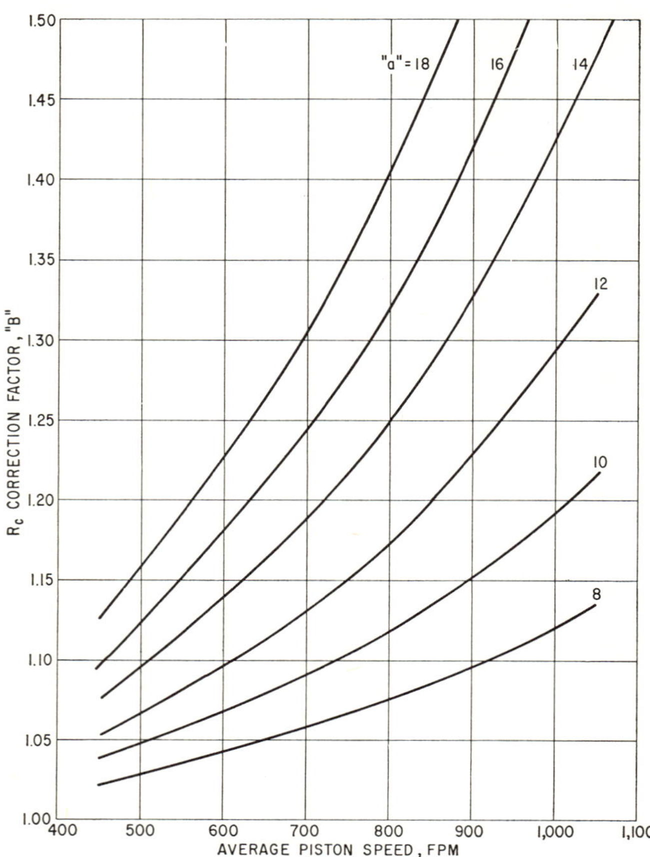

Figure 2.4. Correction B *factors to convert line* R_c *to intrinsic* R_c *natural gas, m = 19,* R_c *= 3,* t_1 *= 60°F (a is the piston/valve area ratio).*

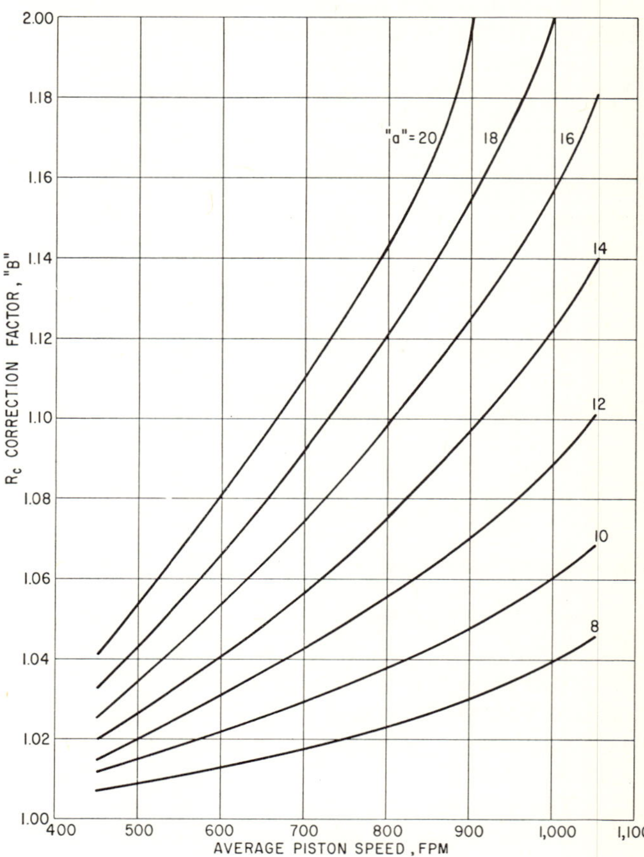

Figure 2.5. Correction B *factors to convert line* R_c *to intrinsic* R_c *for a hydrogen mixture,* m *= 6.5,* t_1 *= 60° F,* R_c *= 3 (a is the piston/valve area ratio).*

factor can be extrapolated as described in Equation 2.12. The suffix *c* represents the operation being changed.

$$B_c = (B - 1)(520/T_c)(m_c/m)(U_c^2/U^2) \qquad (2.12)$$

Compression Efficiency

The term *compression efficiency* has never had a precise definition. The manufacturers confidential guarantee curves have been the prime documents used to rate gas compressors. The issuance of the two adiabatic hp charts in the 1966 NGPSA Data Book has set a precedent in acknowledging the adiabatic power as the basic value. The *compression efficiency* therefore becomes the complement of the valve losses. Equation 2.4 becomes a mathematical expression for *compression efficiency*. It is defined as the ratio of $(R_c^g - 1)$ to $(BR_c^g - 1)$, where R_c is the compression ratio at the cylinder flanges and B represents the algebraic sum of the valve resistances.

The specific heat ratio k values as determined for the proximate mean temperature are entirely adequate for compressor performance ratings. The accuracy of the other factors emphasized in this article is of greater consequence than the k value refinements.

$$k = C_{pm}/(C_{pm} - 1.986). \qquad (2.13)$$

Sample Problem

Given: An 18-inch bore, double acting cylinder with 20-inch stroke has a piston/valve a ratio of 12 and operates at 240 rpm. The cylinder clearance is 11.5 percent. The gas handled has a molecular weight of 17.7 and a k value of 1.275. The suction is 113.5 psia, Z_s is 0.998 at 100° F and compressed to 283.5 psia where Z_d is 0.989.

Determine the hp requirement:

$\sigma = (1.275 - 1.0)/1.275 = 0.216$

$U = 2(20)240/12(60) = 13.3$ fps; $U^2 = 178$

$R_c = 283.5/113.5 = 2.50; R_c^\sigma = 1.22: R_c^{1/k} = 2.05$
$\theta_s = 144(178)17.7/5{,}600{,}000 = 0.081$
$\theta_d = 0.081/1.22 = 0.0665$
$B = (1 + 0.0665)/(1 - 0.081) = (1.0665/0.919) = 1.17$
$BR_c = 1.17(2.50) = 2.92; BR_c^\sigma = 1.260$

Compression efficiency, η_{dy} = 0.220/0.260 = 84.5 percent. The cylinder displacement is determined from

$$Q\text{, cfm} = 0.3275[d^2 - (d_r^2/2)]U. \quad (2.14)$$

An approximate form of Q is $4.25(U/13.3)d^2$ = 1,400 cfm.
The volumetric efficiency is

$$E_v = 100 + C - \Lambda C R_c^{1/n} \quad (2.15)$$
$$E_v = 111.5 - 1.1(11.5)2.05 = 85 \text{ percent}$$

The volume capacity is 1,400(0.85) = 1,190 acfm.

$$v_s = 10.73\ (560)\ 0.998/\ (17.7)\ 113.5 = 2.97 \text{ cf/lb.}$$

The weight flow is 1,190/2.97 = 400 lb/min. The standard volume flow is 400 (1,440) x 379.5/17.7 = 12.35 MMscfd. The adiabatic gas constant is X_σ = 1,545(560)0.993/17.7(0.126) = 225,000 (Equation 2.3). The adiabatic head is L_{ad} = $X_\sigma(R_c^\sigma - 1)$ = 225,000(0.220) = 49,500 ft-lb/lb (Equation 2.2). The adiabatic horsepower is *AD* hp = [(lb/min) *AD*-head]/33,000 = 400 (49,500)/33,000 = 600. Gas hp = 600/η_{dy} = 600/0.845 = 710. Brake hp = 710 + $(710)^{0.5}$ = 737.

The mechanical efficiency η_m = 710/737 = 96.2 percent. If the load was divided between four cylinders, the mechanical efficiency would be 710/[710 + 4(710/4)$^{0.5}$] = 93 percent. This rule has been found most consistent.

Note: The gas conditions used in this problem are similar to Sample Calculation No. 1, NGPSA 1966 Engineering Data Book. The Bhp/MMscfd is 59.7 for both examples. This is more of a coincidence than it is a confirmation of a common objective and technique. The NGPSA example piston speed is 650 fpm compared to 800 fpm for the text example. This factor alone should enhance the NGPSA compression efficiency 8 percent. The NGPSA specific gravity correction Chart E is 0.955. The text gravity correction is 0.61, etc. Fortunately, these inadequacies are fully compensative in this instance.

Equations 2.14 and 2.15 are taken from *Gas and Air Compression Machinery*. Λ is 1.10, which includes a 10 percent valve and piston ring leakage loss affecting only the trapped clearance gas. The heat

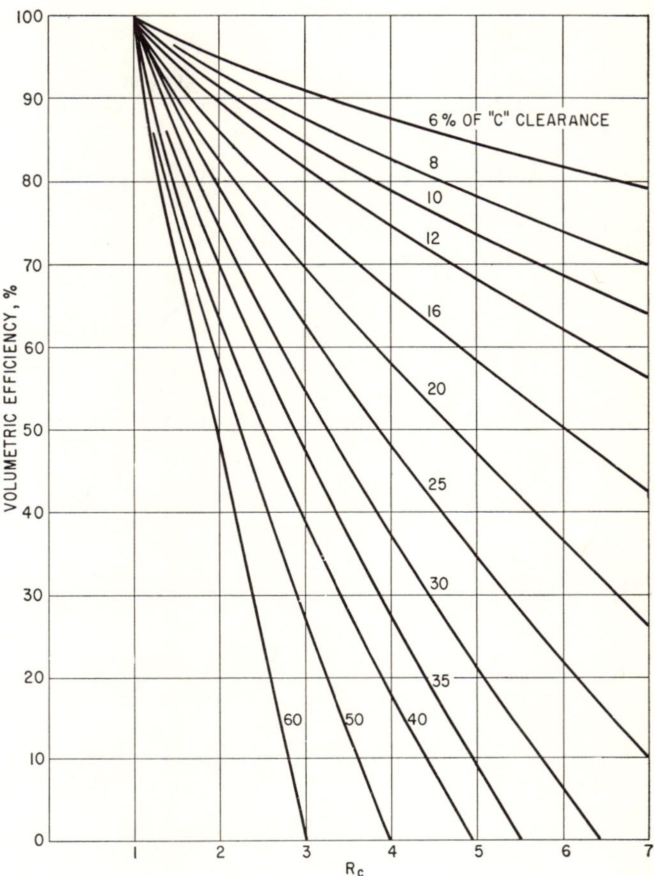

Figure 2.6. Volumetric efficiency for air and diatomic gases, k = *1.40,* n = *1.35,* Λ = *1.10,* η = 1.1, E_v = *100* + C - $C\Lambda R_c^{1/n}$.

rejection from the clearance gas of a typical cylinder has a polytropic factor of 1.10 and reduces the k value of 1.275 to an n value of 1.240 and 85.0 E_v. The extreme range of this polytropic factor is 1.25 which reduces n to 1.23 and E_v to 84.8 percent. The 1.25 factor is suitable for a slow speed, 100-hp cylinder having an abundance of cold water. The cylinder clearance is usually flexible at the purchasing stage. The three volumetric efficiency Figures, 2.6, 2.7 and 2.8, provide quick and accurate E_v selections for preliminary sizing. The foregoing example may be simplified by using Figure 2.4 and the NGPSA adiabatic hp chart. The B factor for 800 fpm and 12 a ratio is 1.174. The actual mole weight and temperature corrections reduce B to (1.174 — 1.0)(17.7/19.0)(520/560) + 1 = 1.151. This compares favorably with the calculated value of 1.154. The intrinsic compression BR_c is 2.88. The hp/MMcfd factors for 2.5 R_c and 2.88 R_c are 44.2

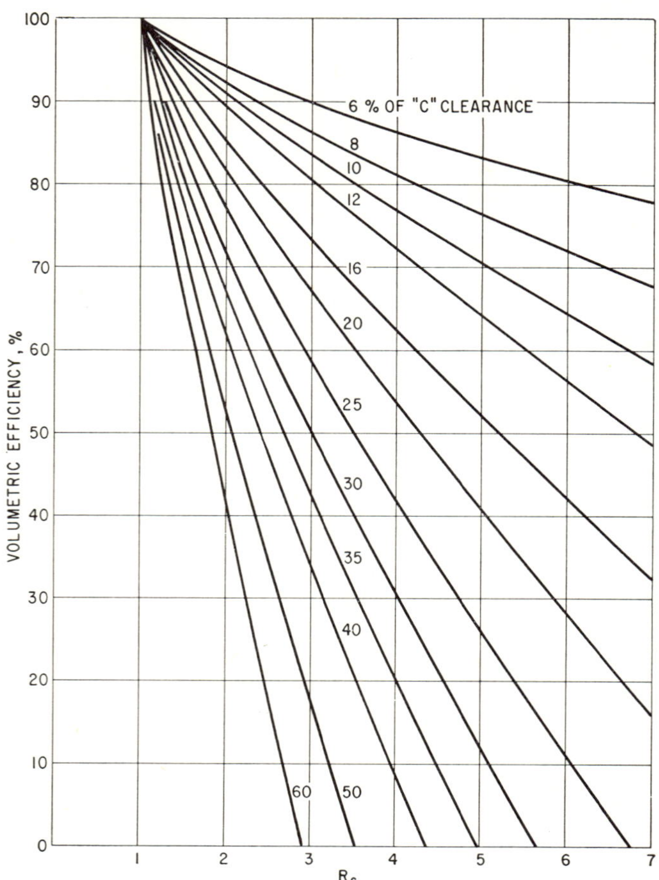

Figure 2.7. Volumetric efficiency vs. R_c for dry natural gas, $k = 1.28$, $n = 1.25$, $\eta = 1.1$, $\Lambda = 1.1$, $E_v = 100 + C - C\Lambda R_c^{1/n}$.

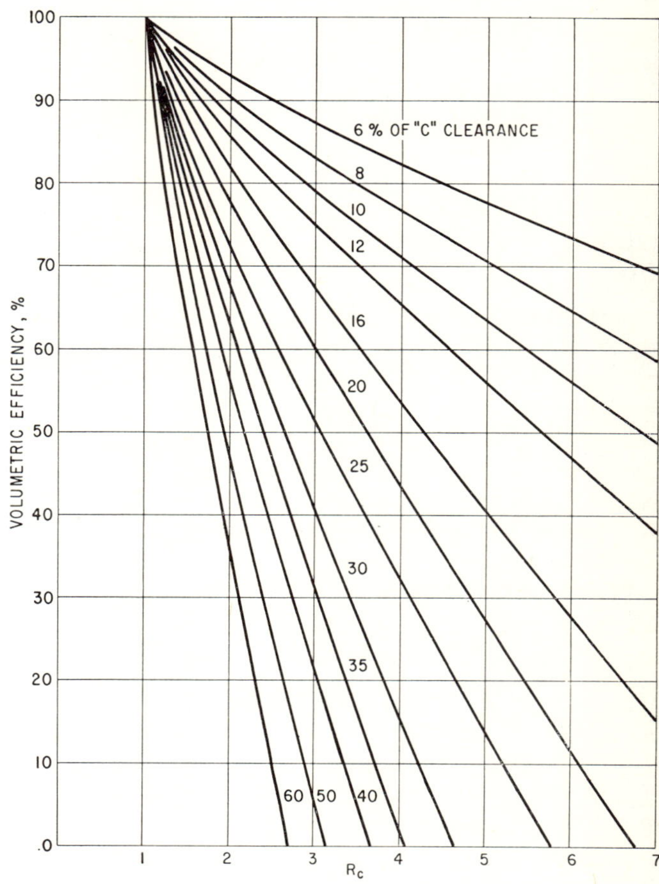

Figure 2.8. Volumetric efficiency vs. R_c for LPG and heavy hydrocarbon gases, Sp. Gr. = 1.55, $k = 1.13$, $n = 1.116$, $\eta = 1.1$, $\Lambda = 1.1$.

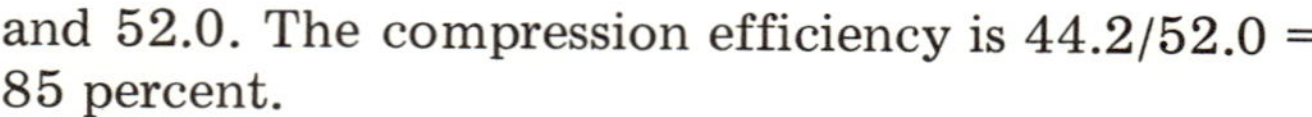

and 52.0. The compression efficiency is 44.2/52.0 = 85 percent.

These hp/MMcfd factors multiplied by 0.547 give the hp/molal ppm. The intrinsic gas hp for compressing 22.6 molal ppm at 100°F is 52(0.547)22.6(560/520) = 690 hp, which checks the more elaborate calculation.

Enthalpy charts can be applied with equal ease. Presume an isobutane refrigeration system takes suction at 24 psi and 40°F. The discharge is 84 psia. The enthalpy is raised from 114 to 137 or 23 Btu per pound. The line R_c is 3.5. The piston speed is 600 fpm and piston/valve a ratio is 10. Figure 2.2 shows B to be 1.243 at 60°F. The intrinsic BR_c is 4.35.

The compression efficiency is

$$(3.5^{\sigma} - 1)/(4.35^{\sigma} - 1) = 0.109/0.129 = 84.5 \text{ percent,}$$

where $k = 1.09$ and $\sigma = 0.0825$.

If exponential calculations are troublesome, the same efficiency can be taken from Chart 12 adiabatic hp values. The power factor is 32.0 for $3.5R_c$ and 38.0 for $4.35R_c$. The compression efficiency is 32/38 = 84.2 percent. The power required to compress 1,000 ppm is 1,000(23)/42.5(0.84) = 643 gas hp. The frictional hp is $(615)^{0.5} = 25$ and the bhp is 668.

Considerable emphasis has been directed to the reader in favor of adiabatic mode of preparing application data. Table 2-1 illustrates what the actual differences are between using the three different modes. For this example, the piston is operated at 900 fpm, and the piston/valve ratio is 12 for all three cases. The compressor handles 100 lb/min of 17.7 mole weight, 100°F natural gas having a k value of 1.275. The pressure is raised from 24 to 72 psia. The adiabatic compression efficiency is 84 percent by the valve loss method previously described. This value is confirmed by Chart 3 in the appendix. The isothermal horsepower is determined from

$$L_{iso} = 144 P_1 Q_1 \, ln R_c / 33{,}000 \tag{2.15}$$

where Q_1 is the acfm of gas handled at P_1 and lnR_c is the Napieran logarithm. The only recourse in establishing the power requirement for an isothermal design is from the empirical records. The service may require a butane to be used in a cylinder which has been completely satisfactory in natural gas service with a 233 hp driver. The volume occupied by 100 pounds of butane is only 30 percent of the volume of natural gas occupied under identical conditions. The isothermal horsepower should be proportionately reduced to 500 hp. The 74.5 percent efficiency was derived from the established gas hp, 162.5/218.

The dynamic efficiency for the polytropic mode can be derived from Chart 21. It shows that the ratio of the polytropic to the adiabatic efficiency is 1.024. This makes the polytropic efficiency of 86 percent equivalent to the adiabatic efficiency of 84 percent and the two gas horsepower values check. In sharp contrast to the reduced power requirement for the butane example, the other two modes require 3.54 times more power by the methods of this text. The NGPSA method requires 2.34 times more power. Every change in a polytropic compressor performance necessitates a new equivalent efficiency and exponent.

Table 2.1
Contrasting Methods of Evaluating Gas Compression Power

Type Mode	Head, ft-lb/lb	Horsepower	Gas hp	Dynamic Eff.
Isothermal	53,600	162.5	218.0	74.5
Adiabatic	60,500	183.5	218.0	84.0
Polytropic	62,000	188.0	218.0	86.0

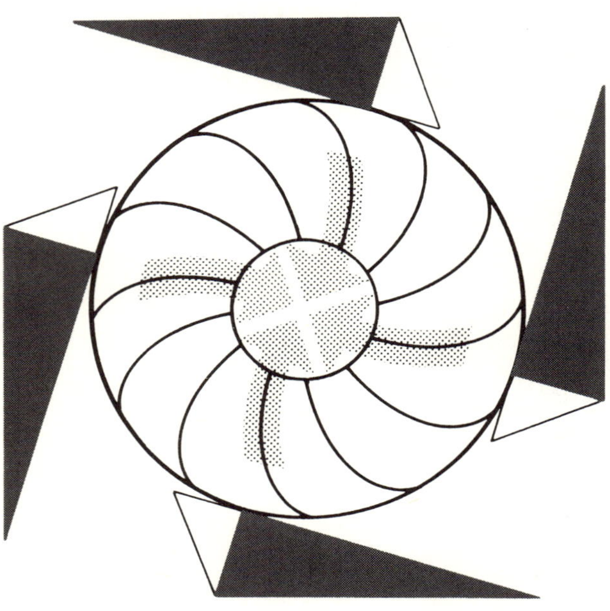

3 Pulse Dampers

One of the earliest sales features that the first commercial pulse damper manufacturer endeavored to establish was that the damper offered a tangible power saving. There have been a few instances where power savings were evident, but there have been a greater number of tests that showed no advantage or some evidence of loss. It is still commonplace to hear glib boasts that a certain pulse damper will save 2 or 3 percent, or even more, of the compression power. The commentary on this subject is that the industry does not have a comprehensive method for evaluating such an economy if it were a reality.

The purpose of this chapter is to examine, discuss and propose certain methods of evaluating the merits and need for pulse dampers and surge bottles, the significance of the compressor valve area and lift and other such elusive intangibles. To avoid a greater confusion, four terms that are pertinent will be defined.

1. *Nominal Piping.* The size of pipe connected to a compressor cylinder is between 40 and 60 percent of the piston diameter. The nominal pipe connections should contain the cylinder displacement within 8 to 10 diameters of length. This is also the distance required to resume normal flow after encountering an obstruction. This arrangement does not include any extraneous reservoir capacity other than the *nominal* piping. The pipe should be sized for a velocity of about 0.035 Mach. Typical inter-facility or interstage piping consumes 0.2 percent of the system pressure or about 1.2 velads.

2. *Surge Bottle.* This is a piping enlargement adjacent to the compressor which provides reserve capacity to receive and supply the cylinder displacement without significant influence on the system pressure. The number of *single acting* (*SA*) cylinder displacements required to fill the surge bottle is referred to as *attenuation quotient* (*AQ*). The pressure drop including a simple low velocity moisture extraction device should not exceed another 1.2 velads and 0.2 percent of the system pressure.

3. *Attenuation.* This means to dilute or weaken. The cylinder charge or exhaust pulse enters and leaves the piping system in a wave form pattern. The full height (or amplitude) of the wave pattern is measured in terms of psi (see Figure 3.1). The degree of amplitude decay is termed the attenuation. The magnitude of the pressure-volumetric decay is referred to as the *attenuation factor* (*AF*), which is $1/(1 + AQ)$. Nominal piping should possess an *AQ* of one or more, and the *AF* should be 0.5 or less.

4. *Pulse Damper.* We shall refer to and respect this category of equipment as a proprietary item. In

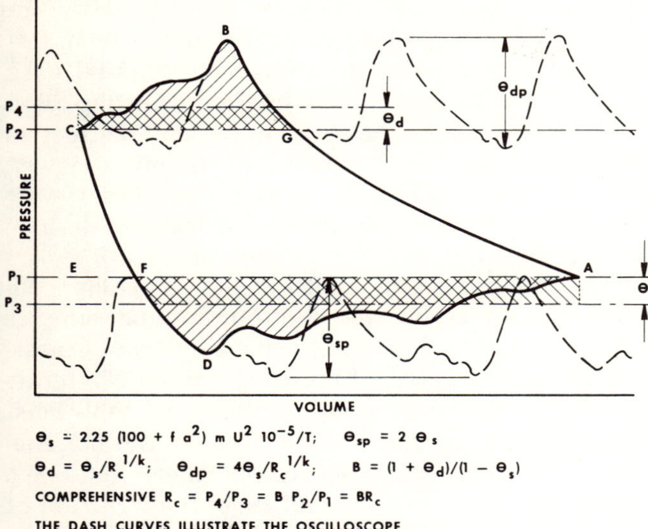

Figure 3.1(above). Compressor indicator P-V diagram illustrating valve loss effect.

Figure 3.2 (below). Attenuation effect on power economy.

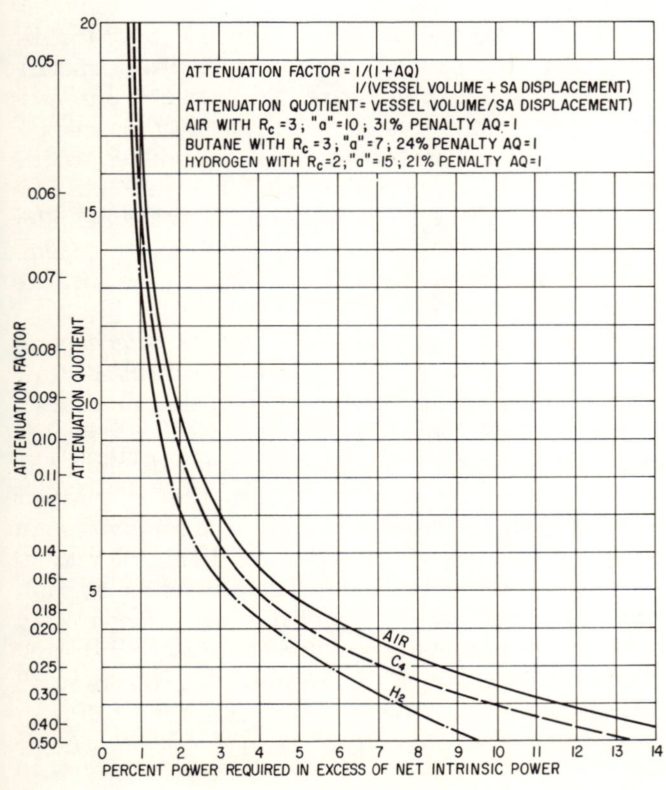

addition to the attenuation volume requirement, a pulse damper contains various choke tubes, sonic filters, etc., designed to reduce excessive pulse amplitudes and disperse harmful wave patterns. Most proprietary designs require two or three compartments; i.e., internal circumferential partitions. Each of these compartments requires the same attenuation volume as a *surge bottle* which makes a *PD* volume two or three times *SB* volume. The overall pressure drop of a pulse damper is between one-half and one percent of the system absolute pressure. The overall resistance of the damper is between one and two velads, based on the maximum velocity of the filter. The latter may vary from 0.06 to 0.16 Mach.

The following example illustrates the pneumatic effect of the *AQ* on the compression power requirement. A surge bottle volume that is nine times the *SA* displacement has an *AQ* of 9 and *AF* is 0.1; i.e., 1/(1 + 9). Presume that the amplitude of the wave front entering the surge bottle is 100 psi, it should leave at 10 psi, insofar as the mass flow is concerned. Figure 2.2 illustrates the effect of *AQ* and *AF* on the power requirement for three different typical gases. Where *AQ* is unity, the power burden can be in the order of 13 percent. When *AQ* is 6, the burden is reduced to 3 percent and when *AQ* is 17, the burden is 1 percent or less. It is also significant to note that the piston/valve lift area ratio *a* is one of the most influential variables that governs the form of the curves on Figure 3.2.

Comprehensive Compression Power

The performance depicted in Figure 3.2 was developed by the following example.

Presume an air compressor, 23 x 14 x 300 rpm, is operating from 14.7 psia and 100° F to 44.2 psia. $R_c = 3.0$, $m = 29$, $k = 1.40$, $\sigma = 0.286$, $a = 10$, $K = 4$, $U = 11.7$ and the *SA* displacement is 5810 in.³ or 1,000 cfm. The weight flow is 107 ppm where the v_s is 14.1 cf/lb and E_v is 75.5 percent. A disk type valve has a resistance of 4.44 velads. The suction valve loss for this type valve is

$$\theta_s = a^2 U^2 m 10^{-4}/T \qquad (3.1)$$

$$\theta_s = 100(136)0.0029/560 = 0.0705$$

$$U = \text{stroke (rpm)}/360 \text{ fps} \qquad (3.2)$$

$$U = 14(300)/360 - 11.65;\ U^2 = 136$$

$$\theta = (k-1)/k = 0.40/1.40 = 0.286 \qquad (3.3)$$

$$\theta_d = \theta_s/R_c^{\sigma} \qquad (3.4)$$

$$\theta_d = 0.0705/3.0^{0.286} = 0.0705/1.37 = 0.0515.$$

The suction valve loss is 14.7 (0.0705) = 1.04 psi. The discharge valve loss is 3(14.7)0.0515 = 2.28 psi.

The internal R_c experienced in the cylinder is a function of BR_c, where R_c is the visual compression ratio and

$$B = (1 + \theta_d)/(1 - \theta_s) \qquad (3.5)$$
$$B = 1.0515/0.9295 = 1.131.$$

The dynamic compression efficiency is

$$\eta_{dy} = (R_c^{\sigma} - 1.0)/(BR_c)^{\sigma} - 1)$$
$$\eta_{dy} = (1.37 - 1.0)/(3.96^{0.286} - 1) = 0.37/0.417 = 88.7 \text{ percent.} \qquad (3.6)$$

The adiabatic head of compression is

$$L_{ad} \cdot (R_c^{\sigma} - 1.0)1545\ TZ_a/m\sigma \qquad (3.7)$$
$$L_{ad} = (1.37 - 1.0)1545(560)1.0/29(0.286)$$
$$L_{ad} = 0.37(104{,}000) = 38{,}500 \text{ ft-lb/lb.}$$

The adiabatic or minimal horsepower requirement without any exterior piping attenuation is (107)38,500/33,000 = 125 *AD* hp. The gas hp is 125/0.88 = 142. The brake hp is 142 + $(142)^{0.5}$ = 154 bhp. The suction line pulse θ_{sp} is $2\theta_s$ = 2(0.0705) = 0.141 or (0.141)14.7 = 2.08 psi. The discharge line pulse θ_{dp} is $4\theta_s/R_c{}^{1/k}$ = $4(0.0705)/3.0^{0.715}$ = 0.129 or 44.2(0.129) = 5.70 psi. These two pressures represent the intensity of the sound waves that permeate the piping system.

The *AF* value is shown by an application to the nominal 10-inch piping for the 23-inch cylinder. The effective attenuation in nine diameters of pipe length is (78.5 x 9 x 10) = 7,065 cubic inches. The *AQ* is $7{,}065/(23^2)0.785(14)$ = 1.215. The *AF* is 1/(1 + 1.215) = 0.452. This immediate volume adjacent to the cylinder dilutes the θ_{sp} and θ_{dp} pulse factors by the magnitude of the attenuation factor, 0.452. The nominal exterior piping pulse θ_{spa} = 0.452.(0.141) = 0.064, and θ_{dpa} = 0.452(0.129) = 0.058. The intrinsic factor as affected by the exterior piping is 1.058/0.0936 = 1.130 and BR_c = 3.39, and when raised to σ power, is 1.418. The compression efficiency for the 10-inch piping is 0.37/(1.418 — 1.0) = 88.5 percent. The total bhp including the inherent valve losses and the 10-inch piping is 154/0.885 = 175 bhp.

A surge bottle having an attenuation quotient nine times the cylinder displacement produces an attenuation factor of 0.1. This reduces the suction exterior peak pulse to 0.1(0.141) or 0.014 θ_{sp}. The discharge exterior peak pulse θ_{dp} is reduced to 0.1(0.129) or 0.013 by the pulse damper. The intrinsic factor is 1.013/0.986 = 1.0274, and BR_c = 3.082, and raised by the exponent 0.286, is 1.38. The compression efficiency as influenced by the surge bottle is 0.37/(1.38) — 1.0) = 97.5 percent. The compressor load when connected to the 0.1 *AF* surge bottles should be 154/0.975 = 158 bhp. This study indicates that the surge bottle could effect an economy of about 10 percent of the input power. Actually, the contributing peak pulse pressures are only active less than 25 percent of the cycle which would reduce the economy accordingly.

The point "*A*" in Figure 3.1 is generally conceded to represent the suction line pressure as accurately as any industrial instrument could measure the line pressure. Likewise, point "*C*" represented the discharge pressure. This concept must further include an *AQ* of 6 or more in the nominal piping or its appendage. Otherwise, there could be a perceptible altering of P_1 and P_2 in the direction of P_3 and P_4.

Mass Flow vs Sonic Flow

The foregoing data only relates to the pneumatic behavior. More frequently than not a pressure wave sensing instrument would show a significant wave pattern being transmitted into the piping system. *PD* manufacturers claim as one of their proprietory skills the ability to equate the pressure drop and the pulse intensity with the *PD* size. A recent bid analysis showed *AQ* values varying from 17 to 42 for the same identical service and pressure drop specifications. This only emphasizes the popular consensus that pulse damping is more an art than a science. The application of the attenuation bottles to the compressor cylinders is illustrated in Figure 3.5. The proximity of the gas reservoir bottle to the cylinder minimizes the pressure surge in the piping to accommodate the charging and exhaust effect of the reciprocating piston.

We would prefer to think of the gas entering and leaving the cylinder in the classical turbulent flow pattern, much as the blustering wind that precedes a Texas "Blue Norther." Accompanying such storms are a pyrotechnic thunder and lightning display. The energy released by the opening of the compressor valves projects pressure waves into the piping system much the same as the thunder and lightning permeates the Texas environment. The blustering 50 mph wind and the flash of lightning are as distinctly separate phenomena as are the compressor gas flow at nominal velocity of 50 fps and the pressure wave front traveling at sonic velocity at rates of 900 to 4,500 fps, depending largely upon the gas molecular weight.

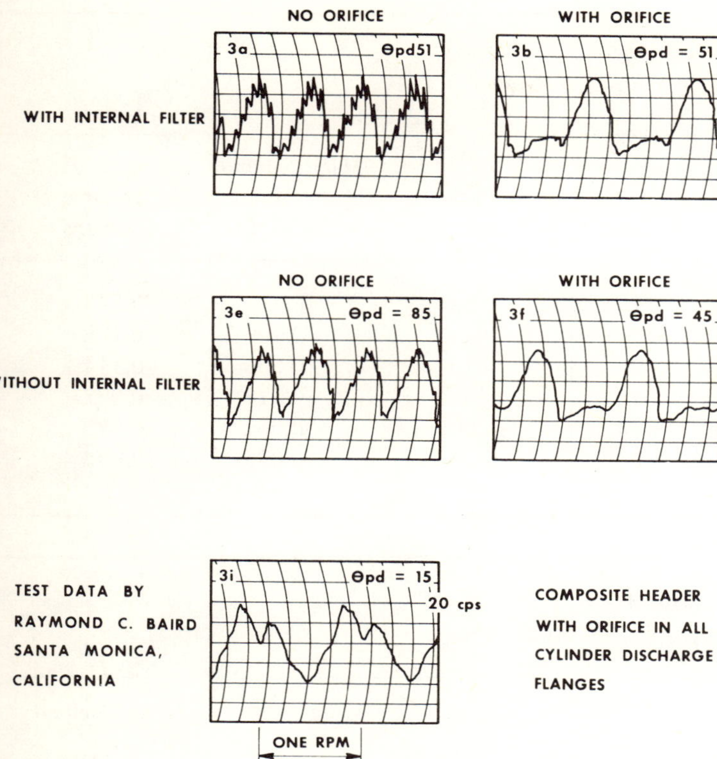

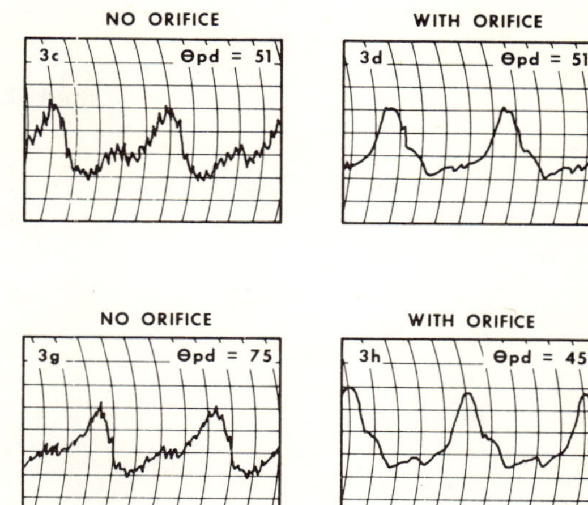

Figure 3.3. Oscillogram of 2,000 psi hydrogen compressor discharge with both ends loaded are shown to left of line. Oscillogram of 2,000 psi hydrogen compressor discharge with only crank end operating is at right. Composite header with orifice is in all cylinder discharge flanges.

It is not possible at this time to provide a similar mathematical solution to pressure wave behavior. But we can report a recent instance where a troublesome standing wave in a 2,000 psi compressor discharge was responsible for an average valve life of three weeks. An orifice 3.5 percent of the piston area was installed between the surge bottle and the compressor cylinder. It completely dampened the troublesome ninth harmonic standing wave, reduced the primary pulse amplitude from 92 to 47 psi and the header residual pulse was reduced to 15 psi (see Figure 3.3). The valve breakage ceased. The orifice velocity was 606 fps or 0.142 Mach, and the resistance was 18.2 psi or 0.92 percent of the absolute pressure. Figure 3.3 shows the oscillograms taken.

While the orifice completely diffused the minor *SA* and *DA* pulse, it had no noticeable effect on the major pulse of the *SA* and *DA* cylinders equipped with internal filters. The internal filters reduced the major pulse intensity by one-third. The deresonator orifice reduced the major pulse intensity by 44 percent in the *PD* not equipped with internal filters. The orifice decay continued to consume the major pulse intensity until at a distance of some 250 feet, less than 20 percent existed in the common discharge line.

Mach Number Velocities

Thus far we have proposed that there are three categories of compressor cylinder appendages concerned with volumetric attenuation and that the pressure wave pulse is a separate phenomena. Now we shall explore further into the gas flow behavior. The effect of various pressure loss heads on the flowing velocities are shown in Table 3.1 for 2.7 mole weight hydrogen mixture, air and butane, all at 240° F and having a common specific volume of 1.39 cf/lb.

The velocity of the 2,000 psia hydrogen experiencing a 1.1 percent pressure drop is determined by converting the 22 psi into the hypothetical gas head; $144v_s = 144(1.39) = 190$ ft per psi, for 22 psi the head is 4,180 ft. The velocity is $(2gL)^{0.5} = (64.4 \times 4{,}180)^{0.5} = 533$. Where the sonic velocity of hydrogen is $224(1.4 \times 700/2.7)^{0.5} = 4{,}260$ fps, the Mach number is $533/4{,}260 = 0.125$.

The effect of various gases on the size of deresonator orifice for an 8-inch cylinder operating at 800 fpm with 0.9 percent pressure drop is shown in Table 3.2.

In Table 3.2, hydrogen is at 2,000 psi and 215° F. The rest of the gases are at 300 psia and

Table 3.1
Pressure Drop Velocity Study

Loss % of System Pressure	Velocity of Hydrogen, 2,000 psia	Velocity of Air, 185 psia	Velocity of Butane, 93 psia	Mach Number Hydrogen, 4,260 fps	Mach Number Air, 1,390 fps	Mach Number Butane, 805 fps
1.1	533	162	115	0.125	0.117	0.143
0.9	482	147	104	0.113	0.106	0.128
0.7	425	129	92	0.100	0.093	0.114
0.5	360	109	78	0.084	0.078	0.097
0.3	278	85	60	0.065	0.061	0.075
0.1	161	49	34	0.038	0.035	0.042

Table 3.2
Sizing Orifices for Equal Resistances

Gas	Mole Weight	Orifice Dia.	Velocity	Mach Number	Nominal Velocity Factor*
Hydrogen Mix	2.7	1.50	606	0.145	1.57
Hydrocarbon	16	2.49	215	0.153	1.72
Hydrocarbon	18	2.56	202	0.148	1.72
Hydrocarbon	20	2.65	190	0.146	1.72
Air	29	2.83	166	0.148	1.57
Butane	58	3.57	105	0.150	1.96

*The nominal velocity factor is the correction between the arithmetical average piston speed and the actual effective velocity, corrected for the simple harmonic motion and volumetric efficiencies. A pressure drop of 0.9 percent of the system pressure is allowed for each case. The system pressure produces a common specific volume of 1.39 cf/lb.

60°F. This table shows the fallacy of sizing flow facilities on a constant velocity for all gases. While the practice of using an orifice plate as a deresonator is not common, neither is it novel. We propose the following equation to size a deresonator orifice:

$$d_o = 0.25 d_c \, (m/kYT)^{0.25} \, (U/13.3)^{0.5}. \tag{3.8}$$

The theme of this chapter presumes that the pressure waves are the product of the energy released with the valve action. Some time ago a method of evaluating valve losses was presented. It was observed that the amplitude of the suction wave front θ_{sp} is approximately twice the suction valve loss θ_s (see Figure 3.1). The pulse intensity equations were developed from the oscillograph study which is typified in Figure 3.3.

Pipe End Thrust

Piping tie-downs, anchor intervals and straight runs of pipe should avoid dimensions of one-quarter gas wave lengths. Pressure waves can induce resonate vibration and end thrusts in the pipe line and cause material damage. Alternate ends of pipe runs that measure close to one-quarter wave lengths or even

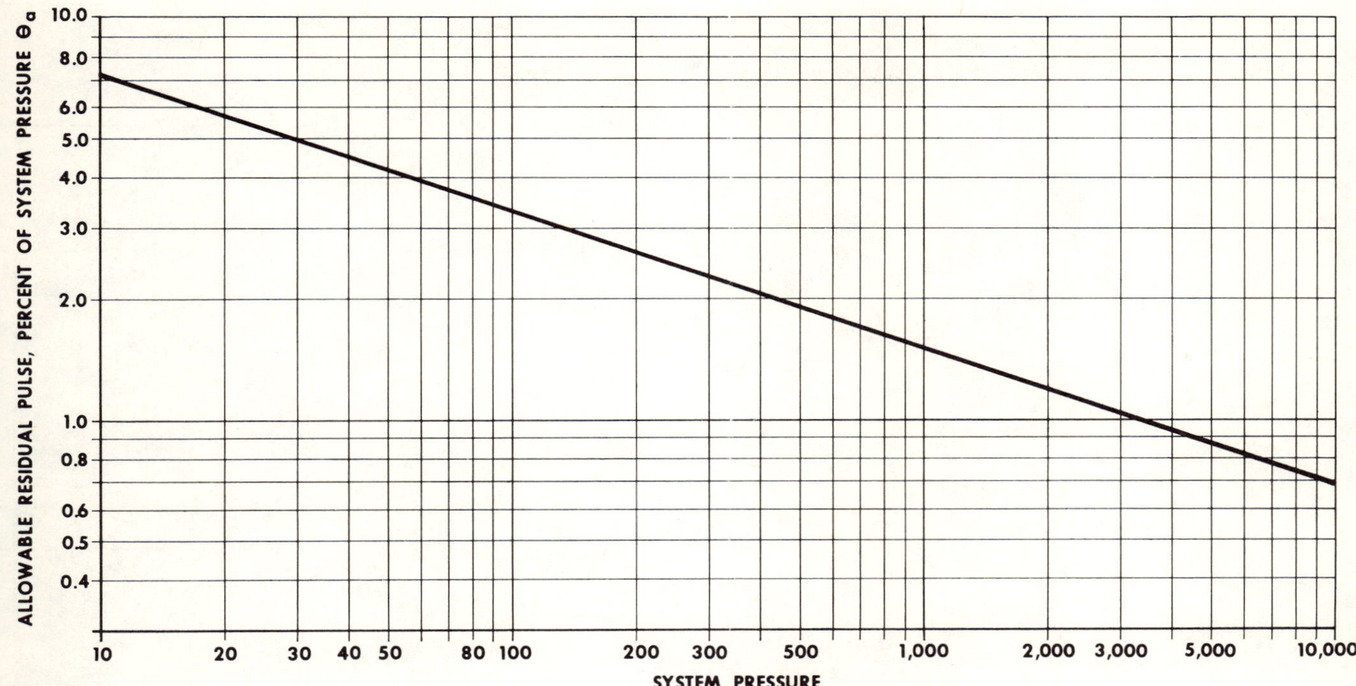

Figure 3.4. Allowable residual pulsation amplitudes θ_a *for compressor installations after Shell Chemical Co. Curve equation % pulse* $\theta_{ap} = 1.5/(0.001\ P)^{0.34}$.

fractions thereof can register the high and the low amplitudes of the pressure wave thereby setting up a substantial vibrating force.

For example, an 8-inch pipe has an internal area of 50 square inches; a wave front amplitude of 100 psi could impose a 2.5 ton end thrust or vibrating force. The fear of damage that might result from such large wave fronts has brought about the commercial development of the pulse damper. The Shell Development Company has given the industry a chart (Figure 3.4) of acceptable residual pulse levels which they have found to be both acoustically attainable and physically possible to anchor. The values given in Figure 3.4 are not absolute but serve as a very valuable reference guide. Pulse levels above the line are likely to give more vibration problems than those points under the line. A simplified expression for the residual pulse is

$$\theta_a = 1.5/(0.001P)^{0.34}. \tag{3.9}$$

Wave Length

The Mach number of sound is the common denominator of a gas flow in reference to the respective sonic velocity. The sonic velocity of gas is

$$\tau = 224\ (kT/m)^{0.5} \text{ in fps} \tag{3.10}$$

The sonic velocity divided by the frequency (cps) gives the wave length. The compressor speed is usually about 300 rpm, which gives the primary frequency of five cps for single action cylinders. When both ends of the cylinder are operating, the secondary frequency is 10 cps, and the wave length is reduced by one-half. The sonic velocities of natural gas is about 1,120 fps, and air is 1,230 fps. Velocities of this magnitude would have a Mach number of one. The closeness of these sonic velocities has created a laxity in presuming that common velocities produce equal resistances when corrected for density. The data given in Tables 3.1 and 3.2 refute this fallacy.

Compressor Valves

Now that the range of gas mole weights being compressed has been extended from 16-30 to 3-200, we must change our velocity reference index to the Mach number. The compressor manufacturers have had difficulty in the past providing sufficient valve area for the common gases. The event of low mole weight hydrogen mixtures has reversed the situation

Figure 3.5. Typical pulse traps application. (Courtesy of Pulsation Controls Corp.)

and now is exposing the effect of too much valve area. This has caused a rash of excessive valve maintenance in hydrogenation service. A valve with nominal lift of 0.100 inch will shatter the disk and cut the seat in a course of several days. The light density and nonviscous character of hydrogen provides a relatively small amount of cushion to dampen the flutter and sonic pulsations.

A valve innovation has been successfully applied to 1,500 psi hydrogen mixtures. It is essentially a damper channel valve, with the pneumatic dampening in the guard recess, rather than lifting against heavy coil or leaf springs. It is illustrated in Figure 1.9.

Valve lifts as small as 0.015 inch were demonstrated to be effective. The resistance was somewhat abnormal, but became consistent with our Equation 3.1 at 0.025-inch lift. Another remedial correction was to provide a laminated phenolic seat facing and valve lifts of about 0.045 inch. Various stainless steels and laminated plastic of 0.188 and 0.25-inch thickness were used for disks. Light resilient materials appeared to have the most merit. Surface should be 15 RMS or less to minimize leakage.

Compressor Leakage

Any compressor service where R_c exceeds 1.9, leakage can be expected to flow at the rate of 0.5 Mach (1). This is a velocity of 2,130 fps for a 240° F, 2.7 mole weight hydrogen gas mixture. An 8 x 16-inch, 300 rpm compressor cylinder will have a valve area of 2.0 square inches per quadrant, with a disk lift of 0.040 inch and an "a" factor of 25 (8-inch piston area is 50.26/25 = 2 square inches). The peripheral edge of such valves will total 2.0/0.040 or 50 inches per quadrant or 100 inches per cylinder end. A 25 RMS surface finish on the valves creates an aperture of 100(0.000025) = 0.0025 square inch. This passage will permit a flow of 2,130(0.0025)60/144 = 2.22 cfm. The cylinder displacement is 4.25(8)8 = (272)0.82E_v = 223, making a valve leakage of one percent (2).

Figure 3.6. Illustrating the application of pulse dampers and completely integrated, two-stage, skid mounted gas engine driven gas compressor. (Courtesy of White Motor Co.)

Admittedly, the actual valve exposure to such a generalized flow condition is subject to modifications. On the same premise, the RMS measurements are not absolute but subject to equal or greater correction. The point is that a minute but perceptible flow does exist under the valve elements. By the same logic, another one percent leakage can flow between the cylinder walls and the piston rings where the fit tolerance is 100 RMS. This combined leakage can raise the suction temperature 15°F, which represents five percent more power of compression, 2 percent for recirculation and three percent for added thermal burden.

The cylinder wall and the piston rod should be given the same high degree of finish as the valve seats. The piston and rings should be equipped for nonlube service and then lubricated with a minimum quantity. Bolted sectional type pistons, using a one-piece carbon filled *TFE* rider ring plus step-cut, snap type carbon or glass-filled *TFE* piston rings, are being used with success. Piston rod packing also made of carbon or glass-filled *TFE* with bronze back-up rings are satisfactory for hydrogen service in excess of 1,000 psig.

Damper Evaluation

The best argument in favor of purchasing a pulse damper system is that it places the responsibility on the *PD* manufacturer. The possibility of a menacing header vibration, a bouncing control system or a rapid sequence of compressor valve failures are a few of the events that a construction contractor and his client are most anxious to avoid, particularly at the time of start-up. They are prone to pay the modest price of approximately 0.1 percent insurance for this peace of mind. Figure 3.6 illustrates the application of pulse dampers.

On the other hand, there are thousands of compressors operating quite satisfactorily without

Table 3.3
Qualifications for Compressor Piping Appendages

Category of Appendage	Nominal Piping	Surge Bottle	Pulse Damper
Discharge pressure, psig	<900	$<1{,}300$	$>1{,}300$
Power classification, bhp	<150	<500	>500
Acceptable, pulse, θ_a	$<1\theta_a$	$<2\theta_a$	$>2\theta_a$
Average attenuation factor	0.4	0.1	0.03
Approximate cost index	1	20	60

the benefit of a *PD* or even a surge bottle. The cost, though nominal with reference to the total plant cost, can be 10 to 15 percent of the compressor cost which occasionally is the largest single expenditure in the plant. There is a need for some guide posts or criteria to classify the performance conditions and the requirements for each of the three categories of cylinder piping appendages (see Table 3.3).

The *nominal piping* category includes the bulk of the field gas compressors built prior to 1950. Few units were operated above 1,000 psi, and they were usually equipped with some form of piping enlargement. Most of these units were operated at pressures below 500 psi. The schedule 40 piping used would be operating below 10 percent of the minimum yield strength and have great resistance to fatigue failure. Having established a method of evaluating the pulse intensity, the magnitude of the end thrust can be determined. The acceptable residual pulse θ_a is taken from the Shell curve, Figure 3.4.

An operation which qualifies for all three of these checks should not experience any serious piping problems. The *surge bottle* is required for cylinders carrying over 150 hp. The cylinder pulse should not exceed $2\theta_a$. When these three qualifications are exceeded, the compressor is a candidate for a *pulse damper*. These criteria, however, should not be judged as absolute and infallible.

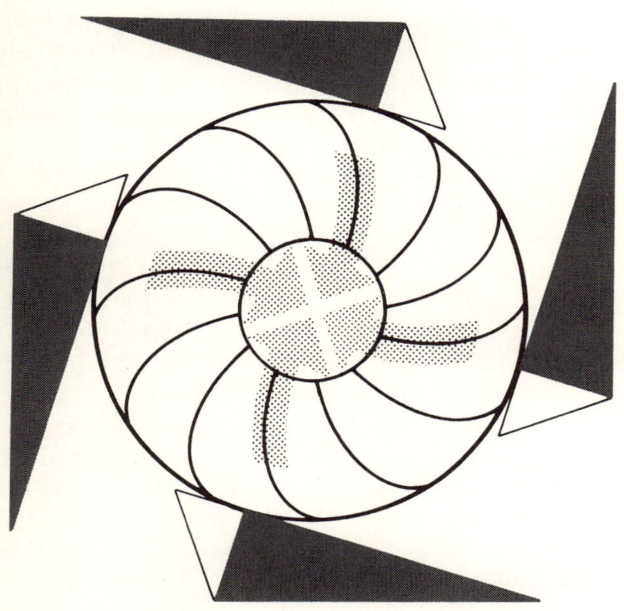

4 Fans and Blowers

Boilers, fired heaters, materials handling, air purifications and the ventilation of large motors are a few of the services which have created endless needs for low pressure air fans and compressors. The standards and application procedures for doing this are well regulated by the Air Moving and Conditioning Association Inc. (AMCA). The performance tables of all sustaining members are based on a standard air density of 0.075 pounds per cubic foot at 70°F dry bulb and 29.92 inches Hg, 760 mm Hg or 14.696 psia, commonly stated as 14.7. This standard is applied throughout this chapter unless other conditions are stipulated. This is slightly lighter than the standard density of 0.0763 pounds per standard cubic feet (lb/scf) at 60°F and 760 Torr as recognized by the American Gas Association and the petrochemical industry. The chemical industry prefers to measure volume flow in terms of moles per hour. In this reference, every 379.5 scf contains a weight of gas equivalent to its respective molecular weight. Each 380 scf of air weighs 29 pounds; each 380 scf of hydrogen weighs 2 pounds, etc.

There are two other standard unit volumes that are used in international operations. The European normal cubic meter weighs 2.846 pounds at 760 Torr ($1.0333 Kg/cm^2$) and 0°C. All European technical calculations are based on $1 Kg/cm^2$, 736 Torr or 14.22 psia. The British standard cubic meter weighs 2.700 pounds at 60°F and 14.73 psia. The AMCA also uses 0.078 lb/scf or 30.2 as the molecular weight for the standard density of flue gas.

It is argued that the error in power calculation is negligible between handling the gas as an incompressible fluid and a compressible fluid. Actually, the former system requires 5 percent more power, which is not negligible at higher pressures and with large volume flows.

An arbitrary line between fans and blowers is set at 12.25 inches of water, which is the maximum pressure sponsored by the AMCA Class III fan. This is a compression of 3 percent; the density change is 2.14 percent with discharge temperature of 74°F. The machinery used for greater pressures than 1.03 R_C are referred to as centrifugal blowers, sliding-vane blowers, axial screw blowers, etc. The power requirements for these units are usually less than 100 hp. The larger powered units operating above 1.25 R_C generally require a piston compressor or a centrifugal or axial compressor.

Static Pressure

The energy density of the air system is measured by the *static pressure* (*SP*) in inches of water, as

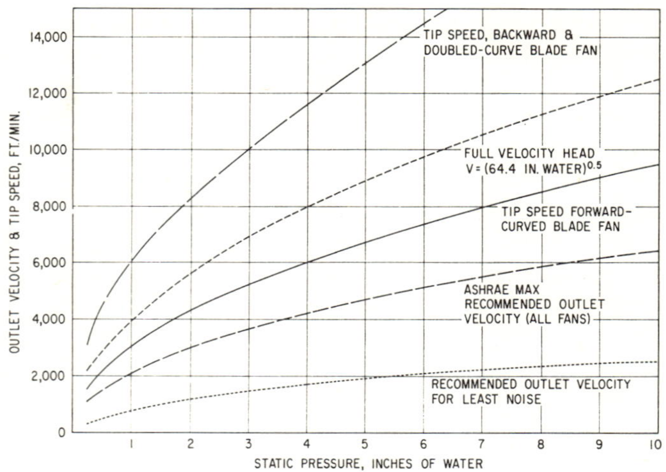

Figure 4.1. Fan tip speeds and outlet velocities for various static pressures.

indicated by a manometer on the discharge duct. The velocity head for fans operating above 6 inches *SP* is about 6 percent of the total head (*TP*). In lower pressure systems, the discharge velocity represents 7 to 20 percent of *TP*. Figure 4.1 gives the fan discharge velocities and tip speeds that are acceptable to the American Society of Heating, Refrigerating and Air Conditioning Engineers (ASHRAE). The lower dotted line gives the ASHRAE recommended outlet velocities for quietness considerations. The second line from the top gives the velocity head in inches of water for the fan discharge velocity.

The velocity in air ducts is converted to equivalent inches of water pressure by the equation

$$VP = (\text{Duct Vel}/4160)^2, \text{ inches } H_2O \quad (4.1)$$

The constant 4160 is the gas head per inch of water pressure. It is derived by dividing 144 square inches per square foot by the AMCA standard air density of 0.075 cf/lb and 27.7 inches/psi. The frictional flow resistance can be estimated in inches of water by $0.02(L/D)VP$, where L and D are in feet of circular pipe. Equivalent hydraulic diameters for rectangular duct sections may be applied where the aspect ratio is less than 8.

For example: A double inlet fan with air-foil vanes on a 33-inch rotor has a capacity of 36,000 scfm at 1,730 rpm and 11.5 inches of water *SP*. The discharge velocity is 3,200 fpm and the power load is 78 bhp.

The velad is $VP = (3{,}200/4{,}160)^2 = 0.6$. The *TP* is 11.5 + 0.6 = 12.1 inches. The standard air hp is

$$\text{Air hp} = \text{cfm } (TP)/6{,}356 = 36{,}000(12.1)/6{,}356 = 68.5 \text{ hp.}$$

The fan efficiency is 68.5/78 = 88 percent.

Flue Gas Example

The air required for stoichiometric combustion of any hydrocarbon fuel is one scf per 100 Btu of the high heating value. One scf of natural gas requires 10 scf of air; one scf of LPG requires 27 scf of air; 190 scf of air is required for complete combustion of one pound of Number 5 fuel oil having a gross calorific value of 19,000 Btu per pound or 150,000 Btu per gallon.

For example, presume that a boiler has an evaporation rate of 110,00 pounds per hour at 900 psig and 800°F, using 200°F feedwater. The enthalpy of the steam is 1,394 Btu/lb. Deducting the feedwater enthalpy of 168 Btu/lb gives 1,226 Btu/lb as the enthalpy absorbed or 134.9 MM Btu/hr. This output divided by the anticipated boiler efficiency of 79.5 percent gives the input requirement as 170 MM Btu/hr.

The natural gas fuel rate is (170,000 M Btu/hr)/1,000(Btu/scf) = 170 Mscf/hr. The fuel oil requirement is (170 MM Btu/hr)/19,000(Btu/lb) = 8,948 lb/hr or 1,133 gallons per hour. The air requirement will be the same for any hydrocarbon, namely 10(scf or air/scf of gas)170 Mscf(1,000/M/hr) ÷ 60(min/hr) = 28,400 scfm. This air requirement checks the fuel oil rate: 190(scf of air lb of fuel oil) × 8,948(lb/hr) ÷ 60(min/hr) = 28,380 scfm. It is advisable to provide about 10 percent excess air or supply 31,200 scfm to the boiler. The resistance of the furnace, tube baffle channels, controls, air and flue gas ducts is presumed to be 3.5 inches of water.

Selection of Fans

The various fan manufacturers catalog their fan performance data in the following manner. For an "AF-DI-40" Class I and II model fan, the performance is shown in Table 4.1.

The asterisk (*) in the table indicates the maximum efficiency selection. Adjacent values are of lower efficiency. The further removed, the lower is the efficiency. The light broken line indicates the demarcation for the Class I and II weight frames. The steel frame and sheet iron plates are relatively light for the Class I fans and limited to 3.75-inch total pressure. Class II fans have a heavier frame and wall plate which make them suitable for 6.75 inches.

Table 4.1
Class I and II Fan Performance

Volume Inlet cfm	Outlet Velocity fpm	Static Pressure inches of water: 3		3½		4	
		rpm	bhp	rpm	bhp	rpm	bhp
26,830	1,600	771 *	15.0	819	17.8	867	21.0
28,500	1,700	783	15.9	830 *	18.5	874	22.5
31,860	1,900	815	18.4	856	20.9	897 *	23.6
33,530	2,000	833	19.7	872	22.4	910	25.1
36,890	2,200	872	22.6	908	25.4	943	28.5

Figure 4.2 (above). Section of air foil for fan vane. (Courtesy of Bayley Blower Co.)

Figure 4.3 (below). Air foil welded in fan impeller. (Courtesy of Bayley Blower Co.)

Class III has still a heavier frame and plate. It is rated to 12.25 inches of water pressure. The fan selected just qualifies for a Class I frame at 3.75 inches of total pressure.

The fan description is contained within the model identification. The fan model "AF-DI-40" refers to an *air foil* (*AF*) type vane in a *double inlet* (*DI*) cage, and the nominal diameter of the rotor is 40 inches. The actual diameter in this instance is 40.25 inches. The air foil fan has a streamlined section welded between the disk and shroud. Figure 4.2 illustrates a foil section which is welded between the disk and shroud. Figure 4.3 shows the complete impeller weldment. Peak efficiencies as high as 88 percent are common for this design category. Its performance curves are similar to the backward-curved (*BC*) vane rotors shown in Figure 4.4. A "BC-DI" rotor mounted on its bearing pedestals and journals is shown in Figure 4.5.

A *Forward Curved* rotor with *Double-Inlet* (*FC-DI*) having a spoke supported peripheral rim is shown in Figure 4.6. This type of rotor has a peak efficiency of 66 percent as shown in Figure 4.7. It has a relatively low tip speed and head potential. A radial bladed fan mounted in a dynamic balancing stand is shown in Figure 4.8. Its performance curve is shown in Figure 4.9. Typical axial vane blower characteristic curves are shown in Figure 4.10. Figure 4.11 illustrates a sophisticated axial blower design detail. The heavier class fan frame is usually specified for industrial applications where weight is not a consideration and continuous dependability is paramount. In a marginal case for the above boiler installation, a Class I fan just qualifies with 3.75 inches *TP*. The total pressure on the fan is $(1{,}860/4{,}005)^2 = 0.22 + 3.5 - 3.72$ inches of water. Should the furnace resistance require 4.25 inches, the speed of a Class II fan can be increased 10 percent to 936 rpm to satisfy the new condition with confidence.

The limiting tip speed for a Class I fan is 9,500 fpm, 13,000 fpm for a Class II and 16,000 fpm for a Class III fan handling air from minus 20°F to 400°F. The temperature range can be extended to 800°F by reducing the tip speed by 30 percent and installing a *cooling wheel*. The latter is mounted be-

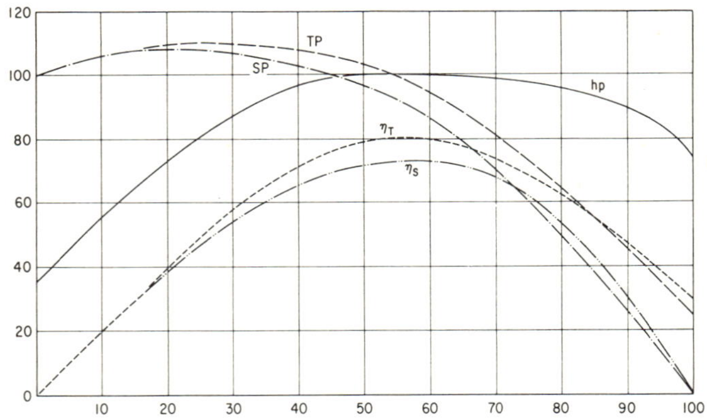

Figure 4.4. Performance curves of backward curved fan.

Figure 4.6. Forward curved impeller having spock supported periphery. (Courtesy of Bayley Blower Co.)

Figure 4.5. Backward curved vanes in double inlet welded rotor on pedestal bearings. (Courtesy of Bayley Blower Co.)

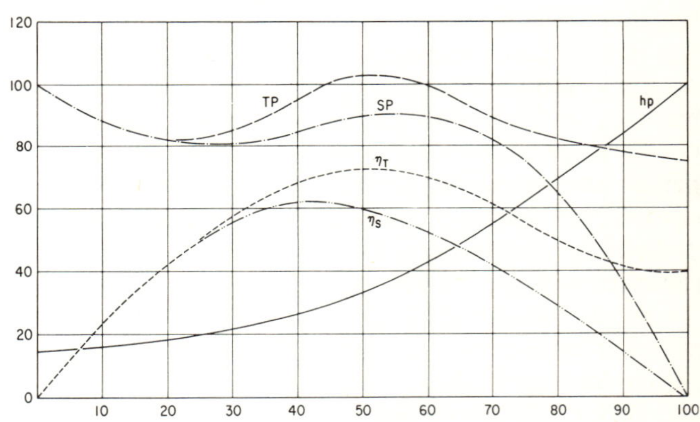

Figure 4.7. Performance curves for forward curved vane fan.

tween the inboard bearing and the fan case. It is a heat dissipating element which serves as an insulator and shields the bearings from the excessive temperature. It is possible to handle gas at 1,000°F by providing separate bearing pedestals, shaft cooling and the cooling wheel with outside air circulation.

The speed for the AF-DI-40 fan previously reviewed is extrapolated to be 851 rpm and the power to be 22.1 bhp. The tip speed is 40.25(3.14)851/12 = 8,960 fpm. The static air horsepower is 31,200(3.5)/6,356 = 17.2 hp. The fan static efficiency is 17.2/22.1 = 78 percent. The efficiency should be high as it was adjacent to the asterisk figures, which denotes peak efficiencies. It is not uncommon to make compromise selections having efficiencies in the low 60 percentages.

The effect of applying an induced draft where the flue gases are extracted from the boiler can be appreciated by presuming that the flue gas is 500°F and that the elevation is 2,700 feet. The volume flow is increased by the ratio of the absolute temperature and inversely as the absolute pressure. The local mean barometric pressure is 27.12 (see Figure 4.11). The new volume is

$$(960/530)(29.92/27.12)31{,}200 = 62{,}500 \text{ cfm}.$$

Figure 4.8. Radial vane rotor in dynamic balance stand. Rotor has no shroud suitable for handling materials. (Courtesy of Bayley Blower Co.)

The performance tables for a Model AF-DI-60 show this capacity at an outlet velocity of 1,800 fpm, 565 rpm and 43.8 hp for standard air. The resistance remains the same at 3.5 inches *SP*. The air horsepower is 62,500(3.5)/6,356 = 34.5 hp. The fan efficiency is 34.5/43.2 = 80 percent. The hot flue gas has a molecular weight of 30.2, which compensates for the 4 percent weight of fuel added to the air supply. The induced fan power must be increased by this amount or 43.2(1.04) = 45.0 hp.

Table 4.2 gives the performance characteristics of various types of fan vanes. Axial fans conform to the pattern of the axial compressor, having a low pressure coefficient. The pressure coefficient (q_{ad}) is 0.12. This factor is applied to the next equation to derive the head potential (L) of any centrifugal compressor. Having this low a coefficient means that it can only develop one-fourth of the pressure head that the air foil or *radial* fan can develop at the same tip speed. The forward curved vane can develop 10 times as much head as the axial fan can develop at the same tip speed, but it is not stable at tip speeds over 65 fps.

$$L = U^2 q_{ad}/g. \quad (4.3)$$

$$N_s = \text{rpm}\ (Q^{0.5}.)/L^{0.75} \quad (4.4)$$

$$D_s = \text{Dia.}\ (L^{0.25})/Q^{0.5}. \quad (4.5)$$

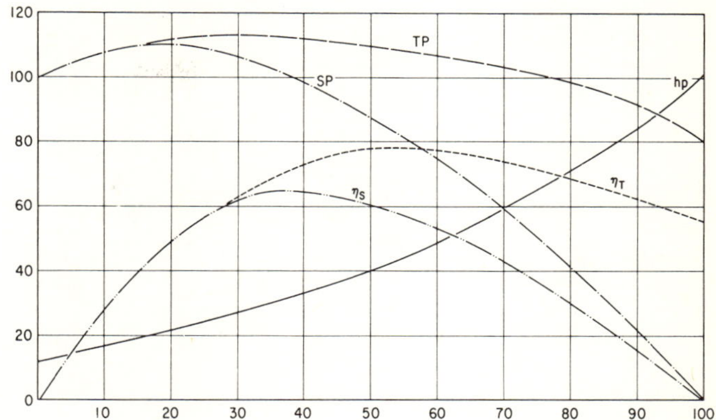

Figure 4.9. Characteristics of a straight radial bladed fan.

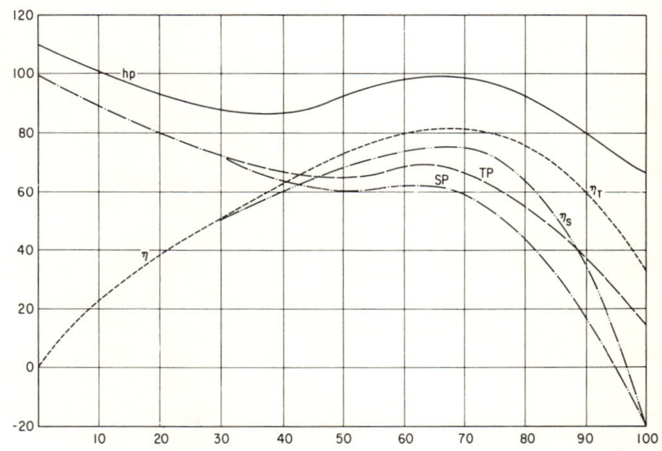

Figure 4.10. Performance curves of an axial proller fan.

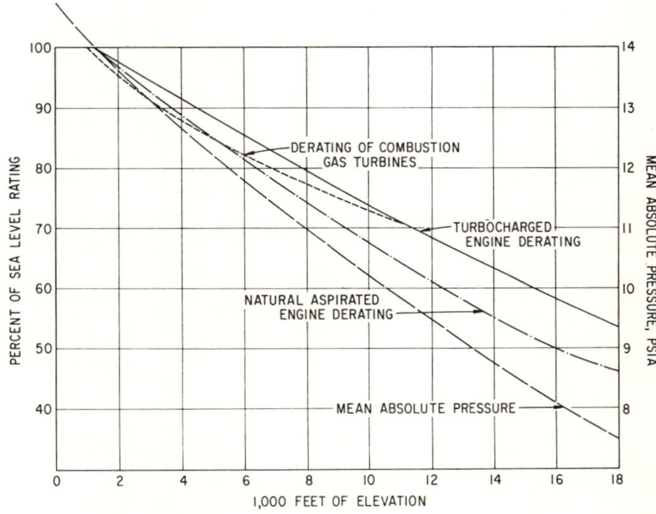

Figure 4.11. Effect of elevation on the mean absolute pressure and engine power potential.

Table 4.2
Fan Performance Characteristics

Description	Quantity 1,000 acfm		Head Inches Water	Opt. U fps	Max. q_{ad}	Diameter Inches		N_s	D_s	Peak Eff.
	Min.	Max.				Min.	Max.			
Axial Propeller	8	20	10	410	0.13	23	27	470	0.63	77
Axial Propeller	20	90	8	360	0.12	27	72	500	0.60	80
Axial Propeller	6	120	2.5	315	0.10	27	84	560	0.50	84
Radial Air Foil	6	100	22	250	0.45	18	90	190	0.85	88
Radial BC	3	35	18	260	0.63	18	90	100	1.35	78
Radial Open MH	2	27	18	275	0.55	18	66	97	1.45	56
Radial MH	2	27	18	250	0.55	18	66	86	1.53	71
Radial LS	2	27	18	250	0.55	18	66	86	1.53	66
Vane BI Flat	1	10	12	250	0.43	10	30	210	0.81	70
Vane FC	1	10	2	65	1.15	10	30	166	0.65	66

These specific speed and specific diameter equations includes a volume flow (Q) in terms of cubic feet per second. Fan manufacturers usually use cfm, and their N_S is $(60)^{0.5}$ = 7.75 times larger than the values given in Table 4.2 and in Figure 4.12. The quantity flow for an axial propellor fan is greater than that of any other category. The ultimate capacity of the axial design is often referred to as being infinite. The pressure range extends from 0.5 to 10 inches of water. This pressure measurement is converted to feet of head (L) by multiplying the inches of water by 69.3 feet. This type of fan operates at tip speeds 50 percent greater than radial inflow impellers. The efficiencies are relatively high, especially if the fan selection offers a N_S value between 300 and 600.

The air foil vane offers the highest efficiency and pressure heads. The heads shown on Table 4.2 represent the performance of commercial fans. It is possible to develop 65 inches of water by running tip speeds to 550 fps. This would require manufacturing techniques equivalent to those used in making centrifugal compressors.

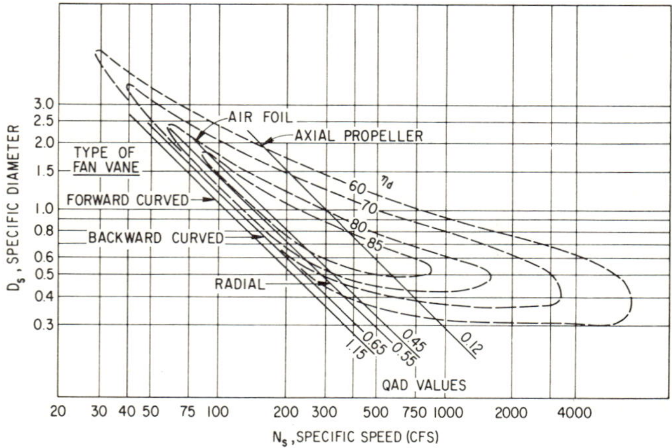

Figure 4.12. Fan characteristics projected on a Baljé Optimum Performance Chart. Total efficiency and pressure coefficients referenced to specific speed and specific diameter. (Reference: ASME paper 60-WA-231.)

The capacities given in Table 4.2 are for single inlet. If the fan has doubled suction inlets, the rated volume should be twice the quantities given in the table. The maximum volumes shown are not limiting. They represent the range of values used to attain the N_S-D_S data presented. For example, fans having backward inclined vanes are available for 750 Mcfm, and axial fans can handle 2 MMcfm.

Figure 4.12 is a plot of the various fans efficiencies and pressure potentials superimposed upon a Baljé chart. The latter is a method of resolving the optimum geometry of turbomachines by the application of specific speeds (N_S) and specific diameters (D_S). The product of these two factors is a constant for a given impeller design. The mystery of selecting the most efficient fan diameter and speed is easily established with this tool. For example, there is need to handle 60,000 cfm at 36 inches of water.

$$Q = 60{,}000/60, Q^{0.5} = 31.6, L = 36(69.3) = 2{,}500 \text{ ft}, L^{0.75} = 350 \text{ and } L^{0.25} = 7.1.$$

Selecting an air foil for maximum efficiency at $200 N_S$, D_S = 0.80, then $D_S N_S$ is 160. The optimum rotor diameter is

$$D = D_S Q^{0.5}/L^{0.25} = 0.80(31.6)/7.1 = 3.56 \text{ feet or } 42.75 \text{ inches.}$$

The optimum speed is

$$N = N_S L^{0.75}/Q^{0.5} = 200(400)/31.7 = 2{,}222 \text{ rpm.}$$

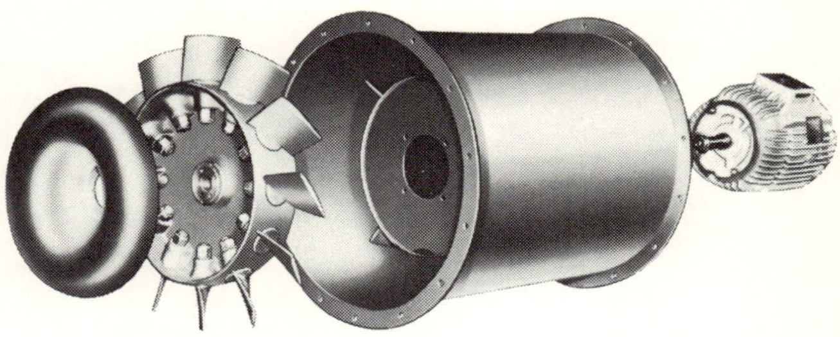

Figure 4.13 (left). A discharge view straight through an axial propeller type fan mounted on the motor shaft in the pipe system. (Courtesy of Joy Mfg. Co.)

Figure 4.14 (above). Illustration of an exploded view of an in-line axial propeller type fan. The propellers are adjusted to a suitable pitch and secured by the radial draw bolts. (Courtesy of Joy Mfg. Co.)

This speed calls for special design features. It exceeds the 12.25-inch limitation for Class III fans and is therefore rated as a *centrifugal compressor*. The tip speed is 2,524(43.6)/229 = 482 fps, which is nearly twice the ratings of Table 4.2, but in every regard, it is a feasible machine. The effect of speed changes can be readily obtained from Figure 4.12. The effect of half speed reduces N_S to 100 and the efficiency to 75 percent. If the speed was doubled, $N_S = 400$ and the efficiency drops to 65 percent. The diameter must also change for the D_S change, 170/100 = 1.7D_S, and the diameter is 6.19 feet or 74 inches at 1,267 rpm.

Figure 4.13 gives a discharge view through an axial propeller type fan mounted on the motor shaft in the pipe system. Figure 4.14 gives an exploded view of an in-line axial propeller type fan.

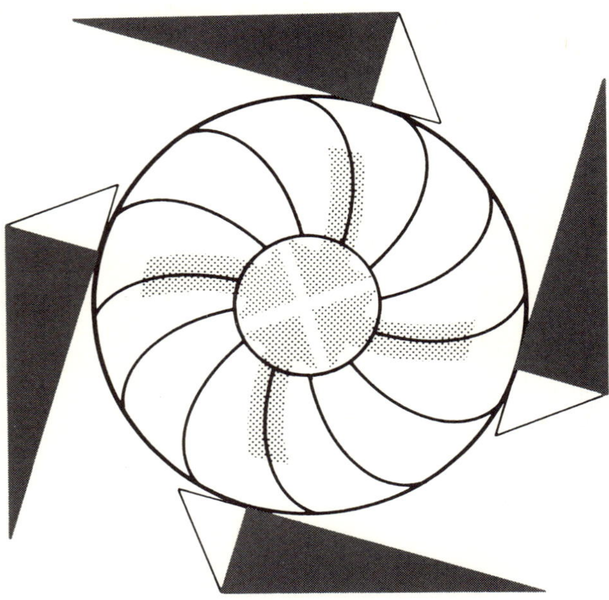

5 Rotary Compressors

The rotary compressor is the least known and least popular type of gas compressors. Rotary compressors have been proposed as the answer to the compressor dilemma—a machine suitable for variable compression ratios and a wide capacity range, one that operates at electric motor speeds and has a potential maintenance history equal to that of a centrifugal compressor.

This chapter considers the rotary compressor to be an adiabatic or isentropic machine. It has no valves. It operates by confining a volume of gas in a rotating pocket (Figure 5.1). The confined gas pockets converge into the discharge chamber. The action is that simple. The speed of the gas is nominal and not great enough to cause polytropic effects such as are experienced in centrifugal compressors when near sonic gas velocities impinge against the vane and volute obstructions. Why then does the discharge temperature exceed the adiabatic rise?

The premise of this chapter is that this type of machine has three aerodynamic losses. The first loss is that incurred in charging and exhausting the pockets which increases the compression ratio within the pocket. The second loss concerns the leakage between the rotors and between the rotors and the case. The third loss is the thermal burden caused by the intake warm-up.

The pocket charging losses are considered to be a function of the rotor tip speed. The gas must follow in the wake of the pocket tip which is continuously opening in the inlet chamber. It is necessary to expend a certain head to match the rotor speed. This is equivalent to the Net Positive Suction Head (NPSH) required to charge a centrifugal pump.

Table 5.1 shows the pressure differential that must be sacrificed to attain a velocity equivalent to the rotor tip speed. The sonic velocity of the air considered herein is 1,130 fps at a suction temperature of 60°F. These velad losses are applied to Equation 5.1 to produce the "B" correction factor. This modifies the visual, line pressure R_C to the more realistic intrinsic pressure differential as it exists in the converging screws. This example considers two cases: one where $R_C = 2$ or $P_2 = 29.4$ and the other where $R_c = 3$ or $P_2 = 44.1$ with $P_1 = 14.7$.

$$B = (1.0 + \Delta P_2/P_2)/(1 - \Delta P_1/P_1) \tag{5.1}$$

$$\eta ad = (R_C\sigma - 1)/(BR^\sigma - 1) \tag{5.2}$$

These equations apply the "B" correction to R_C and give the compression efficiency in reference to the minimal adiabatic power requirement. The compression efficiencies are given in Table 5.2 for the

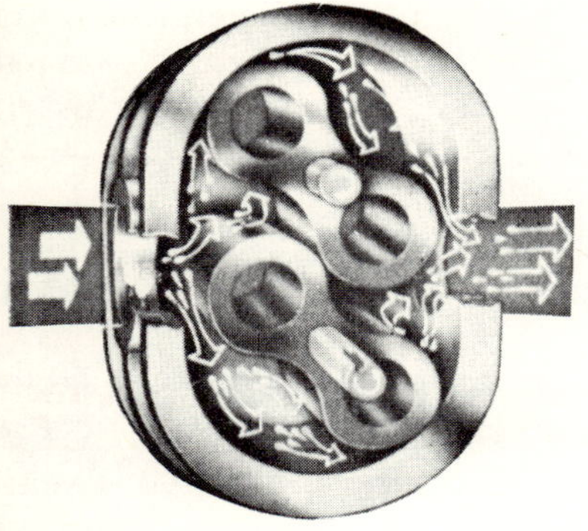
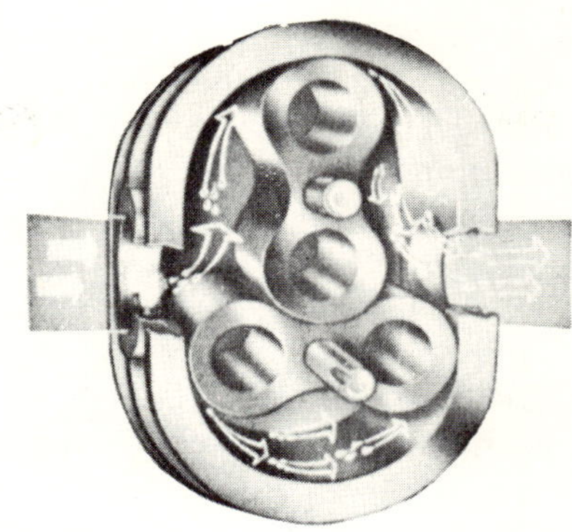
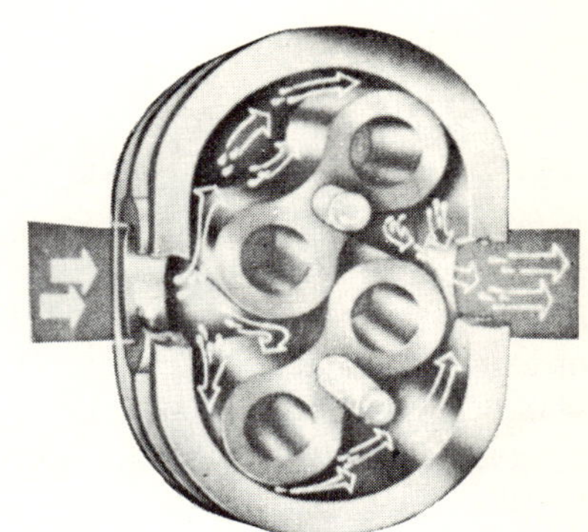

Figure 5.1. The above three positions of a two-lobe rotary compressor give an animated view of the gas flow through the machine.

two cases and for a single K and for two K of resistance. The full gamut of tip speeds is shown in reference to the sonic velocity of the gas handled. Most rotary compressors operate well below 0.1 Mach as shown in Table 5.3. The helical screw compressors are rated at 0.3 Mach; some may go to 0.4 Mach. Certain tests have been made with tip speeds as great as 0.7 Mach.

The compression efficiencies produced from one K of resistance are too optimistic. Those produced from two K are too conservative. A compromise that uses 2 velads of suction resistance and one velad of discharge resistance appears to be realistic. This resolution of compressor performance does not have the supporting test data that should be allocated to such a project. It does present the rudiments and suggest the outline for such a test program. In the interim it offers the most comprehensive procedure for projecting the performance of high speed rotary compressors.

The velad hypothesis used in developing the compression efficiencies in Table 5.2 is best understood by an example application. Using the foregoing data, with rotor tip speed of 0.45 Mach which is 510 fps, the velad is $V^2/2g$ = 510(510)/64.4 = 4,025 ft/K. The specific volume of air at the given standard conditions is 13.1 cf/lb. The gas head per psi is 144 (13.1) = 1,870 ft/psi. Each suction velad represents a pressure drop of 4,025/1,870 = 2.15 psi/K.

Presume that one velad is consumed in overcoming the resistance of the compressor chamber, entering and dead heading into the rotor pocket. The absolute velocity of the air must also match that of the rotor tip. This totals 2 velads for the charging operation or 2 (2.15) = 4.30 psi. This loss is

Table 5.1
Charging and Exhausting Pressure Losses for Rotary Type Compressors

Tip Speed		Velocity Head	Differential Pressure		
			Suction	Discharge	
Mach	fps	$U^2/2g$	PSI	R_c = 2	R_c = 3
0.71	807	10,000	5.35	8.69	11.70
0.45	510	4,025	2.15	3.50	4.70
0.38	430	2,875	1.50	2.46	3.32
0.26	296	1,360	0.71	1.16	1.57
0.14	162	400	0.21	0.34	0.46
0.10	108	180	0.10	0.15	0.21

Table 5.2
Compression Efficiencies for Helical Screw Type Compressors

Tip Speed Mach No.	R_c = 2.0		R_c = 3.0	
	K = 1.0	K = 2.0	K = 1.0	K = 2.0
0.71	44%	22%	58%	30%
0.45	68	50	78	62
0.38	76	61	84	71
0.26	88	72	92	80
0.14	95	93	97	95
0.10	98	96	99	98

expressed as a decimal fraction, 4.3/14.7 = 0.293 or 29.3 percent of the suction system pressure.

It is further presumed that the *built-in* clearance releases the trapped air at or very close to the discharge chamber pressure, so that no overcharging or undercharging exists in the exhaust pocket. The absolute velocity of the gas leaving the rotor and entering the discharge chamber represents another velad expenditure. The analytical difference between the suction and discharge velad is the change in the gas specific volume. The discharge v_{sd} is $v_{os}/R_c^{1/k}$ = 13.1/2.19 = 5.95 cf/lb, and gas head per psi is 144 (5.95) = 857 ft/psi. The gas is dispelled into the discharge chamber without the benefit of any convergence devices to recover the velocity energy of 4,025/857 = 4.7 psi per velad. This energy is wasted and represents 4.7/44.1 = 0.106 decimal fraction of the discharge system. This data is applied to Equation 5.1 to evaluate B = (1 + 0.106)/(1 − 0.293) = 1.106/0.707 = 1.565. This makes the intrinsic BR_c = 4.7. This value applied to Equation 5.2 makes the adiabatic compression efficiency η_{ad} = $(3.0^\sigma - 1)/(4.7^\sigma - 1)$ = 0.37/0.555= 66.5 percent.

This procedure can be resolved into the following simplified equations:

$$\theta_i = 2.5\underline{m}(U^2)/T(10)^5 \tag{5.3}$$

where m represents the molecular weight of the gas, U is the male rotor tip speed in fps and T is the gas temperature °R. This equation includes two velads of resistance. The exhaust loss equation below includes one velad of resistance.

$$\theta_e = \theta_i/2\,(R_c)^\sigma \tag{5.4}$$

The two resistances, θ_i and θ_e are used to evaluate the intrinsic pressure correction factor "B",

$$B = (1.0 + \theta_e)/(1.0 - \theta_i). \tag{5.5}$$

Leakage

Leakage is the next step to evaluate. The rotary compressor configuration offers three longitudinal lines of escape. Two lines follow each rotor and the casing, and the third line is the addendum of the converging rotors. The internal leakage for rotary machinery follows the equation:

$$W_L = 23L\,P_2G/(T_2)^{0.5}\ \text{lb/min} \tag{5.6}$$

The displacement of a rotary machine is:

$$QD = d\,LUX,\ \text{cfm} \tag{5.7}$$

The capacity of the machine is dependent upon the charging pressure. Equation 5.3 has established the percentage of the suction pressure that is lost in the charging operation. The volumetric efficiency simply follows as:

$$E_{vr} = 100 - (\theta_i + W_s/R_c^\sigma). \tag{5.8}$$

Where the suction charging loss is not available, use the approximate expression, E_{vr} = 98 − $V\,R_c$. Table 5.3 gives an average V factor for each type of rotary compressor.

The displacement multiplied by the volumetric efficiency gives the capacity of the compressor. The capacity divided by the specific volume gives the capacity in terms of pounds per minute:

$$W = 0.093\,dm\,E_{vr}\,LUX\,P_1/T_1. \tag{5.9}$$

The *slip leakage* is the ratio of (Equation 5.6/Equation 5.9) which reads:

$$W_s = 30.7\,(0.577G + 0.0038)\,R_c^{1.9}\,d/m\,E_{vr}\,LXU^{0.5} \tag{5.10}$$

The manufacturer usually determines the volumetric efficiency in a unique manner. The speed required to sustain the design pressure at zero flow is known as the *slippage*. The decimal fraction of the slippage speed divided by the rated speed is the volumetric efficiency. For example, the static slippage for an 8-inch, straight-lobe compressor is 120 rpm at 3 psig and 144 rpm at 10 psig. The percent slip in reference to the rated speed of 1,200 rpm is (120/1,200) or 10 percent. The manufacturers' volumetric efficiency is 90 percent for the 3-psig static system. The slippage, N_s*, for the 10-psig system is 12 percent and the manufacturers' volumetric efficiency is 88 percent.

This method does not consider the dynamic loss which is evaluated by Equations 5.3, 5.8 and 5.10. The tip speed loss in most other types of rotary compressors is so low that it is negligible. Table 5.4 shows that the tip speed must exceed 0.12 Mach before it is significant. These *zero-flow* speeds provide valuable checks on the clearance gap and a means of estimating the *zero-flow* pressure. Having the percent of slip speed, N_s*, the likely discharge pressure is

$$R_c^{1.9} = N_s^*m\,L\,E_{vr}\,X\,(U_s^{0.5})\,/30.7d\,(0.577G + 0.0038).$$

Using G as the actual gap = 0.0133 and G^* as the effective gap = 0.0095, U_s = rpm(d)/229 = 144(8)/229 = 5, N_s^* = 0.12, E_{vr} = 0.98 assume L =

Table 5.3
Summary of Rotary Displacement Compressor Performance Data

	Helical Screw	Spiral Axial	Straight Lobe	Slide-Vane	Liquid-Liner
Configuration, Features (Male x Female)	4 x 6	2 x 4	2 x 2	8 Blades	16 Sprockets
Max. Displacement, icfm	20,000	13,000	30,000	6,000	13,000
Max. Dia., in.	25	16	28	33	48
Min. Dia., in.	4	6	10	5	12
Limiting Tip Speed, *y* Mach	0.30	0.12	0.05	0.05	0.06
Normal Tip Speed, *y* Mach	0.24	0.09	0.04	0.04	0.05
Max. L/d, Low Pressure	1.62	2.50	2.50	3.00	1.1
Normal L/d, High Pressure	1.00	1.50	1.50	2.00	1.0
V Factor for Volumetric Eff.	7	3	5	3	3
X Factor for Displacement	0.0612	0.133	0.27	0.046	0.071
Normal Overall Eff., %	75	70	68	72	50
Normal Mech. Eff. at +/− 100 hp, %	90	93	95	94	90
Normal Ratio of Compression, Rc	2/3/4	3	1.7	2/3/4	5
Normal Blank-off, Rc	6	5	5	7	9
Displacement Form Factor, A	0.462	1.00	2.00	0.345	0.535

Note: Mach = 1,250 fps air at 185°F.

d; X = 0.265, $U_s^{0.5}$ = 2.23 and m = 29:

$$R_c^{1.9} = 0.12\ (29)\ 0.98\ (0.265)\ 2.23/30.7\ (0.0095) = 7.00$$
$$R_c = 2.80,\ P_2 = 2.80(14.7) = 41.2 \text{ psia or } 26.5 \text{ psig.}$$

Having the speed and the discharge pressure, it is possible to solve for the clearance gap G in the same equation.

The average clearance gap G for small (3 to 4-inch) rotors is 0.010, 0.020 for 13-inch rotors and 0.030 for 20-inch and larger rotors. A_e is a geometric coefficient based upon the open-end area divided by the square of the male rotor diameter, d is the diameter and L is the length of the rotor, both in inches. The A_e and X coefficients for the various rotary configurations are given in Table 5.3 ($X = 0.1235A_e$). The exponent, 1.9, is an average value suitable for diatomic and heavy polyatomic gases. The complete exponent is $1 + (k + 1)/2k$, where k is the ratio of specific heat at the mean compression temperature.

Table 5.4 gives the percent leakage for tip speed variation of 0.1 to 0.35 Mach in reference to the total weight flow for a typical 6-inch helical type rotary compressor having a nominal clearance gap of 0.0133 inch. The leakages were calculated from Equation 5.10. A variation in the diameter or the gap increases the leakage in direct proportion. The R_c is the most significant factor that affects the leakage.

Table 5.4.
Adiabatic Slip and Thermal Efficiencies

Mach No. (Tip Speed)	Efficiencies	Rc 2.0	2.5	3.0	4.0
0.35 M* (392 fps)	η_{ad}	60.5	66.3	69.5	73.4
	η_s	96.0	93.8	91.3	84.8
	η_t	99.0	98.3	96.7	92.4
	η_o	57.5	61.0	61.4	57.5
0.26 M* (300 fps)	η_{ad}	73.3	78.0	80.5	83.5
	η_s	95.9	93.7	91.2	84.5
	η_t	98.9	98.0	96.7	92.2
	η_o	69.5	71.7	71.0	65.2
0.12 M* (136 fps)	η_{ad}	93.5	94.5	95.6	96.2
	η_s	94.5	91.4	88.0	79.1
	η_t	98.5	97.2	95.4	89.5
	η_o	87.1	84.0	80.3	68.1
0.10 M* (113 fps)	η_{ad}	96.0	96.8	97.3	97.6
	η_s	94.1	90.9	87.2	77.6
	η_t	98.7	97.1	95.6	88.8
	η_o	89.2	85.5	81.2	67.3

Note: η_o is the over-all efficiency.

Table 5.5
Anticipated Leakage Percentage of Total Weight Flow

Mach No. (Tip Speed)	R_c 2.0	2.5	3.0	4.0
0.35 (392 fps)	4.0	6.2	8.7	15.2
0.26 (300 fps)	4.1	6.3	8.8	15.5
0.12 (136 fps)	5.5	8.6	12.0	20.9
0.10 (113 fps)	5.9	9.1	12.8	22.4
Average	5.0	7.5	10.6	18.5
(a) Temp Rise	7°F	12°F	22°F	48°F
Power Loss	1.3%	2.3%	4.1%	8.9%

Note: (a) The above leakage raises the suction temperature by the amount shown.

Table 5.5 also shows the amount of *warm-up* resulting from bypass leakage to the incoming flow. The percent of power loss is directly proportional to the increase in the Rankine temperature. This leakage percentage is also shown in Table 5.5. A simple expression for evaluating the leakage *warm-up* effect for each percentage of leakage is

$$X_t = 0.12(R_c) + 0.02. \qquad (5.11)$$

Equations 5.10 and 5.11 offer a means of evaluating the leakage and *warm-up* losses. Neither are, in reality, dynamic losses and should not be corrected by a gas head alteration. The leakage and the thermal burden directly affect the horsepower and can be applied as an efficiency factor in the denominator of the power equation. The leakage effect is given in Equation 5.12. The thermal effect of the suction *warm-up* is given in Equation 5.13.

$$\eta_s = 1 - W_s. \qquad (5.12)$$
$$\eta_t = 1 - W_s[(0.12\,R_c) + 0.02] \qquad (5.13)$$
$$\text{Adhp} = 0.0468(R_c^{\sigma} - 1)\,W\,Z\,T\,/m\sigma. \qquad (5.14)$$

The basic adiabatic horsepower is the same for all types of compressors, namely:

Where W is the gas flow in pounds per minute, Z is the compressibility factor and σ is $(k-1)/k$.

$$\text{Gas hp} = \text{Adhp}/\eta_{ad}\,(\eta_s)\eta_t. \qquad (5.15)$$
$$\text{Bhp} = \text{Gas hp} + \text{friction (gas hp)}^{0.4} \qquad (5.16)$$

Table 5.4 gives the magnitude of the three principal losses for a 6-inch helical screw rotary air compressor having an average clearance gap of 0.0133 inch. The composite efficiencies (η_o) compare favorably with catalog data. It is essential that the clearance gap be reduced in order to maintain a reasonable efficiency at R_c values in excess of 3. The penalty effect of the greater tip speed is amplified by the charging and expelling loss which are included in the adiabatic efficiency.

It is possible to extrapolate these efficiencies for other conditions. For example, determine the efficiency of a 10-inch straight-lobe compressor having a clearance gap of 0.030 inch operating at a tip speed of 0.05 Mach and at 2.0 R_c. The adiabatic efficiency tends toward an asymptotic minimum value of 98.5 percent. The slip loss varies as the effective gap, G^*, and the square root of the change in tip speed: $([0.030\ (0.577) + 0.0038]/[0.0133\ (0.577) + 0.0038])(0.05/0.10)^{0.5}$ times 5.9 percent leakage, which is the closest approach given in Table 5.5. The leakage for the new condition is $(0.0211/0.0115)0.707(5.9) = 7.65$ percent and the new η_s is 92.35 percent. The new *warm-up* effect is $7.65(0.24 + 0.02) = 2.00$ percent thermal leakage loss, and $\eta_t = 98.00$ percent. The new overall efficiency is $0.985(0.9235)0.980 = 89.15$ percent.

The overall efficiency as given in Table 5.4 is the product of the adiabatic efficiency η_{ad} (see Equation 5.1), the slippage efficiency η_s (see Equation 5.12) and the thermal *warm-up* efficiency η_t (see Equation 5.13). The latter two efficiencies are developed in the preceding sentences. The adiabatic efficiency is assumed to qualify for the minimum asymptotic value of 98.5 percent.

Helical Screw Compressors

The helical screw compressor is the only rotary unit that operates at tip speeds in excess of 0.12 Mach. Table 5.3 shows five distinctly different categories of rotary compressors. The spiral axial operates at 0.12 Mach and the other three operate at or about 0.05 Mach. The latter group can be categorically assigned a composite efficiency of 90 percent, except for the sliding-vane and liquid-liner unit, operating above 4 R_c.

The helical screw compressor is the most versatile form of rotary compressors. An exploded view of the unit is shown in Figure 5.2. A profile of the four-lobe male and the matching six-lobe female configuration is shown in Figure 5.3. The optimized performance of the helical compressor can be projected on a Balje' chart (see Figure 5.4). The optimum performance of a centrifugal compressor lies along a D_sN_s slope of 150 and a pressure coef-

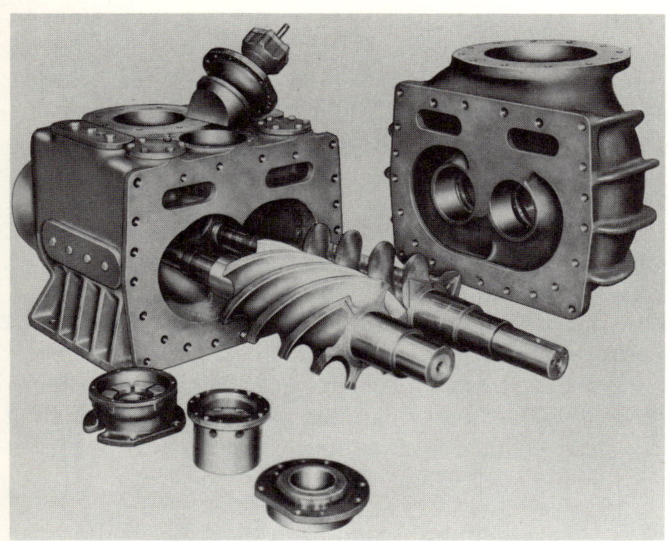

Figure 5.2. An exploded view of a helical screw rotary compressor. (Courtesy of Fairbanks Morse Power Systems Div., Colt Industries.)

ficient of 0.55, with the island peak at $D_S = 1.5$ and $N_S = 100$ as shown in Figure 5.4. The similar optimum performance of a helical compressor has a D_SN_S slope of 26 with the peak performance island at $D_S = 2.0$ and $N_S = 13$. The D_SN_S slope of 26 is identified as having a pressure coefficient of 4.

This means that the helical machine is capable of attaining the same head as a centrifugal compressor at (0.55/4) = 14 percent of the centrifugal compressor's tip speed. In other words, the helical compressor can develop 7.3 times the head of a centrifugal compressor operating at the same tip speed. This *low specific speed* characteristic is highly desirable for compressing small volumes of gas to high pressures. The helical screw compressor has performance characteristics and flexibilities that approach the piston compressor. It operates at moderate speed more comparable with electric motors. It has compression efficiencies comparable to centrifugal compressors. It should have maintenance costs that compare favorably with the centrifugal machines which are less than one-third the cost of maintaining piston machines. With these basic advantages favoring the rotary machine, it might be expected to take over a stronger position in the market. There are deterrents that impede such a take-over, like noise, intrinsic leakage, rotor deflection, casing strength and the technological deficiencies described earlier.

The most successful helical screw compressor applications are delivering air for general con-

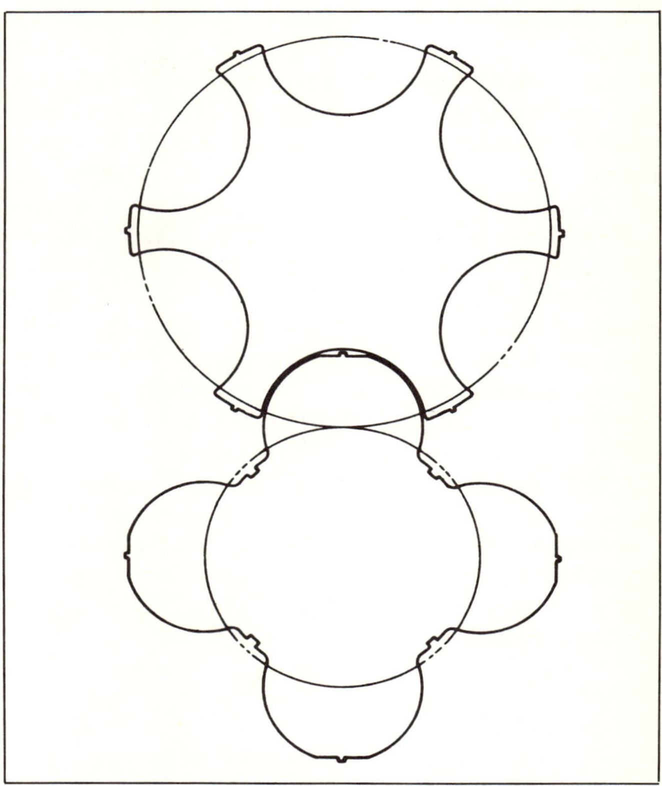

Figure 5.3 (above). Rotor profiles of the four-lobe male and the six-lobe female. Built-in clearance exists between the two rotors. (Courtesy of Fairbanks Morse Power Systems Div., Colt Industries.)

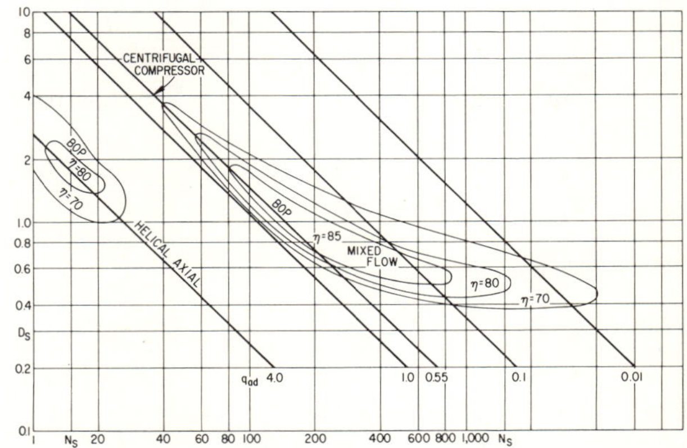

Figure 5.4 The optimized performance of the helical compressor can be projected as seen on the Baljé chart. The optimum performance of a centrifugal compressor lies along a D_SN_S slope of 150 and a pressure coefficient of 0.55, with the island peak at $D_S = 1.5$ and $N_S = 100$.

struction and road building purposes, mainly for operating pneumatic tools. The size of these portable units run from 600 to 1,200 icfm at 125 psig and require 250 to 500 hp. The key to their success is the moderate tip speed of 0.11 Mach and lower. The second factor is the technique of internal oil flooding. The intake air is sprayed with oil at a ratio of one part per thousand, or 7.5 gallons per thousand cubic feet. This flooding absorbs about two-thirds of the heat of compression, reducing a normal 350° F discharge temperature to about 100° F above ambient or 170° F. This eliminates the need for an after-cooler. The oil is recovered from the separator, cooled and returned to the compressor suction. The sheep-skin filters are quite efficient, less than 0.1 percent of the oil is lost. This is equivalent to one gallon per 60,000 hp-hr of operation. The oil also functions as a lubricant and a sealant to minimize the bypass leakage. The latter is perhaps the most important function of all in that it silences the shrill bypassing noise.

The helical unit was originally considered to be suitable only for dry gas. Condensate and grit particles could not be tolerated. These restrictions have been relaxed in recent years. The helical machines are being used to handle the waste gas from a hydrogenation process. This is a most difficult service where occasional lumps of a very viscous coagulate must pass through the machine. The helical compressor is not without its problems, but the helical units perhaps have less difficulties than those which would be experienced from a piston machine in the same service.

The helical machine can be operated without a driving gear. The lobe contact does not experience a locking interference which can occur in other lobe models. Timing gears are used to drive and also avoid this interference. The male lobe is the driver. The female rotates in the opposite direction (see Figure 5.3). The containment wrap requires 300° of rotation. The pitch of the helical screw is about 45°.

Built-In Clearances

The transition of the Lysholm design helical screw unit to the present Svenska Rotor Maskiner (SRM) form included the *built-in clearance* feature. The space between the male lobe and the female recess, as illustrated in the cross-sectional view of Figure 5.3, is the built-in clearance. When the gas retained in this clearance volume is returned to the suction chamber, the respective pressures should balance. When the clearance volume is oversized, there is an excessive pressure and volume returned to the suction chamber. The results are equivalent to another form of bypass. Conversely, when the clearance volume is undersized, the trapped gas pressure is less than the suction chamber pressure. This reduced pressure is undoubtedly an aid to the charging operation.

Perhaps the most suitable and neglected application is its likely potential as a gas motor or still better as a cryogenic expander. Referring to the Baljé chart where the helical compressor had a D_sN_s characteristic slope of 26 and q_{ad} factor of 4, the equivalent rotary expander key performance factor (tip speed/spouting velocity) should be 0.08. This is equivalent to an 8.5-inch rotor running at 3,600 rpm matching the peak efficiency of a 4.5-inch radial in-flow impeller turning at 51,000 rpm.

Spiral Axial Compressors

The spiral axial compressor has a top speed equal to the slowest helical screw machine. A sectional view of the spiral machine is similar to Figure 5.2. The envelope volume is more than twice the capacity of the helical machine having the same d, L and N dimensions. The machine has a timing gear drive and runs dry without the rotors making contact. It provides 20 to 30-psig air for manufacturing operations.

Straight-Lobe Compressors

The straight-lobe rotary compressor operates at the relatively low tip speed of 50 fps. This is equivalent to a 10-inch rotor driven by a six-pole motor or a 13-inch rotor running at eight-pole speed. An animated view of gas flow through the machine is shown in Figure 5.1. This machine can accommodate a considerable quantity of liquid, perhaps as much as 50 parts per thousand. Figure 5.5 is a sectional elevation view.

The public utilities use the straight-lobe compressors as displacement meters, low-pressure pipe line boosters, vacuum gas-well gathering systems and as a supercharger for a gas or diesel engines. This type of compressor offers the most efficient form of vacuum service.

Optimum single-stage performance is realized at 23 inches of mercury and 28 inches for two-stage operation. It is used for desalinization and other evaporative processes where large volumes of vapor are removed at high vacuums.

Slide-Vane Compressors

The slide-vane compressor is illustrated in Figure 5.6. The rotor runs eccentrically within a cylinder.

Figure 5.5. Sectional view of a two-lobe rotary compressor. Note timing gear at left and outside bearings. (Courtesy of Fuller Company, Division of GATC.)

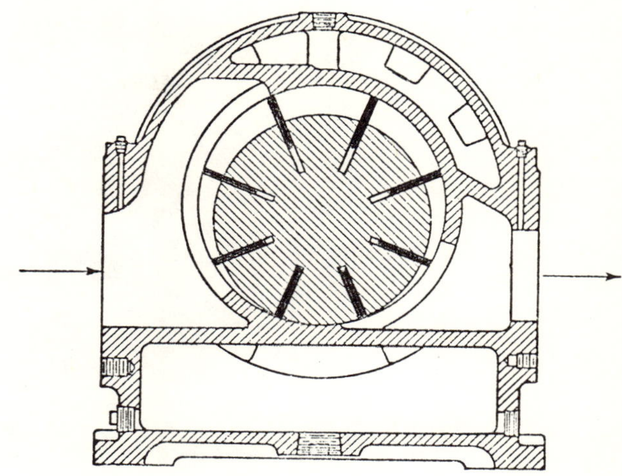

Figure 5.6. End view of a sliding-vane compressor. The rotor turns clockwise, closing the annular cells into the discharge chamber. (Courtesy of Allis-Chalmers Manufacturing Company.)

Radial slots in the rotor carry sliding vanes which form a series of longitudinal cells. The centrifugal spin of the rotor holds the vanes against the cylinder wall. The cell volume diminishes as the rotor approaches the discharge chamber, which is in the lower right-hand corner of Figure 5.6. A single-stage slide-vane compressor can pull a 28-inch Hg vacuum or can pump air to 50 psig. A two-stage unit can compress air at 250 psig. The unit is not suited for handling saturated and super-saturated vapor. It is particularly inapt with cold jackets. This increases the cylinder wall condensation. The best operation requires a warm jacket and the use of a lubricant heavily loaded with a solvent resistant inhibitor such as rapeseed oil, lanolin or tallow. The manufacturers advocate rather generous quantities of cold jacket-water and lubricant. They prescribe 10 drops per minute for a 10-inch and smaller bore cylinder, plus one drop per minute per shaft seal. This is equivalent to 2 to 3 pints per day of SAE 30 lube oil or about 3,000 bhp-hr per gallon.

The oil rate is doubled for the 20-inch cylinder and tripled for cylinders over 30 inches in diameter. The vanes are made of laminated phenolic resins. Their life and cylinder wear are sensitive to the quantity and quality of the lube oil applied. The most successful application of the slide-vane compressor has been with the 100 hp, 360 cfm, portable construction-type air compressor. The rotors are flooded with lubricant using the same quantity and technique described for the larger helical compressors.

These oil rates are on the order of 10 times that required for equivalent piston machinery service. Table 5.6 was compiled from 20 years data accumulated on reciprocating compressors by a major California oil company's gas plants. The gas engine oil consumption varies from 4,000 (two-cycle) to 10,000 bhp-hr per gallon, (four-cycle).

Jacket-Water Requirements

The slide-vane and piston compressors have common difficulties in their lubrication and jacket

Table 5.6
Quantity of Lube Oil Required for Compressor Cylinder Lubrication

Range of cylinder diameters, in.	Nominal discharge pressure, psig	Film thickness, micro-inches	bhp-hr per gal. (thousands)	Compressor lube oil, pints/(cyl.) (100 bhp of load)	
				Pints/day	Drops/min.*
24-36	80	16	13	1.5	13
15-23	200	17	19	1.0	9
10-14	800	23	24	0.8	7
7-9	2,000	24	32	0.6	5
4-7	8,000	47	24	0.8	7
3-5	20,000	57	27	0.7	6

*Based on 12,500 drops per pint of 5,000 SSU SAE 40 lube oil at 75°F with vacuum-type lubricator. Feed includes rod packers, e.g., 300-hp, 16-inch cylinder requires 3 pints/day and 27 drops/minute. Reduce drop count to one-third of above for pressure-type lubricator.

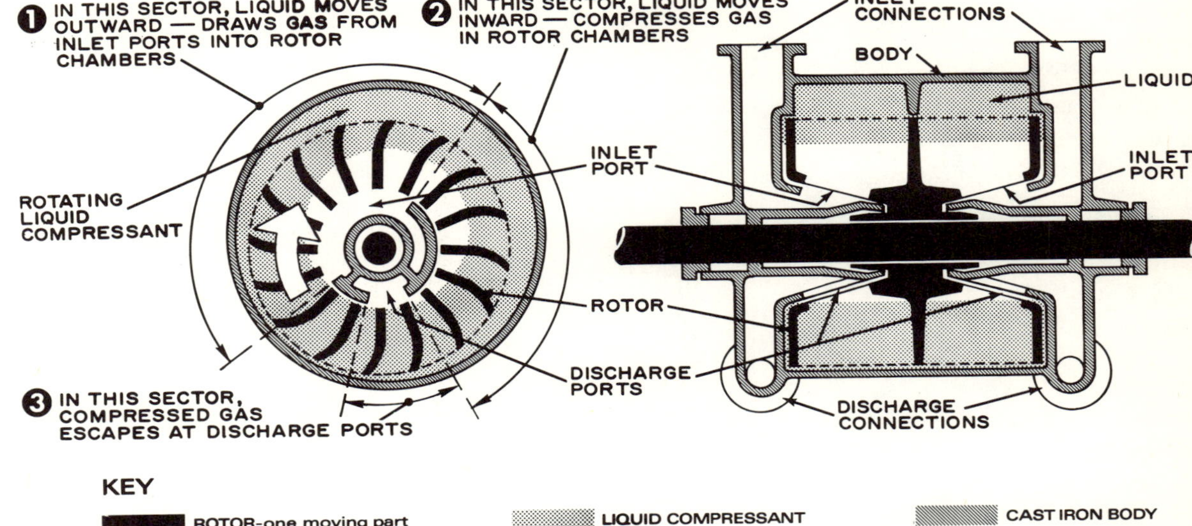

Figure 5.7. A sectional and end view of a liquid-liner type compressor. (Courtesy of Nash Engineering Company.)

temperature. There is a scarcity of credible reference data on both subjects and for that reason these data are included. The quantity of heat absorbed by the jacket-water is best determined from these equations:

$$H_j = 4\ (t_{ag} - t_{aw}) + 100, \text{Btu/bhp-hr.} \quad (5.17)$$

$$Q_{gpm} = H_j\ (\text{bhp})/500\ (\Delta t). \quad (5.18)$$

The letters t_{ag} represent the average of the gas temperature, $(t_{out} + t_{in})/2$, and t_{aw} is the average jacket-water temperature in the cylinder. The heat rejection is 3 to 12 percent of the mechanical heat equivalent, 2,545 Btu per bhp-hr. Statements advocating 20 and even 30 percent of the total power-heat equivalent are absurd. One oil company has operated its entire system of gas compressors with dry jackets for over 15 years (2). They have had no detrimental experience. There has never been a creditable reduction in the power as a result of jacket-water heat rejection (2).

There has been considerable damage experienced as a result of a cold cylinder jacket water system in handling wet gases at or gases close to the saturation line. The cold liner causes the liquids in the gas to condense and wash off the lubricant, reducing the life of the piston rings or the sliding-vanes to a matter of weeks. In instances of this order, it is advisable to reduce the water circulation so that the temperature leaving the cylinder is within 10° F of t_{ag}. The Δt or jacket-water temperature rise in Equation 5.18 should be 2° and 5°F for a warm (140° F) system.

Liquid-Liner Compressor

The operation of the liquid liner rotary compressor is illustrated in Figure 5.7. The sprocket blades throw the sealant (usually water) in a cylindrical form against the internal housing. This cylindrical liner contains the gas flow from the large eccentric entrance to the confinement of the discharge port. This compressor is used to handle the more corrosive and volatile gases in chemical plants. It is indispensible for handling exothermic gases like chlorine, oxygen and acetylene. The liquid-liner compressor is used to handle priming liquids and fibrous material matrix. It is used for vacuum fibrous-plastic molding where material carry-over would interfere with the rotation of the lobe elements.

The liquid-liner compressor offers an absolutely oil-free, 125-psig, single-stage air compression service. Despite a poor nominal efficiency of 50 percent, it is still 10 times more efficient than a steam ejector and is being used as a replacement. It is used as a vacuum pump to remove inert gases from condenser systems and condensate from steam heating plants.

The compressor requires considerable quantities of cold coolant to maintain the discharge within 20° F of ambient or the feed temperature. The manufacturers advocate complete rejection of all mechanical energy expended for the compression. This is an exaggerated requirement and a rather costly auxiliary, especially when the coolant is not permitted to rise above 85° F. A 200-hp liquid-liner compressor requires 100 gpm of coolant at a 10° F differential. Frequently it is necessary to provide a small cooling tower for the coolant service. When an exotic coolant like sulfuric acid is used, it is necessary to provide a special pumping and cooling system.

The liquid-liner rotary compressor has an average composite adiabatic and leakage efficiency of 45 percent at the normal rated tip speed of 66 fps. This efficiency is improved to 63 percent by reducing the tip speed to 30 fps and drops to 40 percent at 80 fps. Judging from the data in Table 5.2, the charge and exhaust loss should not exceed 2 percent. The bypass is estimated to be 3 percent. The mechanical losses should not exceed 5 percent, making a total loss of 10 percent or an ultimate efficiency of 90 percent. The churning power of the sprockets to support the liquid-liner is estimated to be 80 percent of the adiabatic compression power. A rather crude generalized overall compression and mechanical efficiency is

$$\eta_{oL} = 90/[100 + (fps)^{1.10}(\text{Sp. Gr.})^2/(CE)]. \tag{5.19}$$

It has been suggested that the churning loss varies as the square of the specific gravity (*Sp. Gr.*) and inversely as the impeller viscous drag (*CE*) as projected by the Hydraulic Institute. Typical viscosity effects are shown in Table 5.7. The Mach number criterion is not valid for the liquid-liner compressor. The hydraulic resistances control the critical velocities.

Concerning the quantity of coolant required for a liquid-liner compressor, the 90 percent mechanical efficiency factor in Equation 5.19 should be corrected to 100 percent because the mechanical losses are not obvious in the gas temperature. The closure problem after this paragraph deals with a 550 adhp example, which has an overall efficiency of

$$\eta_{oL} = 90/[100 + (71)^{1.1}] = 0.432.$$

The motor load is 550/0.432 = 1,270 bhp, and the motor nameplate rating should be 1,400 to include a 1.10 service factor. The cooling burden is $550/[100/(100 + 71^{1.1})] = 1{,}140$ hp. The coolant required for a 20° F temperature rise is

$$Q_{gpm} = H_j(\text{bhp})/500\Delta_t = 1{,}145\ (2{,}545)/500(20) = 290\text{ gpm}.$$

Table 5.7
Effect of Coolant Viscosity on Sprocket Drag for Liquid-Liner Compressors

Viscosity SSU	Correction Factor "CE"
< 40	1.00
85	0.9
200	0.8
600	0.7
1,000	0.6
1,800	0.5
2,500	0.4

The data given in Table 5.8 require several explanations. All machines are operated at the normal tip speeds given in Table 5.3. The adiabatic losses for charging and exhausting the helical screw compressor are appreciably higher than the other machines. The greater tip speed accounts for this loss. The slippage is controlled by the displacement form factor and the tip speed, which favors the greater displacement, lower speed machines. The overall effect is contrary to the normal overall efficiencies given in Table 5.4. This implies that the clearances for the axial and the straight-lobe machines are considerably greater than 0.025 inch which is nominal for the helical machine.

It is interesting to note that the rotors for the first three types of compressors are almost equal despite a 6 to 1 difference in speed. The power requirement is the most significant criteria and represents the payout for the service rendered. A premium of one-third of the power cost plus perhaps an equal maintenance saving should be worth some consideration.

Figure 5.8 shows the compression efficiencies for four types of rotary compressors as given by catalog data. It offers excellent basis for appraisal.

Rating of Rotary Machinery

The example application presumes a volume of 10,000 icfm of 18 *m* (k = 1.28) natural gas ($Z = 1$) to be compressed from 10 psia and 100° F to 30 psia. What size rotary compressor and motor driver is required using the normal rating criteria given in Table 5.3 and waiving the 1.7 R_c limitation for the

Table 5.8
Various Rotary Compressors Applied to a Common Service

	Type				
	Helical Screw	Spiral Axial	Straight-Lobe	Slide-Vane	Liquid-Liner
(g) Suction Loss, θi	9.35	1.32	0.89	0.90	1.40
(h) Discharge Loss, θe	7.35	1.04	0.70	0.70	1.10
(i) Intrinsic Corr., B	1.185	1.023	1.016	1.016	1.025
(j) Adiabatic Eff., ηad	85.6	97.7	98.5	98.5	97.9
(k) Slippage, Ws, %	28.5	16.6	11.8	11.8	3.0*
(l) Slip Eff., ηs %	71.5	83.4	88.2	88.2	97.0
(m) Thermal Eff., ηt %	89.2	93.7	95.8	95.5	42.5†
(n) Volumetric Eff., Evr	68.0	85.7	89.1	89.9	96.6
(o) Displacement, cfm	14,700	11,650	11,220	11,120	10,370
(p) Rotor Dia., inches	26.6	26.2	27.0	65.0	45.5
(q) Com'l. Size, d x L	25 x 25	22 x 33	22 x 33	46 x 92**	43 x 48††
(r) Speed, rpm	3,500	1,250	593	284	378
(s) Motor, hp	1,100	800	750	750	1,400
(t) Service Factor	1.09	1.11	1.10	1.12	1.10
(u) Disch. Temp., °F	309	270	262	263	120

* Estimate
† By Equation 5.15
**Twin 32.5 x 65 or Triplet 26.5 x 33 (667 rpm) are more realistic
††Twin 32 x 32 (613 rpm) alternate (p) where $L = d$.

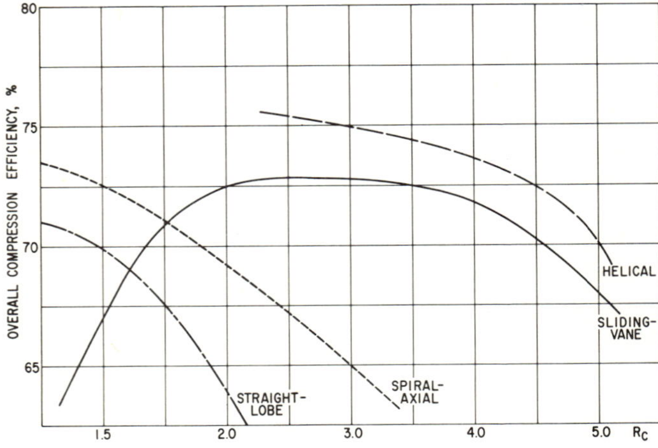

Figure 5.8. Compression efficiency of rotary compressor as compiled from catalog data.

straight-lobe unit? The rotor clearance G is assumed to be 0.024 inch and d is assumed to equal L.

Calculations

(a) Sonic $\tau = 224\ (kT/m)^{0.5} = 224(1.28 \times 560/18)^{0.5} =$ 1,420 fps.

(b) $(yM^*) = 0.24$ (Table 5.3); $y\tau = 0.24(1{,}420 = 341$ fps $= U$.

(c) $k = 1.28$; $\sigma = 0.219$; $R_c = 30/10 = 3.0$, $R_c{}^{\sigma} = 1.273$.

(d) $v_s = 10.73\ ZT/mP = 10.73(560)/18(10) = 33.3$ cf/lb.

(e) $W = icfm/v_s = 10{,}000/33.3 = 300$ lb/min.

(f) Adhp $= 0.0468\ (R_c{}^{\sigma} - 1)\ WZT/(m\sigma) = 0.0468(0.273)$ $300(560)/18(0.219) = 545$ hp (by Equation 5.14).

Note: These calculations are common to all five model applications. The helical case is developed in detail. Only significant variables are shown for the other three dry machines. Highlights are given in Table 5.8 for the liquid-liner model.

(g) $\theta_i = 2.5(18)341^2/560(10)^5 = 0.0935$.

(h) $\theta_e = 0.0935/2(1.273) = 0.0327$.

(i) $B = (1.0 + \theta_e)/(1.0 - \theta_i) = 1.0327/0.9065 = 1.14$.

(j) $\eta ad = (R_c^{\sigma} - 1)/[(BR)^{\sigma} - 1] = 0.273/0.310 = 0.876$.

(k) $W_s = 30.7(0.577G + 0.0038)R_c{}^{1.9}d/m\ L\ X\ U^{0.5}$, where $D = L$, $R_c{}^{1.9} = 8.1$ and $U^{0.5} = 18.5$. $W_s =$ $30.7(0.0176)8.1/18(18.5)\ 0.0612$; $W_s = 0.215$.

(l) $\eta_s = 1.0 - 0.215 = 78.5$ percent.

(m) $\eta_t = 1.0 - W_s(0.12\ R_c + 0.02) = 1.0 - 0.082 = 91.8$ percent.

(n) $E_{vr} = 1.0 - (\theta_i + W_s)/R_c^{\sigma} = 1.0 - (0.0935 +$ $0.215)/1.273 = 1.0 - 0.2625 = 73.75$ percent.

(o) $QD = Q/E_{vr} = 10{,}000$ *icfm*/0.7375 = 13,550 cfm. $QD = d\ L\ U\ X$, then

(p) $dL = QD/UX$ = 13,550/341(0.0612) = 650. If $d = L$, then d = 25.5 inches, or if d = 24 inches, then L = 27 inches. Table 5.3 indicates that a machine of these dimensions should have 13,550 cfm displacement. The maximum displacement of 20,000 cfm is possible for a 25 x 25 helical compressor at 0.37 Mach (525 fps).

(q) The motor load is $545/\eta_{ad}\eta_s\eta_t$ = 545/0.876(0.785) 0.918 = 865 hp. This would impose an 86.5 load factor on a 1,000-hp motor.

(r) The rotor speed is 341(229)/24 = 3,250 rpm for a 24-inch rotor. A two-pole speed motor could be applied to a rotor, 341(229)/3,550 = 22-inch rotor, 29.5 inches long.

(s) The estimated discharge temperature is determined in the following manner: $\sigma' = 0.129/\eta_{ad}\ \eta_t$, = 0.219/0.876(0.918) = 0.272. $T_e = T_i\ R_c^{\sigma'}$ = $560(3.0^{0.272})$ = 560(1.35) = 755 R or 295° F.

(t) A check for the discharge temperature from the adiabatic approach follows. T_{ad} = 560(1.273 − 1.0) = $153\Delta_t$. T_e = 153/0.876(0.918) = 190° F above 100° F at the inlet, makes the discharge temperature 290° F.

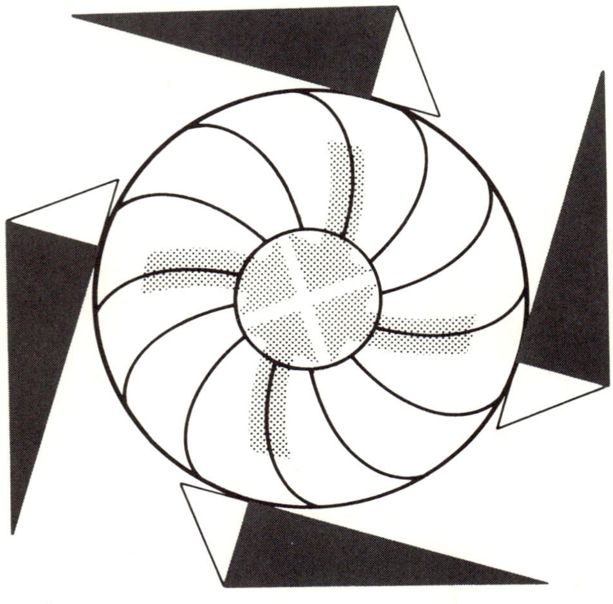

6 Sizing Centrifugal Compressors

There are three basic reasons for using centrifugal compressors instead of reciprocating compressors in hydrocarbon processing plants.

1. Environmental. They occupy less space; they operate with minimum attention and are quieter.
2. Lower Operating Cost. They will run 12 to 30 months without mechanical repair. The maintenance cost is about one-third the cost of piston compressor operation, excluding the drivers.
3. Lower Initial Investment. The lowest cost turbodriver is either a gear-motor or a steam turbine. The cost of such a driver and centrifugal compressor is approximately equal to the least costly type of synchronous motor driven reciprocating compressor in the 2,500 hp category. The centrifugal machinery cost is about two-thirds that of the piston machinery in the 5,000 hp bracket. The cost advantage of the turbomachinery is enhanced as the power demand is increased.

The reciprocating machine and motor drivers are priced on a cost per horsepower basis. Centrifugal machines and steam turbines are sold on a lump sum, plus a small unit cost per hp for the driver. The cost of a 12,000-hp centrifugal compressor may not be much more than the same casing rated for 1,000-hp service. Confronted with such an economy, there is an understandable demand on the part of the hydrocarbon processors for turbomachinery. The purpose of this chapter is to review the prevalent *state of the art* as to the compressors capability and a method of power rating and sizing the machine.

Compression Head

The compression head is the intangible measurement of the energy density imparted to a gas stream by a compressor. It may also be defined as the enthalpy added to the gas by the compressor. It may be observed by the increase in the gage pressure as the gas passes through the machine. Some engineers choose to evaluate the compression head as an isothermal function. A great many European compressors are evaluated on this basis. It presumes that the gas leaves the compressor at the entering temperature. This is an obvious false premise.

Isothermal Head

The isothermal head is determined from the equation:

$$L_{iso} = 144(\text{Log}_e R_c) P_o (v_{so}), \text{ ft-lb/lb.} \quad (6.1)$$

The product of this gas head and the weight flow, pps, divided by 550 gives hp $_{iso}$.

$$\eta_{iso} = L_{iso}(W)/\text{Gas hp}(550). \quad (6.2)$$

The resultant efficiency is related to the reference gas head. The term *gas horsepower* refers to the input shaft power, free of mechanical losses which have no thermal effect on the gas. This value is attained by subtracting the 0.4 root of the *Brake horsepower* from that Bhp value. For example, the Bhp is 5,000; the 0.4 root is 30 hp of friction; the Gas hp is 5,000 — 30 = 4,970. The isothermal head and efficiency serve no particular significance or reference value. It is 6 to 9 percent less than the adiabatic efficiency. It is about 10 percent less than the polytropic efficiency.

Adiabatic Head

The adiabatic head is determined from the equation:

$$L_{ad} = (R_c^{\sigma} - 1)1{,}545(T_o)\ Z_a/\overline{m}\ (\sigma) \quad (6.3)$$

The power developed with the adiabatic head represents the minimum realistic power requirement. Instances where a piston compressor has a generous valve area and is operated at reduced speed, the power tends to approach and almost equal the adiabatic power. The temperature of compression follows the same adiabatic pattern. As the piston speed is increased and the valve area reduced, the power requirement tends to exceed the adiabatic value, reflecting the dynamic losses. The temperature rise generally follows the adiabatic function. The reason is that the gas velocity in a piston machine rarely exceeds 0.2 Mach through the minimum restricted valve lift area. The velocities through a centrifugal machine are seldom under 0.2 Mach and are usually greater than 0.3 Mach. The friction of these high velocities against the obstructions and the resistances that exist in the impeller and the diffuser creates heat. These frictional losses contribute to the abnormal heating above the adiabatic. The loss is directly proportional to the degree of superheating the adiabatic temperature.

Adiabatic versus Isentropic

The adiabatic head and efficiency are often referred to as the isentropic values. The terms adiabatic and isentropic are used interchangeably. The

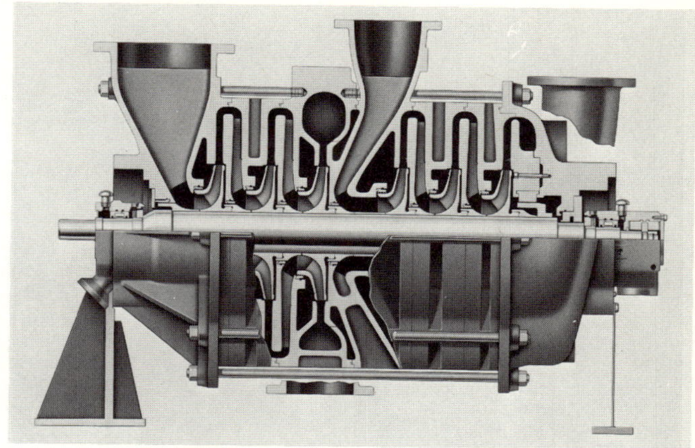

Figure 6.1. A half-sectional view of a six-stage module assembled centrifugal compressor. The radial diaphragms are stacked together over the shaft and impellers. The outer circumference of the diaphragm is the casing shell. Draw bolts lock the diaphragms and the manifold into a unit. The above model shows a mid-point bleeder and return manifold (Courtesy of Allis-Chalmers).

thermodynamic definition of an adiabatic process requires that no heat be added or withdrawn from a facility where a change of state occurs. It may or may not be reversible to qualify as an adiabatic process. If the process is reversible, it is truly an isentropic operation. An isentropic change of state occurs at constant entropy. This identifies isentropic analysis with Mollier charts and tables of gas properties. For lack of a better mark of distinction, we refer to such enthalpy calculations as isentropic and the exponential R_c values as adiabatic processes.

Polytropic Head

The polytropic head is determined from the equation:

$$L_p = (R_c^{\sigma'} - 1)1{,}545(T_o)Z_a/\overline{m}\ (\sigma'). \quad (6.4)$$

The polytropic efficiency can be determined from the Equation 6.6 relationship. The unique and only justification for the polytropic efficiency is its identity with the discharge temperature.

$$T_e = T_o R_c^{\sigma'} \text{ and} \quad (6.5)$$
$$\sigma' = (k-1)/k\eta_p = \sigma/\eta_p. \quad (6.6)$$

The polytropic efficiency can be determined for any operating point, where the suction and discharge temperature and the k value of the gas are known.

Figure 6.2. Versatile model multistage centrifugal compressor in service.

This ratio on the Rankine scale and the knowledge of the gas k value applied to Equation 6.6 give the polytropic efficiency. The usual connotation of the word *efficiency* refers to the degree of perfection.

The adiabatic process represents the ideal, the minimal energy requirement to effect a change of state for any gas. All reference material, Mollier Charts and Gas Tables are so dedicated. The isothermal and polytropic efficiencies are *pseudo* values. The former presumes that the discharge temperature is equal to the suction temperature. There is no experience in nature or by man to sustain this premise. There are numerous industrial and laboratory tests to sustain the adiabatic process.

The polytropic phenomenon is the thermal evidence of dynamic gas losses that are the resultant of the compressors' mechanical deficiencies. It is customary for some manufacturers to present the head-capacity performance curve using the polytropic head. A true presentation requires an established efficiency curve and an individual calculation for each point and each efficiency. This procedure is not always followed. Another disadvantage is individual performance curves are not comparable unless the subject curves have identical efficiencies at each respective flow quantity.

Figure 6.1 illustrates a versatile model multistage centrifugal compressor. The lower discharge flange illustrates the method by which the gas may be withdrawn for intercooling or other process purposes. The gas is reentered through the center vertical flanged manifold. Figure 6.2 shows such a compressor in service. Figure 6.3 illustrates a three-unit train of an ammonia synthesis gas compressor.

Efficiency Conversion Chart

Figure 6.4 presents a simplified correction factor for converting polytropic efficiencies to adiabatic efficiency. It was prepared from ratios of compression of 1.2 to 3.7 for air, natural gas, LPG and hydrogen mixtures. The correction values are not absolute, but are close approximations. The integrated radical $(R_c^{\sigma} - 1)$ is used as the abscissa, and the ratio of the polytropic to the adiabatic efficiency is the ordinate.

The *isoeffic* (constant efficiency) lines cover the normal range of adiabatic efficiencies. For an illustration, an air compressor is operated at $R_c = 2$, the integrated radical $(R_c - 1)$ is equal to 0.22 when $\eta = (1.40 - 1.0)/1.40 = 0.286$. Then if the polytropic efficiency was 65 percent, the correction ratio is 1.057 and the adiabatic efficiency would be $(65/1.057) = 61.5$ percent. If the polytropic efficiency was 85 percent, the adiabatic efficiency would be $(85/1.022) = 83$ percent. These correction ratios are obtained from the equation:

$$R\eta = \eta_p \, (R_c^{\sigma/\eta p} - 1) \, / \, (R_c^{\sigma} - 1). \qquad (6.7$$

The equivalent isothermal efficiency for the above 83 percentage case can be determined by the inverse proportion of the respective heads and efficiencies. The isothermal head is

$$L_{iso} = 144(\mathrm{Log}_e R_c) P_o v_{so}, \text{ ft-lb/lb} \qquad \text{(repeating 6.1)}$$
$$L_{iso} = 144(0.693)14.7(13.1) = 19{,}200 \text{ ft-lb/lb.}$$

The adiabatic head is

$$L_{ad} = (R_c^{\sigma} - 1)1{,}545(T_o)Z_a/\overline{m}\sigma \qquad \text{(repeating 6.3}$$
$$L_{ad} = (0.22)1{,}545(520)1.00/29(0.286) = 21{,}300 \text{ ft-lb/lb}$$

The equivalent isothermal efficiency is $83(19{,}200/21{,}300) = 75$ percent.

Figure 6.3. Three-unit train of ammonia synthesis gas compressor requiring 20,000 hp. This 2,800 psig process has been a most popular application during the past several years (Courtesy of Cooper-Bessemer Co.).

Example Head Calculation

The following example illustrates the derivation and application of the three reference heads and efficiencies. A shop performance test demonstrated that a centrifugal compressor was handling 36,200 cfm of ambient air at 14.7 psia and 60°F to 29.4 psia. The driving motor has an efficiency of 93.5 percent and demands 2,000 kw of power for the test. The discharge temperature is 215°F and the mechanical efficiency of the compressor had been previously determined to be 98 percent. The Gas hp is 0.935(0.98)(2,000/0.746) = 2,450 hp. This value and the discharge temperature represent the vital data pertinent to the plant process. The respective efficiency used to project this information is only incidental to the main objective of fixing the power requirement. The weight flow is

$$v_{SO} = 10.73 T_O Z_O / \overline{m} P_O \tag{6.8}$$

$$v_{SO} = 10.73(520)1.00/29(14.7) = 13.1 \text{ cf/lb.}$$

$$w = 36{,}200/13.1(60) = 46 \text{ lb/sec (pps).}$$

The isothermal head for R_C = (29.4/14.7) was determined to be 19,200 ft-lb/lb in the previous paragraph. Transposition of the fundamental power

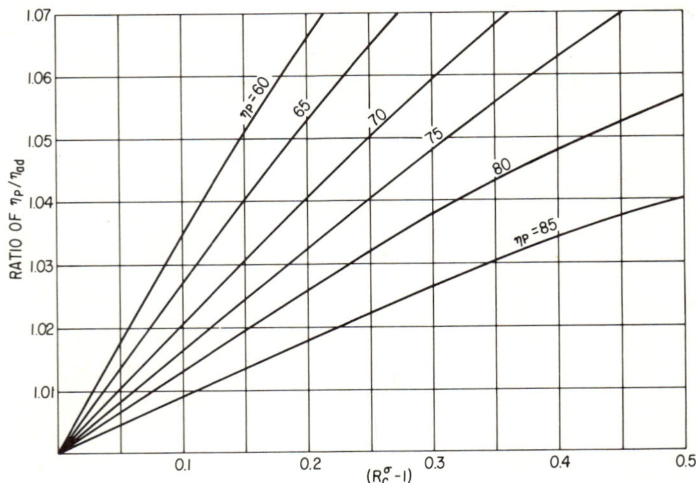

Figure 6.4. Ratio of polytropic to adiabatic efficiency, with reference to the adiabatic exponential function and the polytropic efficiency.

equation produces the isothermal efficiency in the following manner:

$$\text{Gas hp} = wL_{iso}/550\eta_{iso} \tag{6.9}$$

$$\eta_{iso} = wL_{iso}/\text{Gas hp}(550)$$

$$\eta_{iso} = 46(19{,}200)/2{,}450(550) = 65.5 \text{ percent.}$$

The adiabatic head for R_c was determined to be 21,200 ft-lb/lb. The adiabatic efficiency equation has the same form as given above.

$$\eta_{ad} = wL_{ad}/\text{Gas hp}(550) \tag{6.10}$$

$$\eta_{ad} = 46(21{,}200)/550(2{,}450) = 72.3 \text{ percent.}$$

The isentropic efficiency (η_i) should equal the adiabatic efficiency. The enthalpy difference between suction conditions of 60°F and 14.7 to 29.4 psia at constant entropy should be 21,200/778 or 27.2 Btu/lb.

$$\text{Gas hp} = w(H_e - H_o)/2544\eta_i \tag{6.11}$$

$$\eta_i = 46(3{,}600)27.2/2{,}544(2{,}450) = 72.2 \text{ percent.}$$

The fundamental design of the impeller is represented by the vector diagram of Figure 6.5. The Euler equation, Equation 6.12, resolves these vectors into an energy density, gas head:

$$L_{eu} = (V_{u2}U_2 - V_{u1}V_1)/g. \tag{6.12}$$

The Euler efficiency will approach the adiabatic equation if the flows typify the actual flow and the Euler efficiency will be

$$\eta_{eu} = wL_{eu}/550 \text{ Gas hp.} \tag{6.13}$$

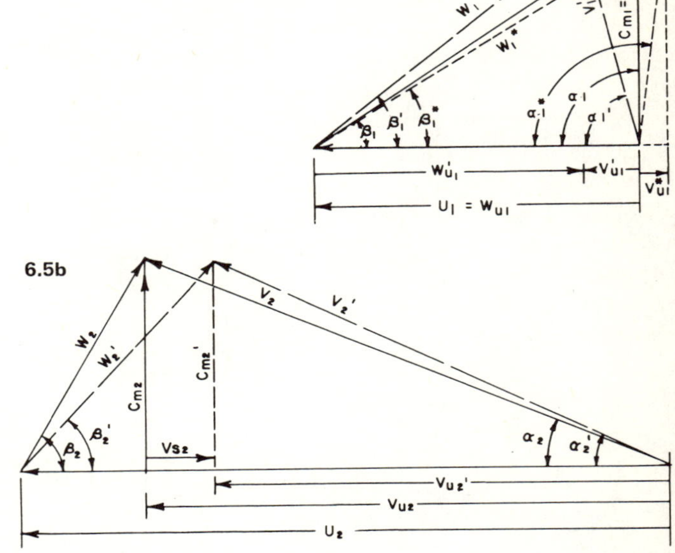

Figure 6.5a. Vector diagram depicting typical suction velocities. C_m is the meridional velocity, U is the peripheral velocity of the impeller, V is the absolute velocity of gas flow, W is relative velocity of the gas flow and α_1 is the eye entrance angle.
Figure 6.5b. Vector diagram depicting typical discharge velocities. β_2 is the backlay angle of the vane at the impeller tip.

In addition to the vector analysis, there are the flow losses through the suction eye, the vanes, the diaphragms, the scroll or diffuser, the guide vanes, the seal leakage and disk friction which contribute to the inefficiencies of the compressor and raising the discharge temperature. The reason for making a shop performance test is to confirm the accuracy of the design premise.

Temperature

The polytropic efficiency is an overall thermal evaluation of the design. The discharge temperature is calculated thus:

$$T_e = T_o R_c^{\sigma'} \tag{6.14}$$

$$\sigma' = (k-1)\,k\eta_p = \sigma/\eta_p \tag{6.15}$$

$$R_c^{\sigma'} = 2.00^{\sigma'} = 675/520 = 1.298$$

$$\sigma' = \log 1.298/\log 2.000 = 0.376$$

$$\eta_p = \sigma/\sigma' = 0.286/0.376 = 76 \text{ percent.}$$

Without the discharge temperature, the power requirement should have reflected the machines efficiency and the polytropic efficiency can be developed from Figure 6.4. The polytropic efficiency is usually 2 to 4 percent greater than the adiabatic efficiency. Taking the abscissa as 0.22 (from previous data) and estimating the polytropic η_p as 75 percent, the $R\eta$ is 1.035 and the resolved adiabatic efficiency is 72.5 percent. The polytropic efficiency has merit in establishing a realistic discharge temperature. It remains constant at optimized specific speeds and specific diameters for any gas providing there is no abnormal leakage and the Mach number and the Reynolds affect are equal or adequately corrected. The polytropic head is as useless and misleading as the isothermal head. But most manufacturers use the term in describing performance of their machinery.

It is determined from the equation:

$$L_p = (R_c^{\sigma'} - 1)1{,}545 T_o Z_a / \overline{m}\,\sigma' \qquad (6.16)$$

$$L_p = (0.298)1{,}545\,(520)/29(0.376) = 22{,}000 \text{ ft}$$

$$\eta_p = 46(22{,}000)/550(2{,}450) = 75 \text{ percent.}$$

Caution

There are two important points to remember concerning the various reference efficiencies.

1. The realistic discharge temperature can only be determined from the polytropic efficiency as applied to Equation 6.14 or to the ΔH in a Mollier chart solution.
2. The realistic power can be determined from any reference efficiency, if these efficiencies are referenced to the same real power data. The adiabatic references are the most professional and useful for this purpose.

Design Criteria

The maximum obtainable efficiencies of turbomachines, together with optimum design geometry, based on the state-of-the-art knowledge can be presented as a function of the similarity parameters, specific speed, N_s, and specific diameters, D_s, for constant Reynolds and Mach numbers. The specific speed and specific diameters are

$$N_s = N(Q^{0.5})/L^{0.75} \qquad (6.17)$$

$$D_s = D(L^{0.25})/Q^{0.5} \qquad (6.18)$$

$$U = d(N)/229 \qquad (6.19)$$

$$d_r = 1{,}980 L^{0.5}/N \qquad (6.20)$$

$$d_b = 1{,}760 L^{0.5}/N \qquad (6.21)$$

$$q_{ad} = g(L_{ad})/U^2 \qquad (6.22)$$

where d_r is the outside diameter of radial vaned impeller. The pressure coefficient is 0.65 for a single stage machine. This coefficient is depreciated about

Table 6.1
An array of gases flowing at 0.2 mach

Gas	Hydrogen Mixture	Natural Gas	Air	Propane	Butane
$\overline{m}$	6.2	18.2	29	44	58
k	1.38	1.26	1.40	1.13	1.09
Sonic, fps	2,400	1,350	1,120	760	700
0.2 M, Velocity	480	270	224	152	140
Velad, ft., ($V^2/2g$)	3,580	1,130	780	360	305
Ft./psi	8,750	3,010	1,890	1,240	945
Δ psi/Velad	0.41	0.38	0.41	0.29	0.32
Equivalent C_m, fps	431	242	202	136	126

2 percent per stage for the resistance of each return diaphragm. The D_sN_s product is 165 for the radial vane. The nominal 45 degree backlay vane has a D_sN_s product of 147 and a pressure coefficient of 0.55 which is subject to the same frictional depreciation. The quick solution for the backlay vaned impeller diameter is given in Equation 6.21. U is the peripheral tip speed (fps); Q is the volume flow in cubic feet per second (cfs); L is the gas head; N is the rpm and g is the gravitational acceleration, 32.2 ft/sec^2. The head coefficient may have numerous reference bases, such as the adiabatic given in Equation 6.21, where L_{ad} is derived from Equation 6.3. The Euler coefficient is sometimes referred to as the geometric or theoretical pressure coefficient, q_{th}. There are design data available for correcting the geometric (vector analysis) to the realisitc q_{ad} values. These factors correct for the number of vanes, the slip, the various inefficiencies of the guide vanes, the rotor, the scroll or diffuser and the discharge velocity conversion.

Mach Numbers

All significant velocities in aerodynamic design are referenced as decimal Mach numbers. Velocities of equivalent Mach numbers have approximately equal resistances in terms of percent of the system pressures. Table 6.1 illustrates the behavior.

The velad is the gas head (ft) required to support the 0.2 M velocity, $V^2/2g$, or 480(480)/64.4 = 3,580 ft. The *head per psi* is determined from the specific volume (144 v_{so}), when divided into the velad, gives the pressure drop, Δp *per velad*. The flow-coefficient (ϕ_2) at the point of *surge* is shown on Figure 6.6 to be about 0.2. If the tip speed U_2 is taken as 0.9 Mach, the meridional suction eye velocity (C_{m_1}) for an air compressor would be C_{m_1} = 0.2(0.9)1,120 = 202 fps. The same equivalent *surge* would be realized if the impeller had a tip speed of 2,160 fps and the *eye* velocity was 431 fps. If the

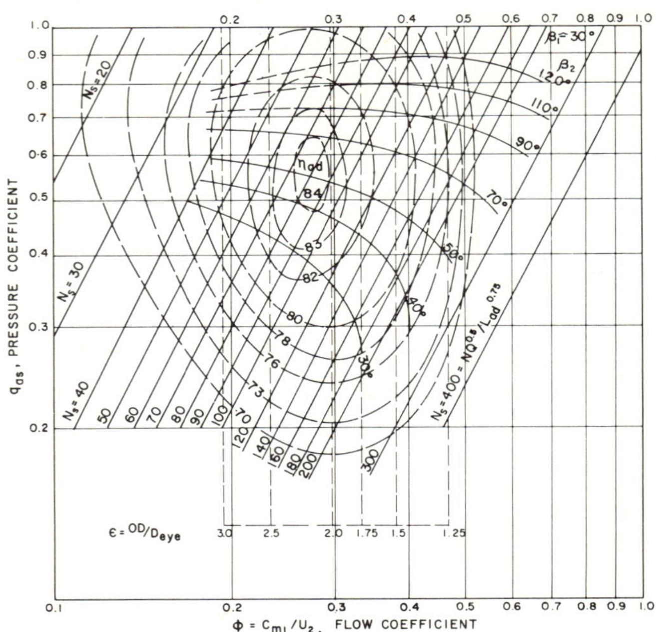

Figure 6.6. Eckert chart giving pertinent adiabatic performance data for impellers having various vane angles.

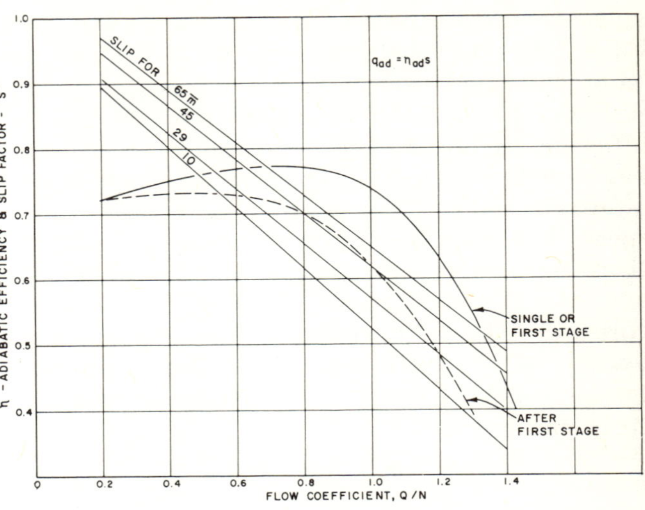

Figure 6.7. Typical basic performance curve of a commercial centrifugal impeller, depicting the pressure coefficient, efficiency and slip plotted against the flow coefficient Q/N. The solid lines represent single stage performance. The broken lines represent multistage factors. The pressure coefficient, qad = SLIP (η_{ad}).

Figure 6.8. Performance curve of highly perfected NACA impeller.

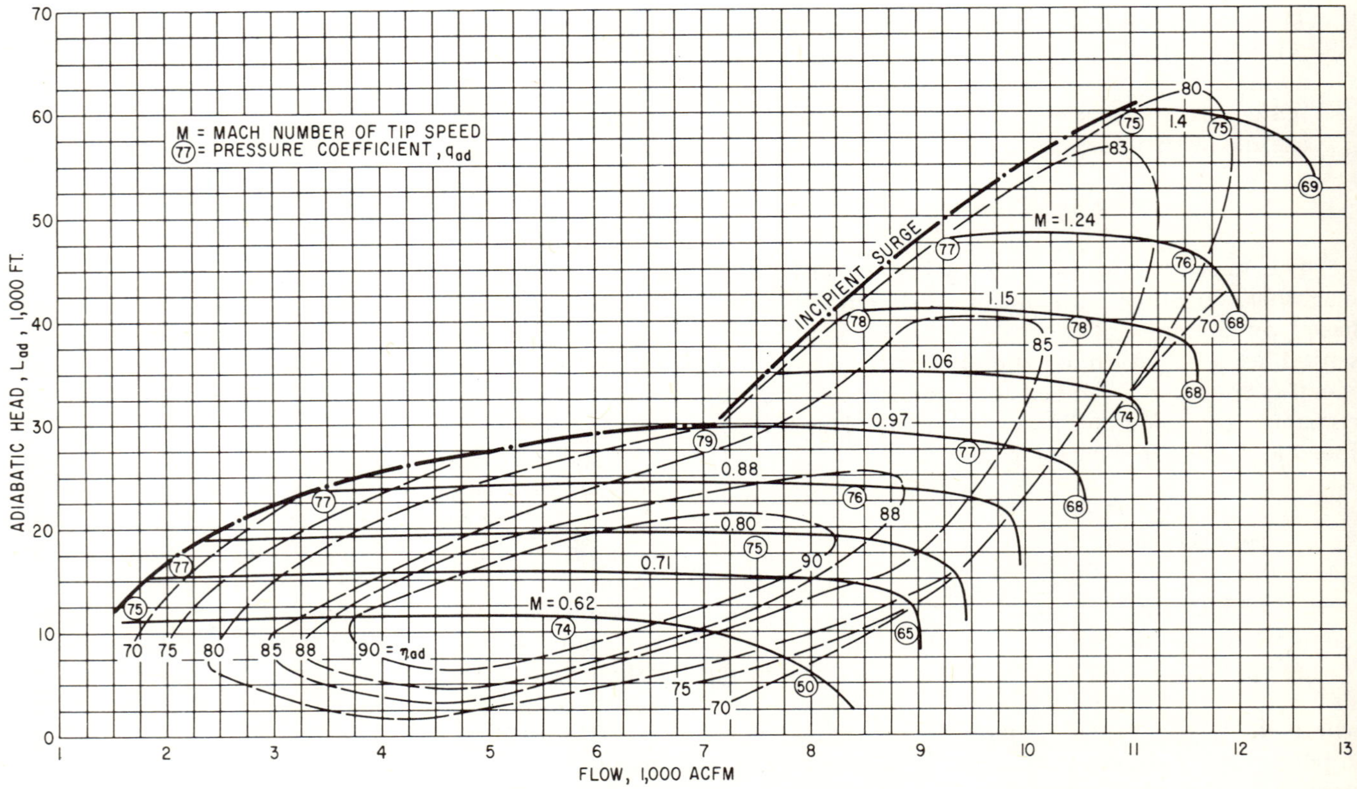

M = MACH NUMBER OF TIP SPEED
(77) = PRESSURE COEFFICIENT, q_{ad}
INCIPIENT SURGE
ADIABATIC HEAD, L_{ad}, 1,000 FT.
FLOW, 1,000 ACFM

tip speed remained the same 1,020 fps as for the air case, the flow coefficient for the greater flow of 431 fps, ϕ_2 would be 0.42, and the impeller would experience the opposite flow characteristic where *choke* exists. The *eye* velocities for the other gases are given on the bottom line of Table 6.1 for 0.2 and U_2 = 0.9 Mach.

Flow Characteristics

The design conditions consist of adiabatic (or polytropic) head and the flow, in actual cfm for a given operation. Figure 6.7 illustrates a typical master impeller characteristic curve which has been developed by thorough testing. The abscissa is plotted as a unit of flow per revolution, usually cfm/rpm, Q/N or other similar flow references. The solid line represents a single-stage performance. The dash line shows the loss from the return diaphragm in multistage operation. The pressure coefficient establishes the head from Equation 6.21. The dotted line is the product of q_{ad} and η_{ad}. The crest of this curve represents the Best Operating Point (BOP). Impellers are selected so that the design point is as close to this crest as practical. The performance represented by the $D_S N_S$ intersection on Figure 6.7 are BOP. The BOP performance for Figure 6.6, 6.8 and 6.9 are obviously the zones of the highest efficiency.

The pump head (and q_{ad} values) will increase some 5 to 8 percent from BOP as the flow is retarded to the point of *stall* or *surge*. The *surge* zone follows the N_s lines of 40 to 60 on Figure 6.6. It is well identified on Figure 6.8. It follows the flow coefficient of 0.10 to 0.15 on Figure 6.9. The other extreme from *stall* is the maximum flow condition of *choke* or *stonewall*. This flow represents the maximum flow that can be drawn through the impeller eye. It is the flow experienced to the right of ϕ_2 = 0.35 to 0.50 in Figure 6.6. The abrupt *knee* breaks on the H-Q curves illustrate the *choke* affect most emphatically. Particular notice should be made of the high q_{ad} values shown in the small circles, ranging from 0.79 at *surge* to 0.77 at BOP and 0.68 at the *knee*. This represents the degree of perfection attained by NACA.

Slip

Gas that is approaching the eye of an impeller will take on a prerotation in the direction of the impeller turning if the casing does not contain inlet guide vanes. The prerotation tends to reduce the compressor capacity and increase the slope of the head curve. Moveable Inlet Guide Vanes (IGV) are sometimes installed to unload the compressor by exaggerating this characteristic. The reverse action of the IGV can create a counter-rotation which has the effect of supercharging each individual impeller so equipped. This tends to reduce the H-Q and the power curve. These effects are illustrated in Figure 6.10. The impact of the gas molecules against the back side of a rotating vane distorts the radial flow pattern. The effect of this distortion on the ideal relative velocity vector, W_2 in Figure 6.5*b* causes the β_2 to be reduced to β_2' and W_2 to increase to W_2'. When this vector is resolved with the tip speed U_2, the resultant absolute velocity is V_2'. The tangential component of the ideal absolute vector is C_{u2} and for the distorted flow is C_{u2}'. The difference between these two vectors is the *Slip* V_{s2}. It is usually about 15 percent of the absolute velocity. The meridional velocity leaving the tip annulus is C_{m2}, which is presumed to equal C_{m2} and C_{m1}.

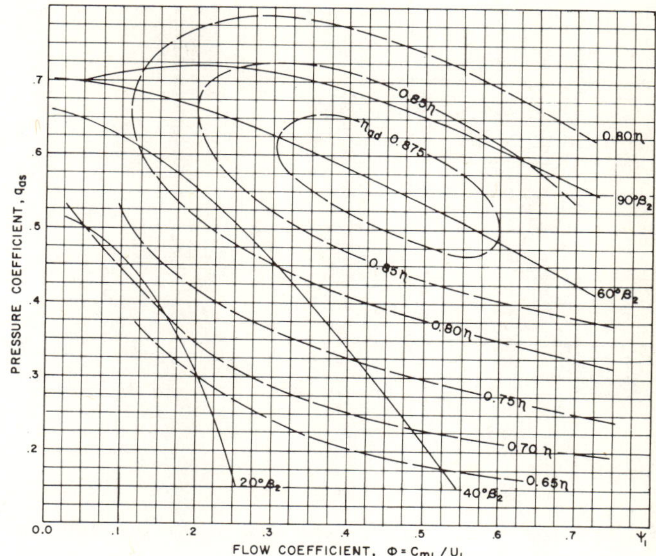

Figure 6.9. Characteristic centrifugal compressor performance. No discharge losses are included.

Figure 6.5*a* shows the same flow behavior for the gas entering the impeller. U_1 is the peripheral speed of the impeller eye. C_{m1} is the meridional vector and represents the average impeller eye velocity. The flow coefficient ϕ is C_{m1}/U_2 and is applied as the abscicca on Figure 6.6. There are occasions when a similar *gulp* factor (ϕ') is taken as C_{m1}/U_1 and is so applied in Figure 6.9. The blade angle at the eye β_1 includes the relative vector W_1

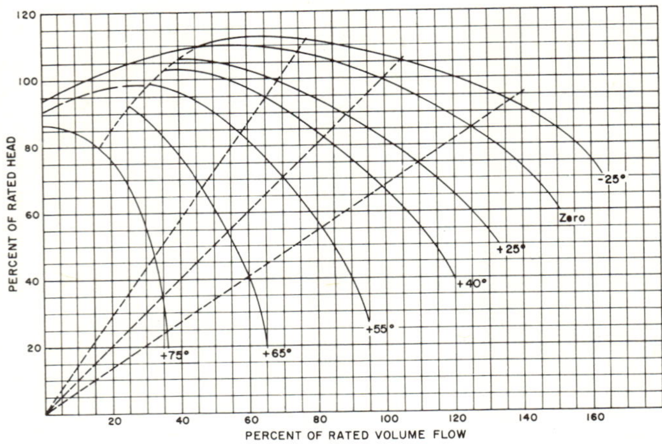

Figure 6.10a. Effect of inlet guide vanes on the pressure head of a centrifugal compressor.

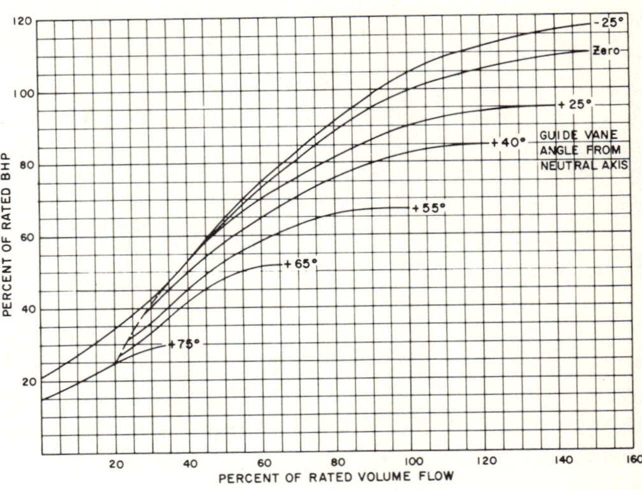

Figure 6.10b. Effect of inlet guide vanes on the power requirement for centrifugal compressors.

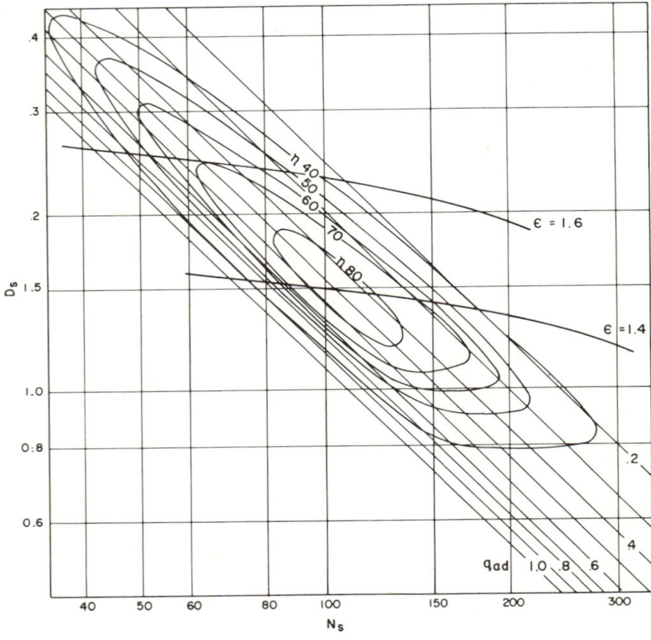

Figure 6.11. Baljé generalized performance curves, orienting the adiabatic-dynamic efficiency and the pressure coefficient as a function of the specific speed and specific diameter.

and the eye velocity U_1. When the guide vanes direct the meridional flow normal to the eye velocity U_1, the horizontal component W_{u1} is equal to U_1. When the inlet guide vanes are adjustable and the gas is given a prewhirl (rotation in the direction of rotation), the relative velocity follows the angle β_1' and vector W_1'. The absolute vector is V_1' and the horizontal component and slip is V_{u1}'. When the incoming gas is given a counter rotation with regard to the impeller, the flow follows angle $\beta_1{*}$ and the relative vector $W_2{*}$. The absolute velocity is $V_1{*}$ and the supercharging head beneficient is $V_{u1}{*}$. The magnitude of the head correction can be evaluated from the Euler equation by applying the following guide angle A degrees of deviation from normal.

$$L_{eu} = [V_{u2}U_2 - 1 + A(U_1 C_{m1} \operatorname{Sin} A)]/g. \tag{6.23}$$

The counter-rotation should be treated as a negative quantity, when corrected by the equation minus sign, it becomes additive.

The *slip* is largely affected by the number of vanes (n). The effect on the Euler equation for an impeller having back-lay vanes is

$$L_{eu} = [U_2\ V_{u2}\ (1 - 2/n) - U_1\ V_{u1}]/g. \tag{6.24}$$

The equation for a radial vane impeller is

$$L_{eu} = (1 - 2/n)\ U_2^{\,2}/g. \tag{6.25}$$

Flow Coefficient

The two heavy lines striking horizontally across Figure 6.11 marked E is 1.4 represent the approximate ratio of the OD to the eye diameter. Where N_S is taken as 100 and D_S as 1.47, E is 1.4 (presuming the impeller is designed for end suction with no shaft extension through the eye and is 24 in. in diameter). The suction eye is 24/1.4 = 17 inches and has 1.585 sq ft area. Presume further that the tip speed U_2 is 1,000 fps and U_1 is (1,000/1.4) = 715 fps. An optimized flow coefficient ϕ_2, value from Figure 6.6 for a 50-degree β_2 back-lay vane impeller is 0.275. The meridional velocity is 0.275 (1,000) = 275 fps. The impeller design capacity is 275 (1.585) = 435 cfs or 26,000 acfm. The flow coefficient ϕ_1 is (275/715) = 0.385. This number is a reasonable value corresponding to a 55-degree back-lay impeller in Figure 6.9.

Application Example Problem

The design flow rate is 18,000 mols per hour of 22 mol weight gas, having a k value of 1.26 and both Z values are unity. The gas is compressed from 85° F and 40 psia to 51.6 psia. Develop the approximate size impeller, connections, speed, efficiency and power.

Solution. The specific volume is

$$v_s = 10.73\ (545)\ (1.00)/22\ (40.0) = 6.64 \text{ cf/lb} \quad (6.8)$$

$$w = 18{,}000\ (22)/3{,}600 = 110 \text{ pps.}$$

$$Q = 110\ (6.64) = 730 \text{ cfs or } 43{,}800 \text{ acfm}$$

$$Q^{0.5} = 27.0$$

$$\sigma = (1.26 - 1.0)/1.26 = 0.206$$

$$L_{ad} = [(51.6/40)^{0.206} - 1]\ 1{,}545\ (545)\ (1.0)/22\ (0.206).$$

$$L_{ad} = 10.000 \text{ ft-lb/lb.} \quad (6.3)$$

An inspection of the Baljé and Eckert charts Figures 6.6 and 6.9) shows that 100 N_s is an optimum design point, and 1.47 D_s is a matching ordinate for a reasonable q_{ad} value of 0.55.

$$L^{0.75} = 1{,}000,\ L^{0.5} = 100 \text{ and } L^{0.25} = 10.$$

$$N = N_s L^{0.75}/Q^{0.5} = 100\ (1{,}000)/27.0 = 3{,}700 \text{ rpm.} \quad (6.17)$$

$$D = D_s Q^{0.5}/L^{0.25} = 12\ (1.47)\ 27.0/10 = 47.5 \text{ in.} \quad (6.18)$$

$$dia = 1{,}760\ (100/3{,}700) = 47.5 \text{ in.} \quad (6.20)$$

The last equation is a simple check of the equation above. The mean tip speed U_2 is 47.5 (3,700)/229 = 769 fps. The flow coefficient is read as 0.32 on Figure 6.5 at the intersection of N_s = 100 and q_{ad} = 0.55. The meridional velocity C_{m1} entering the eye is 0.32 (769) = 246 fps. The eye area is 730 (144)/246 = 427 sq in or 20.6 in in diameter. The A diameter ratio is 47.5/20.6 = 2.31. The effect of the tip speed Mach numbers on the adiabatic efficiencies is shown in Figure 6.8.

The efficiencies (η) in Figure 6.11 relate the L_{ad} to the total inlet pressure and the static outlet pressure. The conditions further presume that the meridional flow entering the impeller eye (C_{m1}) is equal to the velocity leaving the machine (C_{m3}) and that the Reynolds number is 10^6 or greater. The usual practice in multistaged compressors is to relate the compression head to the total discharge pressure (ηt). These values are greater than η, especially for N_s values > 60 and for D_s values < 2.0. The difference between η and ηt is that the former discharge pressure does not include the discharge nozzle velocity head which may exceed the suction nozzle velocity that is presumed to be equal to C_{m1}. The process gas line velocities are usually about 0.08 Mach or 70 to 100 fps for common (20 to 30 $\overline{m}$) gases. The nozzles on multistage compressors are operated at velocities of 85 to 150 fps at design loads. The discharge nozzel is two thirds the diameter of the suction nozzle on multistage casings. The suction and discharge nozzles are usually the same size on single or two-stage cantilevered type compressors.

The discharge temperature for this example at 80% adiabatic efficiency is 545 $(51.6/40)^{0.206/0.8}$ = 582 °R or 122° F. The discharge volume is 730 cfs (1.0677/1.29) = 604 cfs, v_{se} = 5.48 cf/lb.

The process suction line would be sized at 110 fps: 730 (144)/110 = 960 sq. in. or 35 inches ID and 36 inches OD. The discharge line is 0.67 (35) = 23 inches ID and 24 inches OD.

The compressor casing for this volume has 36 by 24-inch nozzles. The discharge velocity is 60.4 (144)/450 = 193 fps. The discharge static pressure correction is calculated from the difference in velocity (velads): $[(193)^2 - (110)^2]/64.4 = 392$. The energy density on the discharge line is 392 ft-lb/lb short of the objective 10,000 ft-lb/lb. Nearly 4 percent of the machine's output is in the discharge nozzle. The conditions for using the efficiency values from Figure 6.10 required C_{m1} to be equal to C_{m3}. The corrected efficiency from conditions of static discharge to total discharge conditions is (10,000 − 392) 80.0/10.000 = 76.6 percent.

This condition can be corrected by providing a 27-inch discharge nozzle on the compressor casing, or by installing a 24 x 32 divergence tube 100 inches long to convert the excessive discharge velocity to static pressure. The velocity in the 32-inch tube is

110 fps and equal to the suction velocity. This restores the dyanmic efficiency to 80 percent. The power required to operate the single-stage compressor is

$$\text{Gas } hp = w\, L_{ad}/550n \quad (6.26)$$

$$= 110\ (10{,}000)/550\ (0.766) = 2{,}610.$$

The mechanical losses from the bearings, etc., which do not affect the temperature of the gas, are not included in Figure 6.11. It is proposed that the mechanical losses are well within the frictional hp of the equation: fph = (dynamic hp)$^{0.4}$.

The frictional hp for this case would be (2,610)$^{0.4}$ = 25. The driver should include this value, plus a 10 percent power tolerance, which totals 2,895 bhp at 3,700 rpm.

Corrections

The data plotted on Figure 6.11 included as much test and operating performance as available. The curves were developed from loss analysis, simplifying assumptions and averaging. The correction for the high discharge velocity is an example of how other corrections may be applied. The Baljé efficiencies include the normal interstage leakage and disk friction. Subsequent parts of this book illustrate how to cope with the abnormal leakage, disk friction evaluation, multistaging losses, materials, ultimate speeds, control features and economics.

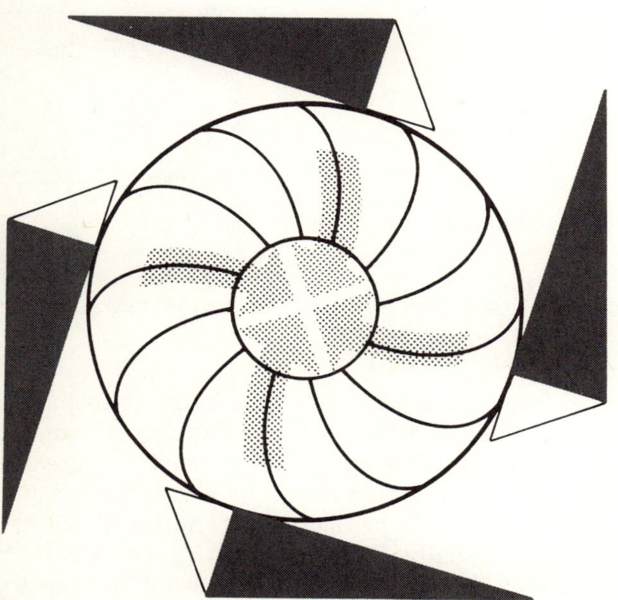

7 Centrifugal Impellers

The high speed rotation of the impeller imparts the vital dynamic velocity to the gas flow therein. The dynamic head is converted to static head by the volute chamber, return diaphragms and other divergent channels. The three equations that express the relationship between the speed, impeller diameter and gas head are

$$L_{ad} = (R_c^{\sigma} - 1)\ 1{,}545\ T_o\ Z_a/\overline{m}\sigma, \text{ ft-lb/lb} \quad (6.3)$$

$$L_{dy} = U^2 q_{ad}/g$$

$$d = 12\ (D_s N_s)\ L^{0.5}/N. \quad (6.20)$$

This chapter is concerned with the physical strength of the impeller. There is general consensus that the impeller stresses are greatest at the hub. This is true for the tangential forces in disks of uniform thickness. The disk is usually tapered, having several times greater cross-sectional area at the hub than is necessary at the tip. The impeller elongation Y that is due to centrifugal forces is determined from the equations:

$$Y = (S_t - S_r/3)\ r/E \quad (7.1)$$

$$F_c = 0.0000142\ WdN^2. \quad (7.2)$$

The elongation of a 7.85-inch diameter aluminum expander rotor is about 0.0023 inch at the tip when the speed was 1,075 fps. The tangential stress (S_t) is about 12,000 psi at the rim and 36,000 psi at the hub. The radial stress (S_r) is 17,000 psi at the rim and only 1,200 psi at the hub. The axial stresses are minor and negligible. The hub elongation is 0.0034 inch under these conditions. The cross-sectional area of the hub is usually enlarged, the shaft bore is undersized by 0.006 inch for a low heat shrink fit and the hub section is finished and fit on the shaft under a compression force. The stress forces are neutral about one inch radially from the shaft. The centrifugal force (F_c) is determined by Equation 7.2, where W is the impeller weight in pounds, N is rpm and d is the inches of diameter.

The integral forged rotor with milled or electronic eroded blading has the greatest structural strength module. The various axial and radial vaned, shroudless impeller rank next in inherent design strength. The conventional end suction impeller with shroud, riveted backlay vanes constitute the weakest design. These appendages have individual centers of gravity and force patterns which produce complex and serious vibration problems. The loosening of rivited components undoubtedly originates from this source.

Table 7.1
Density of Impeller Materials

Material	lb/cu ft
Bronze, Inconel, Monel	550
Stainless and Carbon Steels	485
Titanium Alloys	300
Aluminum Alloys	

Table 7.2
Classification of Impeller Types

Class	Description of Impeller	C	K
1.	Back-lay vanes, with riveted shroud	0.67	7.0
2.	Radial vanes, with riveted shroud	0.50	8.0
3.	Back-lay vanes, with welded shroud	0.40	9.0
4.	Radial vanes, with welded shroud	0.30	10.3
5.	Radial milled vanes, no shroud	0.15	14.5
6.	Axial blades, b/d = 0.25, fir tree	0.55	7.7
7.	Axial blades, b/d = 0.15, fir tree	0.50	8.0
8.	Axial blades, b/d = 0.10, fir tree	0.33	10.0
9.	Axial blades, b/d = 0.05, milled	0.14	15.2

The equation for determining the stress developed in the rotating impeller is

$$S' = \delta C U^2 / g \text{ lb/sq ft.} \quad (7.3)$$

The densities of several likely materials used in impeller design are given in Table 7.1.

The relative effective strength of material for making impellers is best expressed as the quotient of the yield strength divided by the density.

Figure 7.1 lists 16 impeller materials giving this strength-density ratio versus operating temperature. Titanium far outclasses all other materials. Aluminum alloys outclass the stainless steels, except type 410. Equation 7.3 is modified to produce the effective tip speed U:

$$U = K(S'/\delta)^{0.5} fps \quad (7.4)$$

where $K = (32.2/C)^{0.5}$ and $S_{psi} = S'/144$.

An evaluation of the C and K design factors for various configurations is given in Table 7.2.

The b/d in Table 7.2 relates the axial blade length to the outer diameter. The design factors in the table utilize 50 to 75 percent of the minimum yield strength. These factors were generalized from many sources (12, 13, 14, 15, 16). One reference states that "higher admissible tip speeds may be attained with rotor geometries which are carefully stress balanced by rigorous design analysis (12)."

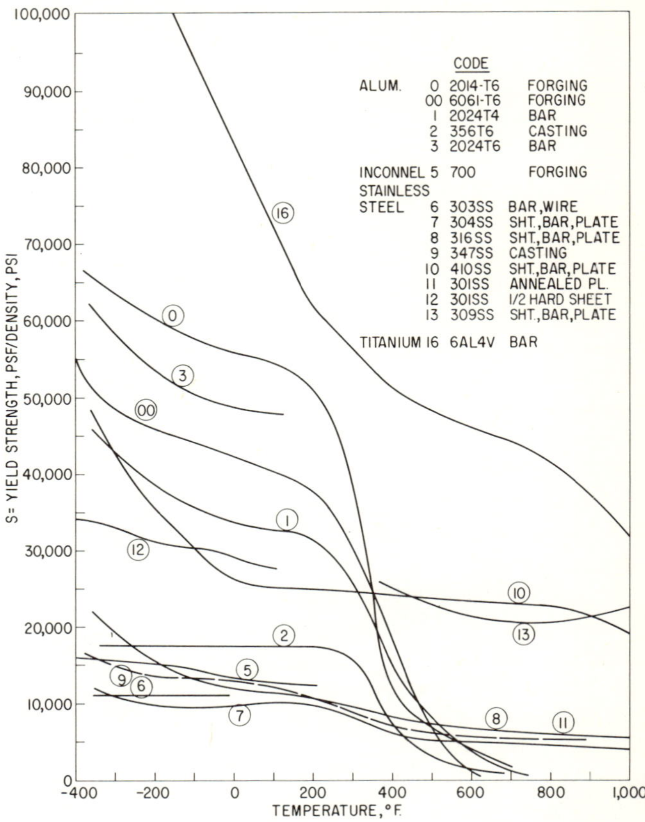

Figure 7.1. Strength-density ratio versus operating temperature for various alloys.

Some 200 aluminum impellers, ranging from 1 to 1,600 pounds and from 12 to 60 inches in diameter were spin tested—some to tip speeds as great as 3,000 fps. Most of the impellers elongated (diametrically) 0.1 to 0.5 percent before failing (17). This elongation usually functions as a brake. When the impeller makes contact with the casing, it drags to a stop.

All industrial impellers are given a *spin-test* of at least 115 (some 121) percent of maximum continuous design speed before assembly. When the compressor is motor driven at constant speed, the impeller integrity is thereby assured to be operating below 75 percent of the minimum yield strength. Steam turbine driven compressors can be operated at 110 percent of design speed. The turbine governor is set to trip at 110.3 percent. The square of this increased speed times the 75 percent minimum yield produces an impeller stress of 91.4 percent of the minimum yield strength. This 8.6 percent is an uncomfortably small safety margin to allow between the *spin-test* stress and the stress that a turbine governor may permit, particularly at startup when the turbine driver has the potential for

Table 7.3
Tip Speeds at 60 Percent Yield Point

Materials	Table 7.2 Class (in fps) 3	5	9
Aluminum 6061-T6 at 500°F	550	650	1,450
Aluminum 6061-T6 at 100°F	1,200	1,450	3,200
Aluminum 6061-T6 at −350°F	1,550	1,850	4,100
Type 316 S S at 100°F	650	800	1,750
Type 410 S S at 100°F	950	1,150	2,500
Titanium 6 AL 4V at 500°F	1,300	1,500	3,600
Titanium 6 AL 4V at 100°F	1,600	1,900	4,200
Titanium 6 AL 4V at −350°F	2,100	2,500	5,500

much greater speed. The spin-test should be 115 percent of the turbine governor trip speed to maintain the working stress about 60 percent of the minimum yield strength. Table 7.3 was prepared to show the tip speed at approximately 60 percent of the yield point for several designs, materials and temperatures.

It is not uncommon to use 80 percent of the minimum yield strength at the coupling end of the shaft. The shaft is not exposed to corrosive conditions at this point. Inside of the casing, the shaft size is fixed by the critical speed rigidity requirements. The internal shaft stress is between 3,000 and 5,000 psi.

Carbon alloy steels and stainless steels of equal strengths have equivalent performance at ambient temperatures. The performance of ferritic type 410 SS is equivalent to aluminum at cryogenic temperatures. Austenitic 18-8 stainless steels are better than aluminum at temperatures above 250°F. Carbon steels are impractical to use at cryogenic temperatures. Martensitic nickel alloy steels should be used for this purpose. We know that some riveted Class 1 impellers as in Figure 7.2, made of ferritic type steels, are spin-tested to 1,000 fps. They experience sufficient distortion in the shroud that the periphery must be remachined and balanced. Class 3 impellers (Figure 7.3) are spin-tested to 1,250 fps without distortion. It was recently reported that a Class 9 ferritic steel impeller withstood 2,750 fps spin test without elongation. We also know that jet engines having Class 6, 7 and 8 impellers operate at blade velocities in excess of 1,800 fps during takeoff in every jet engine powered aircraft. They cruise at 1,200 fps and have a remarkable safe record

Figure 7.2. Sectional view of Class 1, riveted construction impeller having back-lay vanes. (Courtesy of Allis Chalmers Mfg. Co.)

Figure 7.3. Class 3 impeller for twin flow, three-stage impeller for large capacity, low head centrifugal compressor. (Courtesy of Elliott Co.)

Figure 7.4. Modern combination seal and lube oil console for major centrifugal compressor. (Courtesy of Cooper-Bessemer Co.)

during both operations. The metallurgy problem of turbomachines are minor compared to the design inadequacies.

Corrosion engineers of NACE have made an extensive study of the metallurgy problems that have beset some 300 refinery compressors for the past 25 years (15). The mean working stress on the impellers of these units is 37,500 psi, by Equation 7.3. Ten percent of these machines are working on a stress less than 25,000 psi, and 15 percent exceed 50,000 psi. Seventy percent of these machines were relatively trouble free from the startup. One-third of the remaining units, or 10 percent of the total, still have troublesome maintenance problems. There were eight impeller failures. One impeller failed at the periphery and could be classified as a metallurgy failure. It was diagnosed as sulfide stress cracking caused by excessive hardness. Another companion impeller definitely failed at a sharp cornered key slot. The other six failures were the result of similar design inadequacies and showed evidence of stress concentrations. In another instance loose shroud rivets were welded without annealing. This mistake was the cause for three impeller failures.

Clean Seal Oil

Two-thirds of all troubles involved the shaft seals. Twenty-five percent of the trouble was from dirty lube oil. The deluxe lube oil consoles of recent years have undoubtedly reduced this problem. Figure 7.4 illustrates a comprehensive lube oil console. The balance of the maintenance concerns the shaft seals, equally divided between the oil film type and the labyrinth. The latter are usually made of aluminum and set in the casing. The knife edges surround the shaft. Where the temperature exceeds 250°F, a soft stainless steel labyrinth is recommended. Eurpoean manufacturers usually machine the labyrinth edges on the shaft and set the knives on a soft babbit or aluminum half-cylinders secured in the casing. There was only one valid case of corrosion in a fluid catalytic cracking unit which was handling sour hydrogen sulfide gas. In this instance the 11-13 chrome impellers were upgraded to 17-4PH. AISI 4140 is the most prevalent steel alloy used for impellers and shafts for all services. Corrosion engineers recommend that the heat treatment be held to 27 Rockwell G, which still permits a yield stress of 127,000 psi. Figure 7.5 shows the effect of hardness on the yield strength in a hydrogen embrittlement environment (15).

Impeller Attachment

A half-sectional profile of an impeller disk is a right angle triangle shape having a peripheral angle of about 16°. The base profile includes a hub which projects ½ to 1 inch beyond the shaft. The hub design is significant in that it constitutes the impeller suction eye configuration, plus it transmits the driving torque from the shaft, provides the necessary disk rigidity and fixes the impeller balance position. The U.S. manufacturers use square keys staggered 90° to complement the balance of multistage machines. The impeller hubs are reamed to a *push* fit of 0.001-inch clearance or to a *drive* interference fit of 0.003 inch. Cantilever single and dual-stage machines generally do not have a key and are given an approximate 0.009-inch interference fit. The shaft is drilled so as to require a 2-ton hydraulic jack to remove the impellers. European practice usually avoids keys and uses a heat shrink of about 0.006 inch. This can be accomplished by an atmospheric steam hose applied to the impeller which gives 140°F of expansion or 0.001 inch per diametral inch.

The strength of an interference fit is determined from the following equations. The radial pressure P_c on the shaft contact surface is

$$P_c = \Delta E \ (d_s^2 - d_i^2)\ (d_h^2 - d_s^2)/2d_s^3\ (d_h^2 - d_i^2). \qquad (7.5)$$

Where the shaft diameter (d_s) is 5.5 inches and (d_i) is the internal bore or zero when solid, the impeller hub (d_h) is 7 inches and E is the 30,000,000 psi, the elasticity modulus for steel. The interference is delta, Δ (17). The tangential stress S_t is determined from

$$S_t = P_c(d_h^2 - d_s^2)/(d_h^2 + d_s^2). \qquad (7.6)$$

Where the tangential stress is limited to 8,000 psi and applied to the above equation, the surface contact pressure P_c is applied to Equation 7.5, it requires an interference of 0.00264 inch.

The force required to remove the impeller from the shaft is

$$F_s = f\pi d_s L_s P_c \qquad (7.7)$$

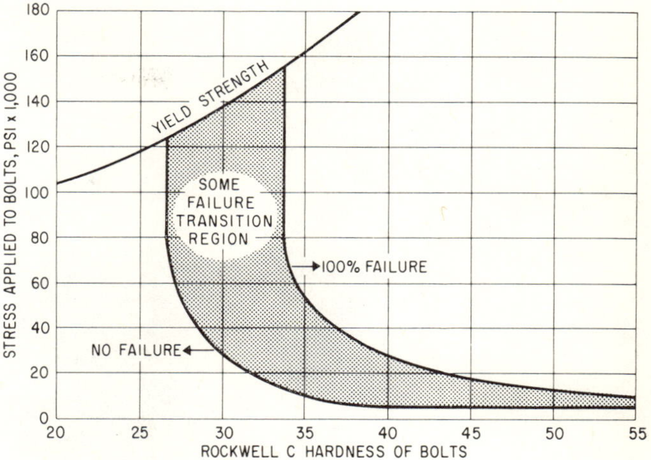

Figure 7.5. Cracking susceptibility of AISI 4140 bolts in H_2S–H_2O for one year at 40°C and 250 psi (from Warren and Beckman, Corrosion, *Volume 13, No. 10, 1957).*

where f is the coefficient of friction (0.12), L_s is 4 inches (the length of the axial contact of the impeller on the shaft), F_s = 0.12(3.14)5.5(4)1,900 = 15,700 pounds.

The torque transmitting capability is 15,700 (2.75/12) = 3,600 ft-lb.

The power potential without slipping is 3,600 (6,000 rpm)/5,250 = 4,120 hp.

The torsional stress in the hub is usually modest and seldom exceeds 10,000 psi. The size of the shaft is predicated upon the critical speed characteristics. The nominal stress is

$$S_t = 5.1(\text{Torque, in-lb})/\text{dia.}^3 \qquad (7.8)$$

For the 5.5-inch diameter shaft just cited, the torsional stress would only be 1,210 psi (16). A sharp cornered keyway with a snug shaft fit can develop stress densities two to three times the nominal stress. A loose fit would compound this stress concentration. The use of two close fitted *feather* keys has merit. The thickness of a *feather* key is one-fourth of the width (peripheral) dimension. All keyways should have well rounded fillets. After studying the corrosion engineers' report on refinery turbomachinery problems and from discussion with the manufacturers, it is the writer's conviction that an impeller attachment specification is needed to guard against loose attachments, sharp cornered and overstressed keyways (14).

A second conviction is that all impellers of sophisticated design should be an investment casting

Figure 7.6 (above). A matched set of Class 6 investment cast 17-4 PH stainless steel impellers for a "Four Poster" air compressor. (Courtesy of Chicago Pneumatic Tool Company.)

or fabricated by electroerosion, arc-welding and/or milled.

The NACA impeller performance curve, Figure 6.8, shows that it is possible to attain pressure coefficients (*qad*) as high as 0.78. This impeller is a laboratory model and is not to be confused with the industrial category. The best examples of high efficiency and high head impellers are found in the so-called "Four-Poster" compressors. These plant utility air compressors have cantilevered pinion shaft driven impellers of which those in Figure 7.6 are typical. The frame for these units is built around a herringbone gear case having dual horizontal or radially oriented pinion shafts (Figure 7.7).

The pressure coefficient for these impellers which have a slight back-lay is about 0.6. They operate at speeds up to 50,000 rpm and tip speeds of 1,200 fps. Figure 7.8 shows one of the larger 15,000-cfm, 3,000-hp air compressors. This type of machine is available in sizes of 1,500 cfm and 400 hp to 15,000 cfm. The bases of these units include the necessary intercoolers, separators, drains and piping.

Figure 7.7 (below). The impeller in Figure 7.3 forms the two first stages of this twin, three-stage compressor. The steam turbine driver on the left illustrates the four categories of axial flow rotors. The Class 6 blade height over the rotor diameter ratio of 0.25 is nearest to the compressor. The Class 9, 0.05 b/d rotor is on the extreme left end of the shaft. Class 7 and 8 have intermediate b/d ratios. (Courtesy of Elliott Co.)

Industrial grades of Class 5 back-lay vaned impellers, which are milled or investment cast without a shroud, have a pressure coefficient of about 0.55. The same quality impeller, Class 1 or 3, may have a pressure coefficient of 0.53 q_{ad}.

Industrial grades of Class 5, radial impellers without a shroud have a pressure coefficient of 0.65 qad. Similar Class 2 and 4 impellers with shrouds average about 0.63 q_{ad}. These coefficients are optimum values applicable to the design point. These q_{ad} values may rise as much as 10 percent greater than the design point as the flow is reduced from the design point to the point of surge or stall. The coefficients will reduce as much as 25 percent when the flow rate is increased to the choke condition or "Stonewall."

Materials for Casings

The API Standard 617 set up the following requirements for steel casings (12).

1. Air or nonflammable gas at a design pressure over 250 psig.
2. Air or nonflammable gas at a calculated discharge temperature over 450°F at any point within the range of the maximum continuous speed.
3. Flammable or toxic gas at a design pressure over 75 psig.
4. Flammable or toxic gas at a calculated discharge temperature over 350°F at any point within the range of the maximum continuous speed.

Metallurgy

1. Cast steel for the above categories operating under 500°F and 1,050 psig are satisfied by the ASTM Spec. A216 Grade WC-8. Weldments are equally acceptable when in full compliance with the ASME Section VIII Code.
2. Operating temperatures and pressures greater than 500°F and 1,050 psig require forged steel which complies with ASTM Spec. A-235 or A-237.
3. For operating temperatures below —20°F, the steel shall have an impact strength of not less than 15 ft-lb (ASTM E 23). ASTM Spec. A-352 Grade LC-B is satisfactory for —50°F, and Grade LC-3 is good for —150°F. These martensitic steels contain 2.25 percent and 3.5 percent nickel respectively.

Figure 7.8. This 15,000-cfm, 125-psig, 3,000-hp air compressor is one of the largest "Four Poster" models ever built. Note the intercoolers and accessories are packaged in the structural base. Space and low installation costs are two additional features. (Courtesy of Joy Manufacturing Company.)

Cast iron is acceptable for all other applications except those specifically defined earlier. The applicable cast iron specification is ASTM-A278. Horizontal-split cast steel or weldment cases are used for gas pressures up to 650 psig, discharging less than 350°F and having a molecular weight greater than 16. Pressures above 650 psig and less than 1,050 psig require a cast steel or weldment *barrel* design. Pressures in excess of 1,050 psig require a forged steel barrel design. Where the partial pressure of hydrogen or helium in the gas handled exceeds 250 psig, the barrel design should be required. The flat joint of the horizontal-split case cannot contain these low molecular weight gases. Weldments are proving popular in lieu of cast steel casings because of the improved delivery. A faulty steel casting can set the schedule back two months. This can happen all too often and accounts for most late deliveries of cast steel casings. The manufacturer has complete control of the production schedule for

Figure 7.9. This large centrifugal compressor is made from weldment sections. The unique design and application won the 1967 Lincoln Welding Award. The unit has capacity in excess of 100,000 acfm and 20,000 hp. (Courtesy of Elliott Co., Div. Carrier Corp.)

weldment casings. They can be fabricated for approximately the same cost (see Figure 7.9).

Nodular (ductile) iron offers an alternative to cast steel and weldment casings at a lower price. There appear to be certain prejudices concerning the use of nodular iron. It can be cast with the same confidence that cast iron can be delivered. Weld repairs can be made as readily as cast steel and better than with cast iron. ASTM Spec. 395-61 Class 60-45-15 provides 15 percent elongation. ASTM Spec. 439-62, Class D-2C and D-5 provide 20 percent elongation. This ductility is equivalent to cast steel and its alloys at normal operating temperatures. At elevated temperatures as experienced in a major fire, the superior strength of cast steel is granted. However, by the time the superiority of the cast steel is evident, the system pressure should be vented by the protective devices or by a ruptured vessel in the system.

Testing

Mechanical Running Tests should be performed as specified in API Std. 617. Some form of performance tests should be made and witnessed. The test procedure proposed by O'Neill and Wickli is recommended to resolve the test data (11). Any head-capacity curves that can be developed from these tests should be useful in establishing the credibility of the machine's performance. These tests are especially valuable when the compressor design exceeds the normal parameters.

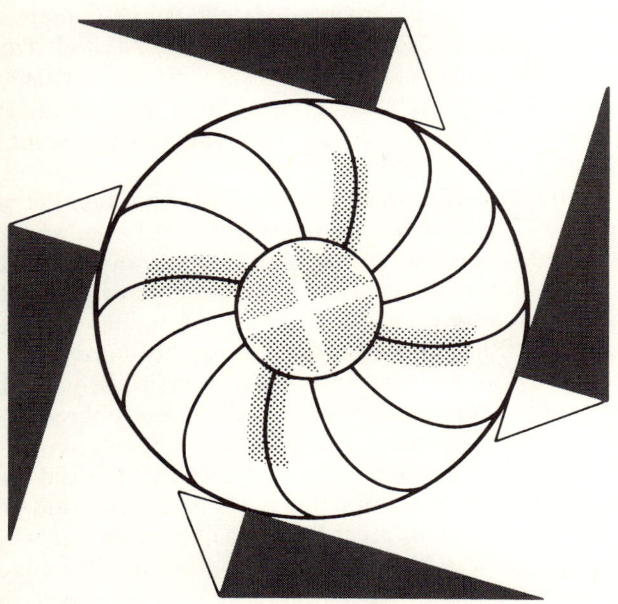

8 Compressor Shaft Seals

The development of centrifugal compressors for high-pressure services has been greatly limited by the capabilities of the shaft seals. This chapter describes the various shaft seals and their performance in typical process compressor service.

The shaft seals may be divided into the following categories: (a) labyrinth, (b) restrictive carbon rings, (c) mechanical face contacting rings, (d) floating bushing, oil film and (e) dynamic or centrifugal pumping chokes. The first two seal categories are usually operated dry, when the casing pressure is less than 100 psig and the outer terminal exhausts to the atmosphere.

Labyrinth Seals

Labyrinth seals are used extensively for interstage shaft and impeller eye sealing. Figure 8.1 illustrates the simplest form of labyrinth seal as it is applied to the shaft. Most labyrinth seals are machined from bronze, babbitt or aluminum and fit in the casing in the form of a horizontal split cylinder. Other designs use aluminum rings cut from coil strip about 0.020 inch thick. The labyrinth rings are caulked into small grooves machined in the casing. An alternate type of coil strip blading has a J-section form. The hook form is pressed into the groove, instead of using the more tedious caulking procedure. When the temperature exceeds 250°F, Monel or stainless steel strips are used. The labyrinth seals are fit as snug as possible or given a slight negative clearance.

The clearance usually provided for turbomachinery shaft seals, using labyrinth blading, is about 0.002 inch. The clearance is considered excessive when it exceeds 0.030 inch. Labyrinth seals on casings operating at pressures in excess of 50 psig have 8 to 20 blades. Lower pressure casings contain 3 to 6 blades. Leakage tolerance is frequently stated as approximately 0.5 percent of the compressor capacity. This is a rather crude and irrelevant gage. The leakage can be estimated with reasonable accuracy by the following procedure. The velocity of the gas through the annulus of the smooth shaft and sharp blades as shown in Figure 8.1 is

$$V = 30.4\, K(\Delta \text{psi} \times v_{sm})^{0.5} \text{ fps} \qquad (8.1)$$

$$K = 5/[7+(n-1)]. \qquad (8.2)$$

The constant 30.4 is the square root of the product of 144 (square inches per square foot) and $2g$ (64.4). The Δ psi is the flow resistance through

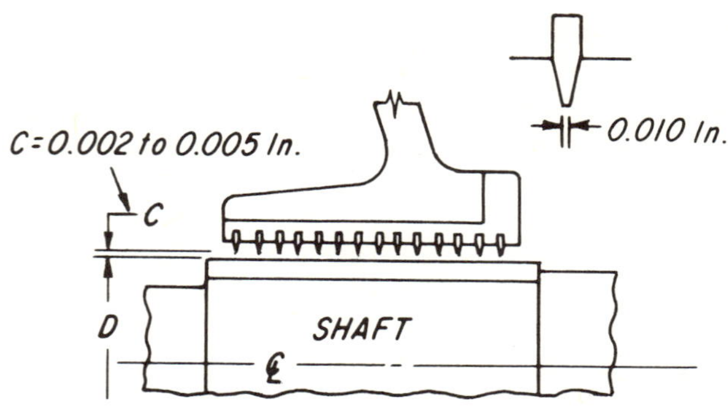

Figure 8.1. Straight labyrinth seal as used on an interstage shaft, on suction eye seals and on low pressure outside seals. Staggered and tapered diameters can reduce the leakage as much as 40 percent.

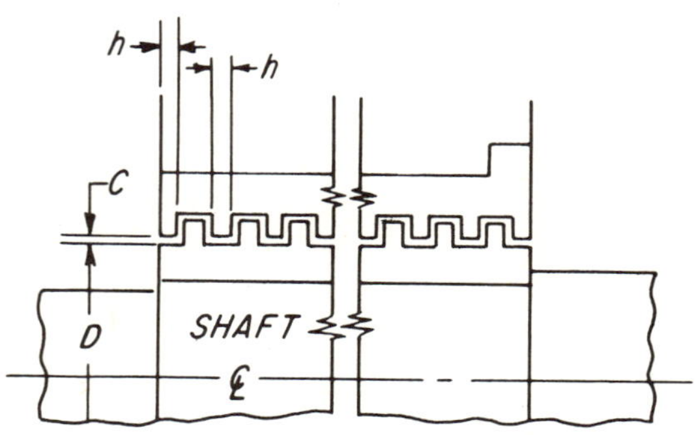

Figure 8.2. Interlocking labyrinth seals require clearances of 0.014 plus 0.004 per diametrical inch. Clearances greater than 0.030 inch are considered excessive.

the shaft seal and v_{sm} is the mean specific volume of the gas flowing through the labyrinth. The flow coefficient K is the ratio of the actual velocity through the seal to the potential dynamic velocity. The small n represents the number of barriers or labyrinth blades included in each seal. An approximate flow coefficient of 0.75 is a reasonable value for a smooth shaft in a plain bushing. A seal composed only of a series of step-cuts or staggered barriers, as shown in Figure 8.2, has a K factor of 0.6. The resistance of a staggered offset and a bladed labyrinth is the product of the two resistances. If the ratio of the shaft radial clearance to the blade spacing exceeds 15 percent, the "carry-over" velocity tends to increase the K factor by the square root; i.e., a 0.49 K factor becomes 0.7 K. Another demonstration of this carry-over velocity has been observed in testing flat topped male thread labyrinth seals. The K factor for a machine thread is consistent with Equation 8.1 where R_c is less than 3. When the pressure ratio is great than 3 and less than 4, the numerator in Equation 8.1 goes to 8; it goes to 10 for ratios greater than 4. The objective is to avoid a straight passage for the gas flow, especially where the clearance is 0.030 inch or more. The more torturous passage and the more turbulent flow that can be created, the less leakage that is experienced. The speed of rotation has a negligible influence on the leakage rate.

Example

The following example will illustrate the application of the above equations. An FCC main air blower handles 133,000 cfm or 167 pps of air from 60° F and 14.5 psia to 46.0 psia. The unit is driven by a 17,500-hp motor at 3,600 rpm and the adiabatic efficiency of the five-stage compressor is 76.5 percent in Table 8.1.

The outside diameter of all impellers is 48 inches. The mean diameter of the eye seal is assumed to be 30 inches, and the clearance is 0.020 inch, making the area A_e = 0.013 square feet. The mean diameter of the shaft seal is 10 inches with the same 0.020-inch radial clearance; the area A_s is 0.00435 square feet. The pressure drop through the interstage shaft seal is presumed to be 50 percent of the static differential, between the respective stages. The gas head is the product of this differential, the constant 144 square inches per square feet and the mean specific volume of the gas through the seal. The gas head across the seal at the impeller eye is determined in the same manner, except that the effective differential across each seal is presumed to be 75 percent of the respective stage pressure rise. The leakage that is expected from such a compressor is listed in Table 8.1 and illustrated in Figure 8.1.

The two seal differential percentages are arbitrary, but reasonable and amenable to more specific reference data. The shaft leakages are all less than 0.02 percent of the total weight flow of 10,000 pounds per minute (ppm). The third stage leakage is slightly in excess of 0.02 percent and could be corrected by adding another barrier to the labyrinth seal. The average leakage through the impeller eye is 0.058 percent of the total flow rate. The outside eye seal leakage is greater than the capacity of the shaft seal (49 − 11), 38 acfm or 5.7 ppm. It is advisable to vent this surplus from port D to the third stage

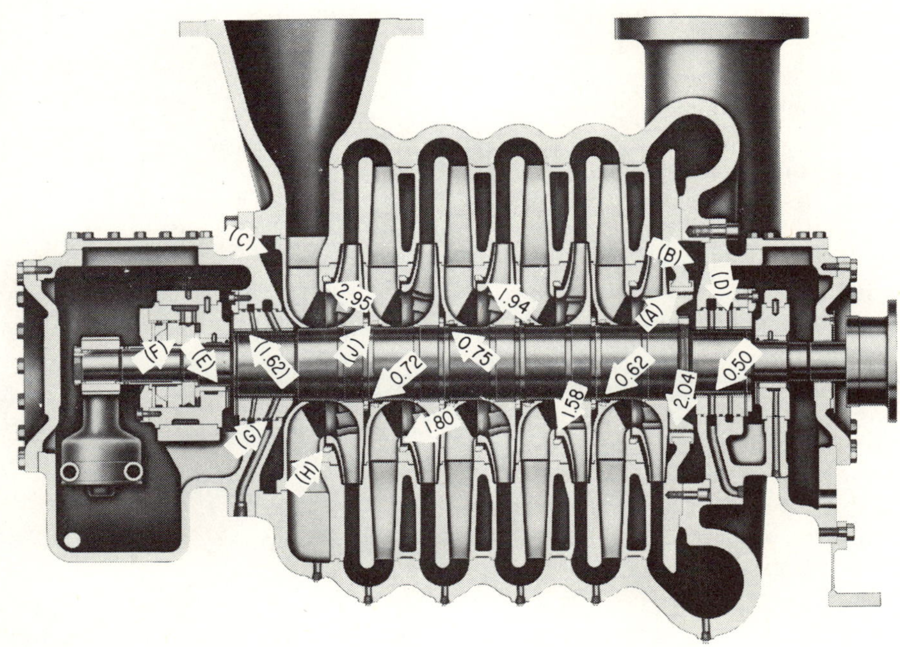

Figure 8.3. Various labyrinth seal applications and leakages (cfs) are shown for a similar five-stage centrifugal compressor.

(A) The seal between the discharge and equalizer chamber.
(B) Leakage is 49 acfm with seal described in Table 8.1.
(C) Suction chamber draws in 18.9 acfm of outside air.
(D) An injector or inductor port. Sweet gas can be introduced or an eductor can withdraw gas from this port insuring against lube oil contamination when handling sour gas.
(E) Radial sleeve bearing.
(G) Can be used as a drain as shown or as a source of feeding sweet gas.
(H) Depicts the suction eye seal where the leakage is 55 acfm.
(J) Shows the shaft seal between the first stage impeller back and the second stage suction, where the leakage is estimated to be 20.4 acfm.

diaphragm. The pressure at the outside seal is 46 − 0.75 (46 − 36.8) = 39.1 psia. The pressure in port D is 28.6 + 0.2 (line loss) = 28.8 psia. The differential pressure is 10.3 psi. The dynamic velocity through the seal is $30.5\ (10.3 \times 7.0)^{0.5}$ = 260 fps. The actual velocity as retarded by the 12-blade labyrinth seal is (260 x 0.24) = 62.5 fps. The area is 0.013 square feet, and the flow is 49 acfm or 7 ppm. Figure 8.3 shows the various seals in a five-stage centrifugal machine. It also gives the anticipated leakage through each seal. Table 8.1 gives similar leakage data.

The temperature of leakage gas is that of the higher pressure condition. It may be considered to be an isothermal flow, or, at best, the flow is at constant enthalpy, where the Joule-Thomson effect may show a small temperature drop. The hot gas leakage will blend with the incoming gas, raising the temperature of the mixture by the mass-heat ratio. For the example cited, the discharge temperature is 345° F, ΔT is 285° F and ΔT_S is 57° F per stage. The temperature of the gas returned from port D to the third stage diaphragm is assumed to be 200° F by reason of the ambient radiation from the small tubing and the small mass flow. The effect of such small leakage is negligible. A leakage of 5 percent through the fifth stage eye seal raises the suction temperature from 272° R to 276° R and the discharge from 337°F to 342°F. From this information, it follows that each percent of internal shaft and eye seal leakage can account for raising the discharge temperature one degree. This generality is only applicable to relatively large diameter impellers (greater than 30 inches). This recycling and abnormal heating directly affects the dynamic or total adiabatic losses. The discharge shaft seal leakage is considered to be a portion of the mechanical losses, along with the gear and bearing friction.

The effect that the labyrinth leakage has on the dynamic efficiency can be appreciated by the following evaluation: The sum of the average internal gas leakage plus the outside seal leakage is divided by the total weight flow of the compressor. For this example, the total power loss due to seal leakage is $[(8.72/5) + (29/5) + 7.0]/10{,}000 =$ 0.00145 percent of the power input. In large process compressors, the discharge seal chamber is balanced with the suction system. There is no mechanical loss involved in permitting the suction gas leakage to the ambient. There is just a commodity loss. The venting loss of the high-pressure gas back to the suction seal chamber is charged to the internal dynamic losses.

Restrictive Carbon Rings

This type of shaft-seal has been used extensively for the past 50 years on steam turbines. The steam encountered by the carbon rings is relatively clean and moist at the exhaust condition.

The latter offers the necessary lubricant to minimize the wear and to sustain the carbon bond. Fifteen years of such service is commonplace. Figure 8.4 illustrates a typical carbon ring seal element. It shows an assemply of the two, three-piece carbon ring elements. The first ring is cut in three segments with tangential separations. These segments are backed with identical size carbon rings having three radial separations. The solid sections of one segment will overlay the alternate cuts and wear gaps. Both rings are secured to the shaft by means of garter springs. These springs also take up the wear. The clearances used for carbon ring shaft are less than used for labyrinth. The rings are given a generous external radial clearance to adjust for shaft *run-out*. The gas leakage can be estimated from Equation 8.1, using $K = 0.75$ and $C = 0.005$.

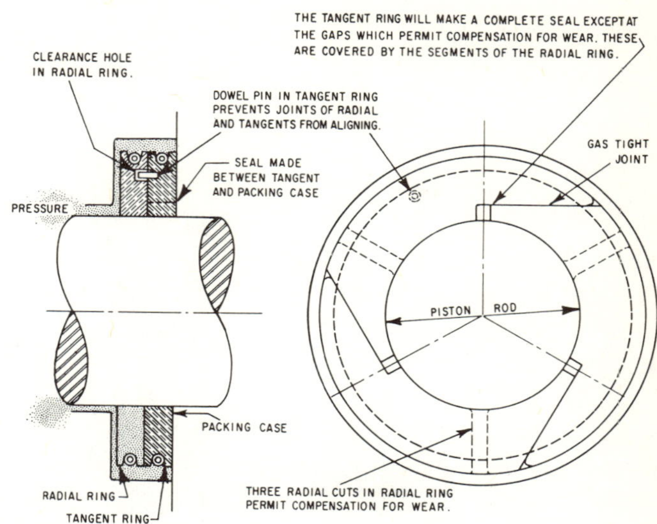

Figure 8.4. Restrictive ring shaft seal side view of the gland is shown at the left. A front view assembly of a tangential cut and a radial cut segmental carbon ring.

Balancing Piston

The leakage and thrust loads reported in Tables 8.1 and 8.2 presume no balancing pistons are applied. There is a surplus leakage of 48 less 11 or 38 acfm that can be returned to the fourth stage volute vent. The disk that supports the 12-barrier labyrinth seal "A" constitutes a *balancing piston*. The mean effective pressure in back of the impeller disk is approximately 75 percent of the head developed for each respective stage. The shroud face is exposed to about 82 percent of the added differential pressure per stage. The effective pressures and thrusts for the 31.5 psig air compressor are shown in Table 8.2. The overall thrust for the air compressor without a balancing piston is approximately 15,000 pounds in the direction of the suction connection or the outboard. The compressor is presumed to have a 16-inch Kingsbury type dual thrust bearing with a net supporting surface of 155 square inches. The thrust load would be 97 psig for the machine as described with a 10-inch hub, 36-inch eye for the first three stages and 30-inch diameter eye for the last two stages.

A 36-inch balancing piston is connected to the discharge chamber. The pressure therein is 1.2 psig and connected to the suction chamber "C". The overall thrust is reduced to 9,250 pounds or 60 psig. These bearing pressures are acceptable and only one-fourth of the ultimate bearing loading capabilities. A 40-inch balancing piston creates a 16,645-pound

Table 8.1
Labyrinth Seal Leakage for 17,500 hp FCC Air Compressor

State	1	2	3	4	5	Out
Discharge pressure, psia	18.2	22.8	28.6	36.8	46.0	46.0
Discharge temp., °*R*	566	617	672	732	797	660
Shaft seal, Δpsi	1.9	2.3	2.9	4.1	4.6	23.0
Shaft seal Pm., psia	15.4	19.9	24.3	30.7	39.1	32.2
Shaft seal Sp. Vol., cf/lb.	13.6	11.5	10.3	8.9	7.6	8.7
Dyn head vel.(*). fps	155	156	166	183	180	136
Number of barriers	4	4	4	10	10	10
Flow coefficient, *K*	0.50	0.50	0.50	0.31	0.31	0.31
Leakage Q(†), acfm	18.9	20.4	21.9	14.8	14.6	11.0
Leakage wt., ppm	1.23	1.77	2.13	1.66	1.93	1.27
Eye seal, Δpsi	2.9	3.5	4.4	6.1	6.9	10.3
Eye seal Pm., psia	14.9	18.7	23.5	29.7	38.0	35.0
Eye seal Sp. Vol., cf/lb.	14.2	12.2	10.6	9.3	7.9	7.0
Dyn head vel.*, fps	196	199	208	228	224	260
Number of barriers	8	8	8	8	8	12
Flow coefficient, *K*	0.36	0.36	0.36	0.36	0.36	0.24
Leakage Q†, acfm	55.0	56.8	58.3	64.0	62.8	49.0
Leakage wt., ppm.	3.9	4.7	5.5	6.9	8.0	7.0

*Dyn Head Vel = $(2g \times 144 \text{ in}^2/\text{ft}^2)^{0.5} [\Delta psi(10.73 \times T_m/\overline{m}P_m)]^{0.5}$; fps.

Vel = $30.5(\Delta psi \times v_{sm})^{0.5}$ fps.
†Leakage Q = AKV60 ft/sec. A = $d\pi C/144$ ft². $K = 5/[7 + (n-1)]$;
$Q = 0.262KV$, where d = 10 in. and C = 0.020.

Table 8.2
Determination of Thrust Load for Centrifugal Compressor*

Stage	1	2	3	4	5
Discharge pressure, psig	3.7	8.3	14.1	22.3	31.5
Suction pressure, *Po*, psig	−0.2	3.7	8.3	14.1	22.3
Stage differential press, psig	3.9	4.6	5.8	8.2	9.2
MEP_s on shroud, psig	3.0	9.2	12.9	20.6	28.6
MEP_d on disk back, psig	2.8	7.9	12.5	20.1	28.0
Thrust to coupling or inboard					
Shroud area, A_s sq in.	811	811	811	1101	1101
Shroud thrust, $A_s MEP_s$ lb	2,430	7,460	10,450	22,700	31,500
Eye area, A_e *sq in.*	932	932	932	630	630
Eye thrust, $A_e P_o$, lb	−185	3,440	7,750	8,900	14,050
Total thrust inboard, lb	2,245	10,900	18,200	31,600	45,550
Thrust to suction or outboard					
Disk (Area - 1731)MEP_d, lb	4,850	13,700	21,600	34,800	48,500
Net thrust to outboard, lb	2,605	2,800	3,400	3,200	2,950
Total thrust	14,955 or 96.5 psi				

*OD of Impeller is 48 inches, Hub = 10 inches, Eye dia = 36 & 30 inches

Figure 8.5. Deep scallops, cut out of the disc between vanes, reduces the back pressure factor to 20 percent.

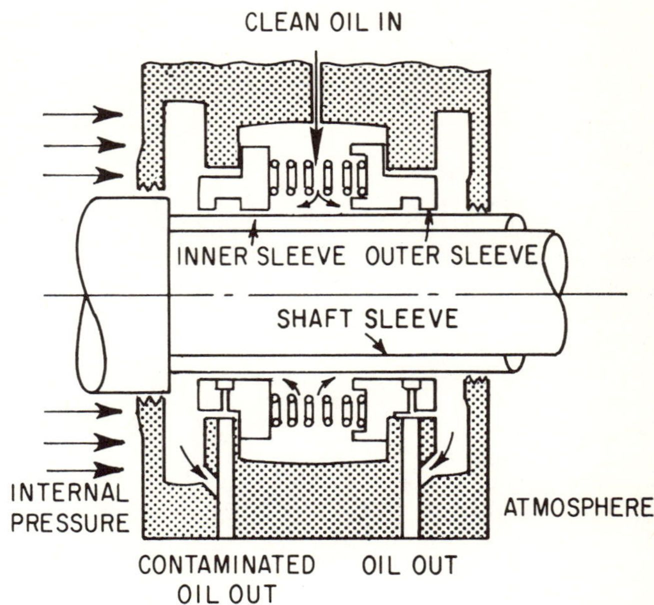

Figure 8.6. The liquid shaft seal contains two "L" section sleeves which are flushed with seal oil at 5 psi above the process gas pressure. The oil is supplied from an overhead reservoir.

thrust toward the coupling. A 30-inch balancing piston reduces the coupling directed thrust to 2,845 pounds or only 18 psig against the bearing.

Another technique of balancing impeller thrusts is to stack the disks back to back. In these five stages, the first three thrusts opposed to the opposite oriented fourth and fifth stages would result in a 1,655-pound outboard thrust. It is usually more economical to apply balancing pistons wherever it is possible.

There are numerous other devices used to reduce the impeller back pressure. A series of 0.200-inch ribs machined on the back of the impeller are so effective as a *pump-out* device that the 75 percent back-pressure factor is reduced to 15 percent. Deep scallops cut out of the disc between vanes reduces the back-pressure to 22 percent (see Figure 8.5). Pressure equalizer holes drilled through the eye hub, about every 3 to 4 inches apart, will reduce the back-pressure to 22 percent. The equalizer hold diameters should be about 2 percent of the impeller diameter. Several thin 0.060-inch ribs can reduce the back-pressure to 25 percent. A shroudless disc has a thrust factor of 30 percent of the differential pressure on respective front and back impeller areas.

Liquid-Film Seal

The shaft seals considered thus far were concerned with relatively low pressure compressors operating at discharges of less than 125 psig, 250° F, and the seals are operated dry. There are exceptions where lube oil is applied to carbon-ring seals by drop-feed lubricators. This is done to reduce the frictional heat of the higher pressure seal contacts and where the lube oil carry-over is not a nuisance. Where the compressed gas is inexpensive and expendable, the pressure range of labyrinth seals is extended beyond this arbitrary limitation; otherwise, the *oil-film* type seals are applied. The casing discharge pressure is reduced with a labyrinth seal to suction pressure and equalized to that suction seal chamber. The *oil-film* shaft seal consists of two sleeves which are supplied 2-3 gpm of seal oil at a midpoint about 5 psi greater than the gas chamber pressure. The inner sleeve facing the gas is shorter than the outer sleeve. It is also fit to within 2 mils clearance which will only permit a leakage of 1-3 gallons per day. This leakage is collected and gas separated by means of a trap draining the internal chamber. When the process gas carries undesirable contaminants, the trapped oil is drained to waste. The bulk of the oil circulation passes through the outer sleeve which is somewhat longer than the inner sleeve to break down the greater pressure drop. The circulation maintains the seal at or about 120° F with a temperature rise of about 5° F. The design of both sleeve cylinders is such that an adequate degree of radial movement is permitted and the sleeves may follow the shaft play. The entire assembly is sealed into the casing with suitable O-rings and gaskets to avoid gas leakage (19)(see Figure 8.6).

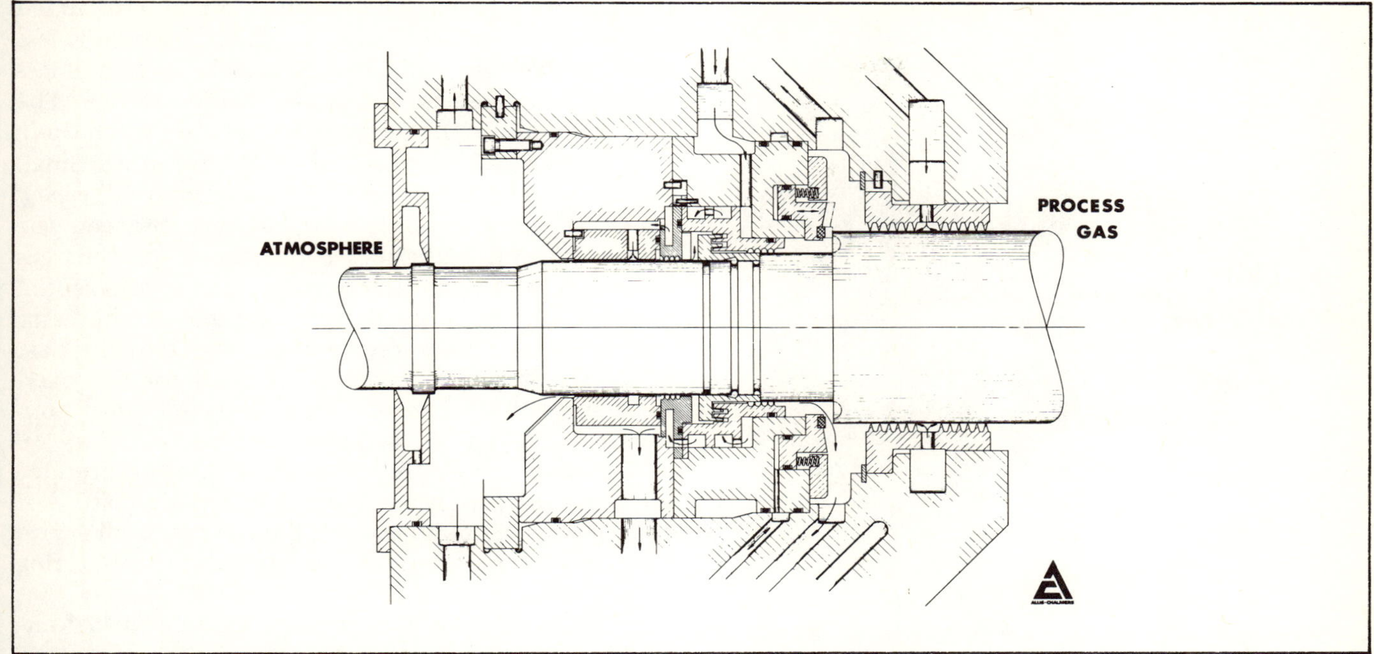

Figure 8.7. The trapped bushing shaft seal is operating on pressure as high as 2,800 psig and test stand pressures of 3,500 psig. The trap impeller sets up a dynamic resistance to restrain the gas passage. (Courtesy of Allis Chalmers.)

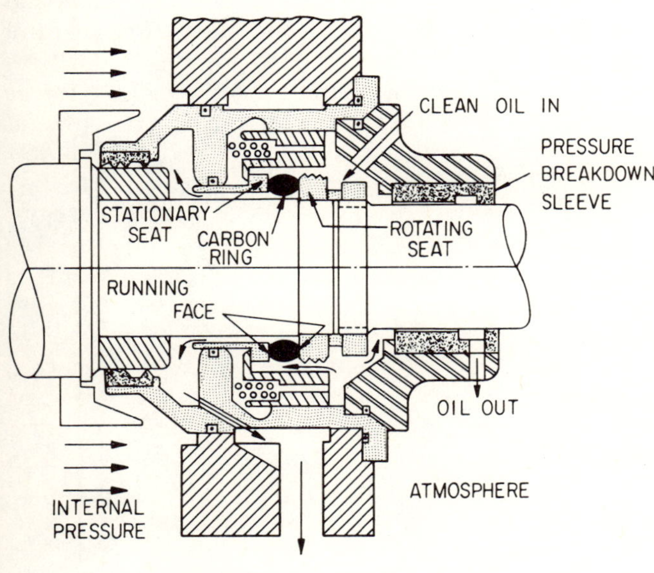

Figure 8.8. Mechanical contact seal with radial clearance reducted to zero, the sealing faces normal to the shaft instead of longitudinal and featuring an unattached floating carbon ring and a less complicated oil supply system with an automatic closing device.

Dynamic Seals

There are variations of the oil-film seal which use the dynamic pumping action of the inner sleeve. Instead of the straight sleeve, a conical sleeve is substituted which has rather large clearances over the shaft. The inner seal oil-flow is impeded by the dynamic pumping action of the cone acting as an impeller. There are other dynamic types of seals developed principally for gas turbines which operate at low pressures, 15 psig to high vaccum and temperatures up to 1,400° F. Their leakage is only a few pounds per day, and they have a life of one to five years. One of the more interesting seals is shown in Figure 8.7. It depends upon the trap action of an impeller containing two interfering rake-like teeth (20).

Mechanical Contact Seal

The difference between this and the oil-film seal is that the radial clearance is reduced to zero and the sealing faces are normal to the shaft instead of longitudinal. The seal requires a greater differential of 30 to 50 psig above the internal casing pressure. The features of the seal are shown in Figure 8.8. The unattached floating carbon ring is presumed to run

Figure 8.9. Mechanical contact shaft seals have no face clearance.. The faces are normal to the shaft and operate at 30 to 50 psig above the process pressure. They do not require an elevated seal oil tank. The seal has an automatic check to prevent back flow.

at half shaft speed. The mating surfaces are lapped to two microns. The inner gland oil leakage is less than the oil-film seal. The oil supply system is less complex and has an automatic closing device. The seal will contain the internal gas pressure when there is a loss of seal-oil pressure or at shutdown. Small auxiliary pistons are activated by internal casing pressure which maintains a contact between the stationary seat and the carbon ring to prevent gas escape to the atmosphere. The higher differential pressure on the seal oil permits higher compressor discharge pressure surges without gas blowing past the seals. It also eliminates the need for the overhead surge tank. The oil-film seal system must contain a 90 gallon (or larger) pressurized auxiliary oil reservoir to supply the necessary seal oil to produce a gas shut-off, equivalent to the automatic feature of the mechanical contact seal. This reservoir oil is also used to provide minimum bearing lubrication during shutdown (19).

Figure 8.9 shows a cartridge module for a four-stage, sunflower type compressor. It illustrates the unique hydrostatic bearing system and the shaft seals.

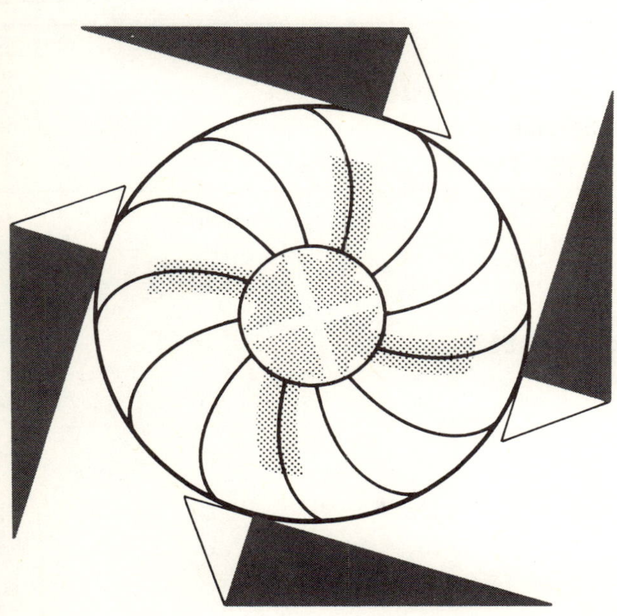

9 Refrigeration

The compression efficiency holds the key for selecting the most economical medium. Where does the economy of the centrifugal overtake the reciprocating machine? An economic analysis of the amortization, energy and maintenance is presented which illustrates an example case for refrigeration.

A brief study contrasting the economy of reciprocating and centrifugal compressors for process refrigeration should be of interest to the hydrocarbon processor. Having previously presented realistic and comprehensive methods for evaluating the compression efficiencies for both types of machines (see Chapters 2 and 6), the cost study that follows shows that these efficiencies very largely control the selection of the most economical machine.

Table 9.1 gives the annual operating costs for centrifugal refrigeration machines ranging from 400 to 5,000-hp nominal rating. Table 9.2 gives operating costs for the same duty using piston machines having 85, 75 and 65 percent adiabatic efficiency. An amortization rate of 20 percent is applied to each case. There are an infinite number of amortization refinements that can be included, depending upon the operator's accounting system. The amortization only represents a minor portion of the *total cost* and is applied equally to each case, so that variations are inconsequential.

The cost of power at 0.7 cents per kwh for 8,750 hours per year, with 98 percent gear efficiency and 93 percent motor efficiency, is $50 per hp-year.

The *centrifugal compressor* is the preference of the hydrocarbon processor because of its ability to operate one to three years without a turnaround. A machine that is in continuous operation obviously requires little or no maintenance.

The repairs required for a turnaround are essentially preventive in nature. This would include a new set of bearings, labyrinth seals and an occasional replacement of the first-stage impeller and a rotor rebalance. A sinking fund of 10 percent of the compressor cost per year would adequately provide for the normal maintenance of centrifugal equipment.

A *piston compressor* may have an operating-time record of 85 percent which is equivalent to 1,300 hours of downtime per year. This means that some maintenance, such as a valve replacement, was performed every third day on an average unit. The cost of such maintenance is on the order of $25 per hp-year. This is a high but realistic cost, as shown by the maintenance costs reported in FCC reports of the *big-inch* pipe line stations.

Table 9.1
Operating Cost of Centrifugal Compressors in Refrigeration Service

Shaft hp	5,000	2,500	1,500	1,000	665
Compression Eff.	81	80	76	60	40
Mechanical Eff.	98	97	96	95	95
Isentropic hp	3,900	1,940	1,100	570	254
Refrig. Tonnage	3,000	1,500	840	440	200
	Operating Costs in Thousands of Dollars Per Year				
Amortization	35	30	25	19	18
Power Cost	250	125	75	50	33
Maintenance	10	9	7	6	6
Total Annual Cost	295	164	107	75	57

Table 9.2
Operating Costs for Reciprocating Compressors in Refrigeration Service (Thousands of Dollars per Year)

Adiabatic horsepower	3,900	1,940	930	570	254
Mechanical Efficiency(*)	96.8	95.5	93.3	91.7	87
Refrigeration, tons	3,000	1,500	710	440	200
	Compression Efficiency of 85%				
Shaft hp	4,740	2,420	1,100	740	343
Amortization, $1,000	53	33	19	13.4	8
Power, $1,000	238	121	55	37	17
Maintenance, $1,000	71	36.4	16.5	8.6	5
Total at 85% η_{dy}	362	190.4	90.5	59	30
	Compression Efficiency of 75%				
Shaft hp	5,370	2,700	1,290	830	390
Amortization, $1,000	60	37	22	15	9
Power, $1,000	269	135	64.5	41.5	19.5
Maintenance, $1,000	81	40.5	19.5	12.5	6
Total at 75% η_{dy}	417	212.5	106	69	34.5
	Compression Efficiency of 65%				
Shaft hp	6,200	3,120	1,490	955	450
Amortization, $1,000	69	42.5	25.5	17.5	10.5
Power, $1,000	310	156	74.5	48.5	22.5
Maintenance $1,000	93	47	22.5	14.5	7
Total at 65% η_{dy}	472	245.5	122.5	80.5	40

*By Equation 1 and using 4 cylinders for each case.

Table 9.3
Effect of Piston Speed and Mol Weight on Power

Three Example Cases	A	B	C
Piston speed, fps	9.75	14.6	9.75
Molecular weight	42.0	42.0	154.5
Piston/valve area ratio	10.0	10.0	10.0

First Stage, Case	A	B	C
Discharge, psia	70.0	70.0	67
Suction, psia	16.3	16.3	16.3
Suction temperature, °F	−35	−35	−35
Ratio of compression, R_c	4.28	4.28	4.125
Ratio of compression, R_c^g	1.208	1.208	1.125
Suction valve loss, θ_s	0.094	0.210	0.346
Discharge valve loss, θ_d	0.078	0.174	0.308
Intrinsic factor, B	1.19	1.47	1.94
Intrinsic R_c, BR_c	5.10	6.27	8.00
Intrinsic R_c, BR_c^g	1.236	1.270	1.189
Compression efficiency, %	0.88	0.77	0.66
Isentropic ΔH Btu/lb*	32	32	11
Circulation, lb/min†	360	360	1,400
Cylinder capacity, acfm	2,300	2,300	2,520
Cylinder displacement, cfm	4,600	4,600	5,040
Cylinder bores, no. x diam.	3 x 22	3 x 18	3 x 13
Isentropic horsepower, Ise hp	270	270	372
Gas horsepower, Gas hp	307	350	565**
Brake horsepower, Bhp	340	380	604

Second Stage, Case	A	B	C
Discharge, psia	245	245	185
Suction, psia	67.3	67.3	64.0
Suction temperature, °F	20	20	35
Ratio of compression, R_c	3.64	3.64	2.90
Ratio of compression, R_c^g	1.183	1.183	1.093
Suction valve loss, θ_s	0.094	0.210	1.346
Discharge valve loss, θ_d	0.079	0.177	0.317
Intrinsic factor, B	1.19	1.47	1.98
Intrinsic R_c, BR_c	4.33	5.35	5.75
Intrinsic R_c, BR_c^g	1.205	1.243	1.157
Compression efficiency, %	89	75	59
Isentropic ΔH, Btu/lb*	27	27	8
Circulation, lb/min†	720††	720	3,550***
Cylinder capacity, acfm†††	1,150	1,150	1,700
Cylinder displacement, cfm	2,000	2,000	2,500
Cylinder bores, no. x diam.	3 x 8¼	3 x 8¼	3 x 9
Isentropic horsepower, Ise hp	467	467	669
Gas horsepower, Gas hp	513	610	1,133
Brake horsepower	550	650	1,190
Total, Bhp	890	1,030	1,794

*Isentropic enthalpy data from Elliott Co. Propylene Mollier Charts and E.I. DuPont T-115 Table and Chart Freon 115.
†For equal duty of 3.2 and 3.67 MM Btu/hr., Stg. 1 & 2.
**Freon 115 at 14.6 fps piston speed Comp. Eff.-44% & 900 Bhp.
††Includes 146 lb/min of Propylene Economizer Boil-off.
***Includes 1332 lb/min of Freon 115 Economizer Boil-off.
†††Clearance 17%, First Stage E_v = 50, Second Stage, E_v = 58% & 68%.

A well-designed plant operating at conservative load factors, with nominal valve velocities and handling a light, clean gas, could have an operating cost of or about \$5 per hp-year. The maintenance costs used in Table 9.2 is the mean of these two rates or about \$15 per hp-year. The same gas-hp used in Table 9.1 is used in Table 9.2. For example, the 4,740 shaft hp in the nominal 5,000-hp category is the result of the 3,900 adiabatic hp divided by the mechanical and dynamic efficiencies of 96.8 and 85 percent. The mechanical efficiencies are determined from the equation:

$$\eta_m = 1 - 1/(\text{bhp/cylinder})^{0.5}\ . \qquad (9.1)$$

Heretofore the compression efficiency of all piston compressors were considered to be equal. The matter of applying a centrifugal compressor was contingent upon its limited minimal capacity and efficiency in contrast with the assumed constant efficiency of the reciprocating unit. The data in Table 9.3 shows that the latter compression efficiency can vary from 93 to 71 percent by increasing the piston speed from 9.75 to 19.5 fps or by reducing the valve lift area by one-half, with all other pertinent dimensions and conditions being equal.

Figure 9.1 shows the annual cost of refrigeration, 200 to 7,000 tons. Forty-five percent of the tonnage load is handled by the first stage at minus 35°F. The 55 percent balance is treated in the second stage at 20°F intake.

Figure 9.2. The two cylinders shown constitute the first stage of a two-stage, 400-ton propylene refrigeration service. The intake is minus 50° F, which accounts for the white frost line. The suction "pulse bottle" is insulated.

The broken line "C" represents the cost line for the centrifugal compressor. Beyond 1,800 tons the efficiency remains constant at or above 80 percent. This value tapers to 67 percent at 200 tons as shown in Table 9.1.

The four straight lines represent the dynamic compression efficiencies, 65 to 95 percent for piston compressors. The intersection of curve "C" with the 65 percent efficiency lines at point A shows that parity exists at 500 tons. Parity tonnage for the 85 percent compression is 870 tons and 1,600 tons for 95 percent compression efficiency.
percent compression is 870 tons and 1,600 tons for 95 percent compression efficiency.

When the piston machine has an adiabatic efficiency of 75 percent, equivalent costs are realized at 620 hp, and the marginal advantage is 68 percent for loads greater than 2,000 hp. When the piston compression efficiency is 85 percent, the equivalent cost is realized at 800 hp, and the marginal advantage is

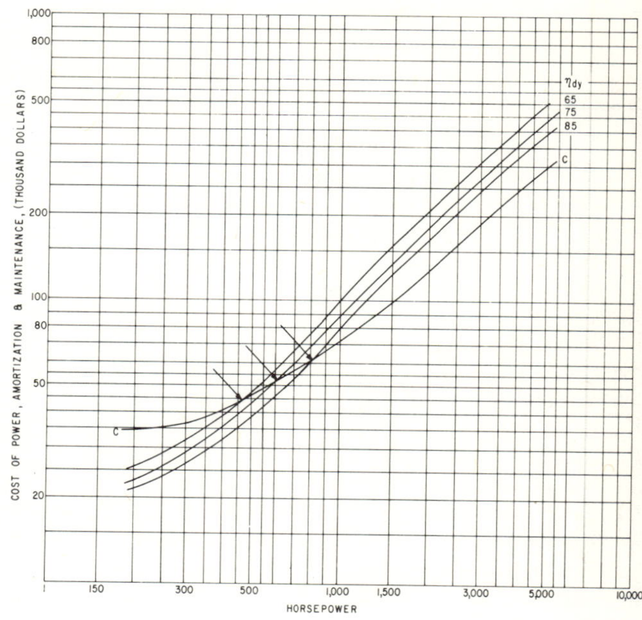

Figure 9.1. Study of refrigeration costs, centrifugal (C) versus piston machine at 65, and 85 and 95 percent dynamic efficiencies. The cost data for curve C are in Table 9.1; remaining costs are in Table 9.2.

76 percent. This question may be raised: "Why not exploit the 90 and even 95 percent efficiencies, since it is *well known* that a piston compressor is more efficient?" The purpose of this chapter is to establish the credibility of the efficiencies assigned to each category. Table 9.3 shows the effect of changing the speed of identical, 6-inch stroke machines from 585 rpm to 1,175 rpm. This is done in reference to the adiabatic efficiency (using the Universal Gas Law) and to the isentropic efficiency (reference to Mollier or accredited enthalpy tables).

The first stage of a two-stage refrigeration service is illustrated in Figure 9.2.

Piston Compressor Efficiency

Table 9.3 outlines the procedure used to establish the piston power requirements for the five load levels presented in Figure 9.1 and Table 9.2. The effect of high piston speed, of small valve area and of heavy mol weight gases are also given in Table 9.3. A piston/valve area ratio of 10 is used for all three examples in Table 9.3. The P/V ratio of 10 exemplifies good design proportions. A ratio less than 7 is rare and difficult to accomplish designwise.

A compressor cylinder having a P/V ratio greater than 13 is only suited for handling air or lighter gases. The piston speed could have been held constant and the same tolerance permitted in varying the P/V ratios. To operate the first case at 14.6 fps and 88 percent, the P/V ratio must be reduced to 6.5. The compression efficiency is quite sensitive to a $m(aU)^2$ value of 500,000. This critical number includes the mol weight of the gas times the square of the P/V ratio and the piston-speed. Smaller $m(aU)^2$ values produce compression efficiencies in the 90 percentages, which are acceptable in the relatively low (< 1,000) horsepower category. The difference between the first and second case is 140 hp worth $7,000 of electric power per annum. This is the price of running 875 versus 585 rpm. In large stations including 20 times this power, the power penalty would be $140,000 per annum.

The operating conditions for the first two cases are identical. The third case in Table 9.2 involves Freon-115 having the same duty as the two propylene examples. All five cases given in Tables 9.1 and 9.2 involve the same refrigeration duty. The flow sheet for the operation is shown in Figure 9.2. The gas compression required 1.31 isentropic horsepower per ton. This is relatively high performance. The weighted average power given in Reference 4 is 2.26 Bhp per ton. This indicates that the dynamic efficiency is 62 percent, and the mechanical efficiency is 96 percent. These values closely resemble those used in the last example in Table 9.2. Figure 9.1 illustrates that such economy can be matched by a centrifugal compressor having a dynamic efficiency less than 40 percent at the refrigeration duty of 365 tons (point A). At or about 500 tons of refrigeration (point B), a reciprocating machine must have a compression efficiency of 65 percent or better to match the operating expense of available centrifugal compressors. At the 870 tonnage level, the compression efficiency of a reciprocator must exceed 85 percent (point D), and at the 1,600 tonnage, the reciprocator must exceed 96 percent. These last two objectives are more difficult to accomplish than it is to exceed 80 percent with a centrifugal machine, where the onus has been on the centrifugal compressor to produce low capacity (< 500 acfm) impellers with plus 60 percent efficiencies. These objectives have been achieved. The onus is now on the reciprocating machines.

Other than the technique presented in this text, there is no other means of making such an analysis. There are several suppliers of centrifugal machines of the 1,000-hp and larger sizes that develop credible efficiencies. There are only a few qualified to offer smaller machines. The efficiencies are somewhat nebulous. But the point is that the centrifugal data is more credible than the reciprocating data.

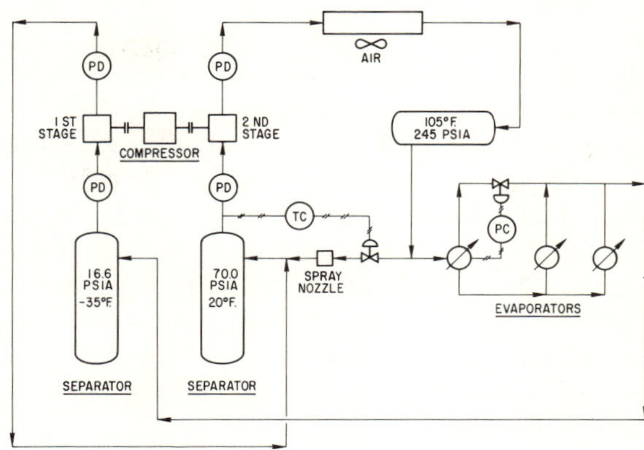

Figure 9.3. Schematic layout of the refrigeration used in the example reported.

The Freon data is displayed in Table 9.3 to show the effect of heavy molecular weight gas in identical machines doing the same refrigeration duty against the similar operating conditions. The effect of a 154.5 versus a 42 molecular weight gas increased the power by 78 percent at a relatively low piston speed of 9.75 fps. The power is increased 165 percent at 14.6 fps piston speed.

The valve losses and pulse intensities were determined from the following equations (1)

$$\theta_s = m(aU)^2/T_o(10^4) \tag{9.2}$$

$$\theta_d = \theta_s/R_c^\sigma \tag{9.3}$$

$$B = (1 + \theta_d)/(1 - \theta_s) \tag{9.4}$$

$$\eta_{dy} = (R_c^\sigma - 1)/(BR_c^\sigma - 1) \tag{9.5}$$

Resistances

The API Standard 618, Section 20c, provides one percent of the system, absolute pressure for the pulse damper resistance. Usually this tolerance can be reduced to one-half of one percent. The same resistance can generally provide ample frictional head for various pipe connections, including a low velocity gas separator between the compressor cylinder and the reference terminal. An intercooler or aftercooler has a resistance usually equivalent to 15 to 20 velads, in reference to the acceptable pipe velocity of 45 fps.

Pressure and Enthalpy

The intrinsic pressure range shifts from an intake of 16.30 and 70.0 psia discharge to 12.87 and 82.2 psia. The enthaply differential is increased from 36 to 42 Btu per pound or a power increase of 17 percent. The second-stage intrinsic pressure range shifts from 67.3 and 245 psia to 53.2 and 288 psia discharge. The differential enthalpy is increased from 29 to 38 Btu per pound. The power is increased 31 percent, for Case B pressures.

The cylinder sizing is accomplished by using the displacement equation:

$$Q(\text{cfm}) = 0.3275[d^2 - (d_r^2/2)]U. \tag{9.6}$$

The gas flow, Q, is a product of 360 lb/min and the 6.4 specific volume at the suction line condition, 16.3 psia and minus 35°F. The capacity, Q, is 6.4 (360) = 2,300 cfm. The cylinder displacement must complement this volume. It is found by dividing the displacement by the volumetric efficiency. A normal clearance volume of the cylinder is 17 percent. The volumetric efficiency would be

$$E_{vs} = 100 + 17 - 17(1.10)R_c^{1/k} = 50 \text{ percent,} \tag{9.7}$$

where the radical is $(70.0/16.3)^{0.87} = 3.57$.

The displacement of a single cylinder or the composite of several cylinders is represented by 2,300/0.50 = Q = 4,600. Then d^2 = 1,480 or a single cylinder must have a 38-inch bore. This is too large to be practical; the thrust would be excessive. Two parallel cylinders would have a diameter of $(1{,}480/2)^{0.5}$ = 27 inches and three cylinders would have a bore of 22 inches. The specific volume of 6.4 of/lb. for −35°F and 16.3 psia is used for the cylinder charging pressure rather than 8.2 of/lb. for 12.87 psia. The effective suction pressure is drawn from the suction line 1,634 psia to 14.86 psia and the discharge line pressure of 70.0 and 75.5 psia insofar as the power is required. The piston speeds drop to zero at each end of the stroke. There is a hypothetical condition of stagnation and zero flow that exists for several milliseconds where the suction pressure inside the cylinder is in equilibrium with the suction line pressure. The discharge pressure is defined by the same stagnation conditions that occur 180° from this position.

Overall Efficiency

The overall efficiency is the ratio of the ideal adiabatic or isentropic (Mollier enthalpy differential) divided by the rated brake horsepower required for the operation. The margin between these two values includes some nine vague concessions. This list includes the following:

1. No negative tolerance, API Std. 618-55.1 (a 3 percent additive),
2. Nonlube blow-by loss of 5 percent or one percent per R_C, whichever is greater,
3. Spring loss, 1-psi suction deduction and 2-3 psi discharge addition, which may be a 2-6 percent burden, at low pressures,
4. Arbitrary mechanical efficiencies, an additive of 2 to 7 percent,
5. Arbitrary compression efficiencies (9),
6. Arbitrary inlet gas temperature *warm-up* corrections (one percent per 5°F) for as much as 25°F,
7. Arbitrary undercharging and supercharging pressure correction to match a pulsating header condition,
8. Electric analog assimilation of piping resonance,
9. Arbitrary exponential corrections.

No Negative Tolerance

This is simply a matter of diffidence or nonconfidence. When a standard method of rating compression power is established, such inadequacies will be recognized as a penalty burden.

Nonlube Blow-by Loss

There is no documented reference literature that proves or states evidence that any additional power is required to operate a *nonlube* machine packed with carbon, TFE or the like. A frictional drag of 5 percent, or 1 percent per R_C, would surely consume the synthetic rings in the course of a few weeks. Failures of this order are usually a progressive deterioration. A 5 percent leakage may continue with a 20°F discharge rise for several weeks until the leakage reached 10 percent, and the discharge temperature would rise another 20°F. The deterioration beyond this point is generally accelerated, and the life of the failing component is only a matter of hours. All manufacturers guarantee their *nonlube* trim for one year. Some *nonlube* machines are experiencing normal service for more than two years.

Spring Loss

There may be certain instances where depressing the suction pressure, in perhaps vacuum service, where spring losses per se′ can be substantiated. It would be precocious to generalize with fixed corrections until more is known about valve behavior (21).

Arbitrary Mechanical Efficiencies

Prior to 1941, when the normal compressor cylinder load was 100 hp, the average mechanical efficiency was well established and acknowledged to be 90 percent (22). During the next two decades the average power packed per cylinder was about 400 hp, and the mechanical efficiency of 95 percent was again well established and accepted. By 1961 the mechanical efficiencies that prevailed were between 97 and 98 percent (16). This reference is based upon the best and most recent test data. This concerned large pipe line units that averaged 1,200 hp per cylinder. The following equation for mechanical efficiency conforms to these three conditions:

$$\eta_m = 1 - 1/\text{bhp}^{0.5}. \qquad (9.1)$$

The bhp should represent the maximum power rating of the cylinder under consideration.

For example, a six-throw balanced opposed frame rated at 2,400 hp would have a mechanical efficiency of $1 - 1/400^{0.5}$ = 95 percent. A similar estimate for a centrifugal machine is attained by substituting an exponent of 0.4 for the exponent 0.5. It is only reasonable to presume that the percentage of frictional load diminishes geometrically with the increase in frame load. There is some question for the need for a 1,800,000-Btu/hr oil cooler and 700-gpm lube oil system to accommodate 7 percent friction fron the new 10,000-hp frame. A system of one-third proportions seems more than adequate.

Arbitrary Compression Efficiencies

The fallacy and shortcomings of generalized compression efficiencies have been discussed at length in earlier chapters.

Arbitrary Inlet Gas Temperature Warm-Up Corrections

One phase of the CNGA tests was to measure the degree of *warm-up*. The compressor cylinders were supplied jacket rates of zero to 25 gpm. The jacket was maintained at 70° to 210°F. A thermocouple probe placed directly into the internal gas flow through the valve guard showed no perceptible rise over the suction channel temperature (22).

Arbitrary Undercharging Correction and Supercharging Pressure

The attenuation of an adequate sized *surge bottle* should reduce the pulse intensity to a negligible figure (see Chapter 3). The matter of establishing an absolute system pressure, much less a correction thereto, is a great challenge for existing instrumentation. A method which has never failed to match the cylinder capacity with the meter flow, within the accuracy of the instrumentation, has been described previously (see Chapter 2).

Resonance

There is a school of thought which holds that once a compressor valve is open, the flow is infinite and without friction. They believe that the system resistance is a function of the resonance present. These resonant waves may be in or out of phase with the valve action at the terminal of the clearance gas expansion or any other compression functions. They believe that these waves may be decidedly out of phase with these functions. They maintain that the intersection of this wave with the head or crank end positions of the piston determines the suction charge and discharge pressure, thus creating an illusion of *under*-and *super-charging* effect. The consequences of pursuing such a hypothesis are most disconcerting.

The pressure system may be visualized as the result of three energy stratus: the static head, the frictional head, including the θ valve resistances, and a nebulous layer of resonant forces, revibrating through the piping system. The latter is a function of natural (Helmholtz) chokes and chambers. Such reactions may be anticipated and avoided by a study of electronic analogies, which constitute the next concession. Before such a resonant condition could alter the system pressure to an identifiable degree, there would be a back-flow situation equivalent to *surge* operation of a centrifugal compressor. The noise and intense vibration associated with a *surge* behavior would leave no uncertainty as to its presence.

Electric Analog Applications

It is possible to establish the flow behavior on electronic analog to represent the gas to and from a reciprocating compressor through a given piping system. The sonic flow in the gas to and from the compressor establishes resonance in certain piping spools and vessel volume combinations.

These high intensity sonic pulse waves of low frequencies of 3 to 11 cps are responsible for the vibration of the compressor, its cylinder valves and piping. The pulse frequencies greater than 40 cps are relatively passive. The intensity of pulse has been correlated with the valve head resistance as noted in Equations 9.4 and 9.5. The velocity under the valve elements is approximately five times that experienced in any other cylinder passage or piping constriction.

The use of any constriction smaller than the minimum lift valve area is only justified as additive resistance. Using the valve port as a resistance criteria is like sweeping the dirt under the rug. It has been proposed that the analog program should include geometric valve losses. This feature should be a part of every compressor analog study program—it just happens to be one of the most significant features in the system. The geometry of the valve controls the intensity of the basic mass flow and amplitude of the sonic waves.

The analog is by no means an answer to the daily need for comprehensive compression efficiency values. The later chapter on computers offers comfort for those who cannot read the compression efficiency on Figures 2.2, 2.3, etc., or who find the Equation 9.5 too complex. The analog study should provide the answers to many of the current baffling problems. Apparently, the existing analysis of pulse intensities are empirical or relative and without geometric substance. The electronic analog is the only device capable of digesting the massive solution of identifying these troublesome sonic waves traps. The analog provides a trial and error method which is extremely efficient in *de-bugging* the system of these hazards. There is a considerable degree of skill and experience in knowing how to circumvent these complexities.

Exponential Corrections

The usual approach to analyze a compressor calculation is to obtain a representative k value, which is the ratio of specific heats. This is accomplished by taking a weighted mole percentage summation of molal heat capacity of each significant component at constant pressure and at the approximate mean temperature.

$$k = mc_p/(mc_p - 1.986). \tag{9.8}$$

There are numerous refinements to solve for a representative k value corrected for the compressibility effects. The Redlich-Kwong equations (4) are suitable for hydrocarbon compressibility factors. It is adapted for computer programming in a later chapter.

The irony of making elaborate k value corrections is these exponents are largely cancelled by the exponent applied in the numerator and the same exponent applied to the denominator of the adiabatic head equation. There is merit and a high degree of reliability in using the simplified k value from the mean temperature in solving for the molal heat capacity. The compressibility correction factors should be applied separately, as proposed by Ridgway (16). The average suction and discharge compressibilities are applied to the adiabatic head equation. The ratio of the suction/discharge compressibility is applied to the last expression in the volumetric efficiency equation.

Another thought to bear in mind concerning the application of exponential corrections is that they only change the slope of the compression and the expansion curve. The valve loss concept extends the intrinsic pressure boundaries of the P-V diagram and enthalpy range. Speaking in behalf of the process and project engineers of the contractural and gas operating concerns, there is no other method to secure a realistic power evaluations for cylinders having undersized valves and for gases heavier than air. It therefore falls upon us to make our own appraisal of the power requirement and judge the vendor's proposal data accordingly.

Figure 9.3 illustrates the type of reciprocating compressor that is applied to foregoing refrigeration service. Figure 9.4 is a typical centrifugal that is being considered for the same refrigeration duty.

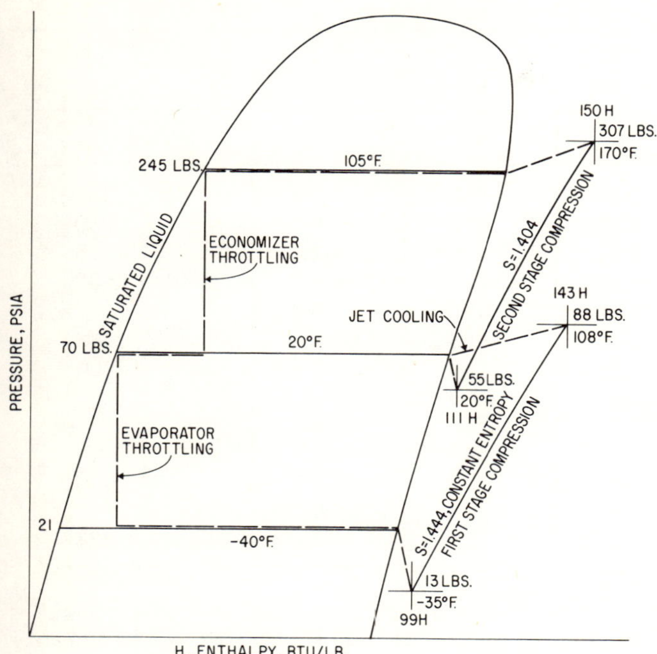

Figure 9.4. Refrigeration cycle used for piston compression on skeleton pressure enthalpy Mollier chart.

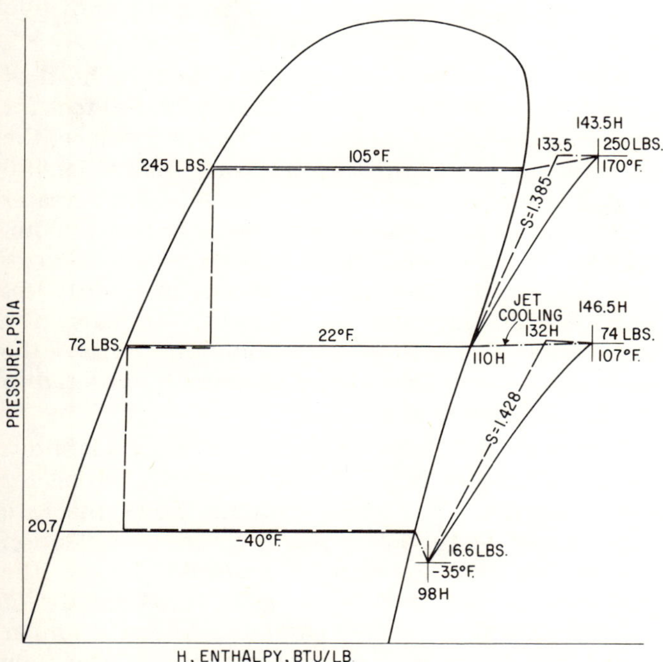

Figure 9.5. Refrigeration cycle used for centrifugal compressor on skeleton pressure enthalpy chart.

Centrifugal Application

The compression cycle for this operation is shown in a heavy solid line on the skeleton Mollier chart, Figure 9.5. The isentropic load is 34 Btu per pound for the first section and 23.5 for the second section. The dynamic losses bring the polytropic load to 48.5 and 33.5 Btu per pound. The comparable isentropic (and total) load for the piston compressor is 36 and 26 Btu per pound.

The centrifugal compressor requires 410 and 450 bhp or a total of 860 bhp as shown on Table 9.4. This shows that an average dynamic efficiency of 70 percent is expected from both sections of the centrifugal compressor.

The respective adiabatic heads for the centrifugal compressor are 28,950 feet and 19,300 ft-lb per pound of gas flow. The calculation method is described in Chapter 6. These heads can be attained with three stages of 9,650 feet potential for each impeller stage. The second section requires two 9,650-feet stages. The technique for developing the impeller program is shown in Table 9.4 and on impeller performance chart, Figure 9.6.

Supporting material for Table 9.4 calculations follow

$k = 1.163$, $\sigma = 0.140$, $m = 42$.

First Section $R_c = 72/16.6 = 4.33$ or $4.33^{1/3} = 1.63$/stage (3 stages).

Second Section $R_c = 250/70 = 3.58$ or $3.58^{1/2} = 1.89$/stage (2 stages).

The incremental volume change is a function of $R_c{}^{1/n}$, where n is the polytropic corrected k value. $\sigma' = \sigma/\eta_{ad} = 0.112/0.7 = 0.160$, $(n-1)/n = 0.20$, $n = 1.25$, $1/n = 0.80$. The first section inlet volumes are found by applying the equation $1/R_c{}^{1/n} = 1/1.63^{0.80} - 1/1.48 - 0.675$. The second section inlet volumes are reduced by $1/1.89^{0.80} = 1/1.67 = 0.60$. The pressure advances by the factor R_c for each stage, and the temperature increases by R_c^{σ} or 1.1025 and 1.136 respectively.

The selection of the impeller diameter and speed are resolved by applying the specific diameter and speed equations. The 3/4 and 1/2 root of the required head, which averages 8,675 feet, are 890 and 92.

The volume for the first stage is 2,300 acfm or 38.4 cfs and $Q^{0.5} = 6.2$.

Taking specific speed, N_s, as 100 for optimum efficiency, the operating conditions are $N = N_s L^{0.75}/Q^{0.5}$ $N = 100\ (890)/6.2 = 14{,}350$ rpm.

Table 9.4
Impeller Selection for Refrigeration Unit

Stage	1	2	3	4	5	Discharge
Intake Flow, acfm	2300	1530	1020	810	480	283
Intake Press. psia	16.6	27.0	43.0	75.0	137	250
Intake Temp. °R	425	455	487	480*	525	523/575
Intake Compr. Z	0.91	0.93	0.90	0.94	0.88	0.86/0.82
Flow Rate, lb/min	354	354	354	564	564	
Flow Rate, Q/N	0.150	0.100	0.066	0.053	0.032	N = 15400
Impeller Number	11-7	11-5	11-4	11-3	11-1	
Impeller Eff., η	0.790	0.615	0.705	0.735	0.665	
Impeller Coef., q_{ad}	0.570	0.405	0.530	0.545	0.495	0.503
Gas Head, L_{ad}	9400	7300	9300	9200	8400	43,600
Dynamic, hp	136	122	135	224	206	
Total, dy. hp		393			430	823

* Refrigerant is withdrawn and cooled from 63°F to 20°F by jet spray between stages three and four. The volume is established for each intake condition. The Q/N is selected from the optimized speed. The latter is corrected by trial and error to balance the pressure coefficient head with the enthalpy head.

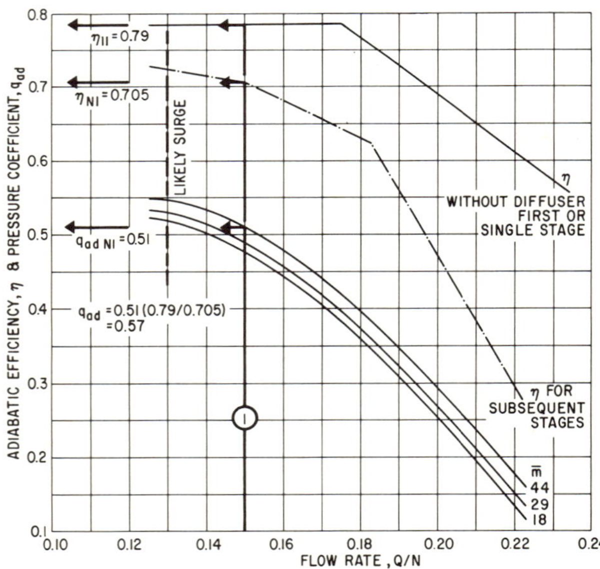

Figure 9.6. The upper broken line of this performance chart gives the single stage or first stage efficiency for the Q/N flow range. The second broken line gives the efficiency for subsequent stages. The lower family of curves gives the pressure coefficient q_{ad} for subsequent stages. The first or single stage q_{ad} value is obtained by upgrading the given q_{ad} values by the respective efficiencys. Table 9.4 shows that in the first stage Q/N is 0.15. The chart gives the η_{11} as 0.79, ηN_1 as 0.705 and $q_{ad}N_1$ is 0.51. It is corrected to (0.79/0.705)0.51 - 0.57 $q_{ad}1$.

Estimating the pressure coefficient as 0.50, the impeller diameter is d = 1,760(0.55q_{ad}) $L^{0.5}/N$ = 1,760(0.55/0.50) 92/14,350 = 12.4 inches. This gives a tip speed of $U = dN/229$ = 12.4 (14,350)/229 = 780 fps.

This tip speed of 780 fps can develop a head of 29,000 feet for the first section and 19,400 feet for the second section (see Chapter 6). For instance, the latter is L_{ad2} = (0.53 + 0.495)780^2/32.2 = 19,400 feet. The sum of these two heads is 6 percent greater than the design requirement which has a comfortable margin of compliance. The economizer pressure appears to be about 68 psia or slightly less than the design value of 70 psia. Actually, the impellers in the first section should out perform the latter stages, and the normal pressure is expected to be slightly greater than 70 psia.

Presume that there is a choice of an 11 and a 15-inch diameter impeller and respective casing for this operation. The small volume flows that are experienced in the latter stages favors the selection of the smaller impeller. Figure 9.6 shows the performance for one of an established family of 11-inch impellers. Each manufacturer has a similar set of curves for each compressor model. The flow rate may appear in various forms. The Q/N or cfm/rpm is the simplest abscissa scale. The ordinate should carry a scale for the adiabatic efficiency and

Table 9.5
Comprehensive Adiabatic Head Evaluation

	1st Section	2nd Section
R_c for section	4.33	3.58
Stages	3	2
R_c per stage	1.63	1.89
R_c^{σ} (σ = 0.14)	1.071	1.093
$T_{o1,\, o2,\, o3,\, o4,\, o5}$	425	480
	470	545
	519	
Z_{o1}	0.96	0.94
Z_{o2}	0.93	0.88
Z_{o3}	0.90	
Z_e	0.86	0.82
X_1	112,000	126,000
X_2	123,000	143,000
X_3	136,000	
Lad_1	8,000	11,700
Lad_2	8,800	13,300
Lad_3	9,600	
Total = ft lb/lb	24,400	25,000

Table 9.6
Summary of Centrifugal Compressor Power Evaluation

Solution Method	Gas hp	Remarks
a) Dynamic head by Equation 6.21	830	Ref. Table 9.4
b) Adiabatic, no Z, no T_o corrections	903	Conventional method
c) Adiabatic, with Z, no T_o corrections	808	$Z \leqslant 1.0$ seldom applied
d) Adiabatic, no Z, with T_o corrections	1,016	No precedence
e) Adiabatic, with Z, and T_o corrections	871	An innovation
f) Isentropic from enthalpy rise	850	Using Mollier Chart

pressure coefficient charts. The performance curves for a centrifugal compressor can be developed from these data. The pertinent data for the impeller selections are shown in Table 9.4.

A summary of the power requirement needed for the various evaluation methods is given in Table 9.6. Method A involves the evaluation used in Table 9.4 and q_{ad} is 0.503. The extent to which this technique is used to prepare a proposal is not known. It was and may be limited to the design stage which precludes the application analyst. These more complex evaluations are better suited for the computer. See Chapter 11. They should gain favor as the speed, accuracy and comprehension of the computer programs are better appreciated.

Applying the individual pressure coefficient for each stage to the U^2/g factor for U at 780 fps, the head potential is 48,050 feet. Table 9.4 shows that the average dynamic efficiency for the impellers in both sections is 70 percent. The gas hp required for the first section is hp = $wL_{ad}/33{,}000\eta$ = 354(26,000)/33,000(0.70) = 400 gas hp. The second section requires 564 (18,000)/33,000(0.70) = 430 gas hp. The total load is 830 gas hp.

The calculated adiabatic head-horsepower method is set-up for comparison in Table 9.5 with the isentropic system.

$$X_\sigma = 1545\ ToZ_{ave}/m\sigma. \tag{9.9}$$

$$L_{ad} = X_\sigma(R_c^\sigma - 1). \tag{9.10}$$

A majority of compressor ratings is made by Method B. This includes a total head based on a single R_c function. The head may be developed by the isothermal, adiabatic or polytropic cycle, so long as the power involved the same cycle efficiency.

A compressibility correction factor that is less than unity is seldom, if ever, applied. Such corrections are applied where the Z correction factor is greater than unity in such instances as involving high pressure hydrogenation operations. This concerns Method C.

The acknowledgement of an incremental temperature rise in Equation 9.9 is without precedence at the project engineering level. Method D should be recognized at the *firm design* level of the manufacturer.

Applying both the compressibility and the corrected intake temperatures constitutes an innovation which has merit. The complexities of Method E are well suited for a computer program.

There is no better definition of state than that offered by a Mollier Chart or from a gas property table.

The enthalpy change procedure of Method F accounts for all the physical conditions and energy measurements. When dealing with pure or near (plus 90 percent) pure gases, the Mollier Chart offers the most realistic and reliable method of power evaluation. Eventually mixed gases will be reduced to their respective partial pressures by a computer and their complex enthalpy changes will be resolved in a matter of seconds.

The Brake horsepower for each case is determined from Equation 9.1, using the 0.44 exponent for each casing power rating. This sum is approximately 20 hp, which is added to Gas hp evaluations. The API Standard 617 requires that the driver nameplate rating be 10 percent greater than the Bhp evaluation. This provides the necessary contengency to cover the discrepancies evidenced in Table 9.6.

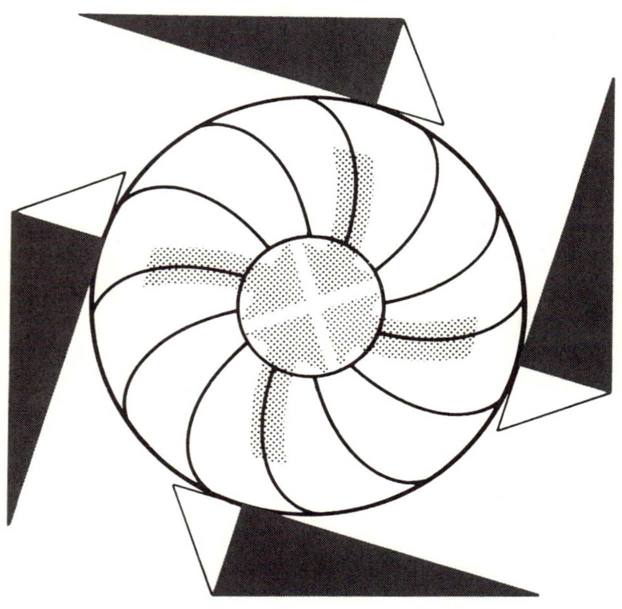

10 Turboexpanders

There is little general information known about gas expanders. Relatively few are used in hydrocarbon processing operations. It may be that we do not use more gas expanders because we do not understand them. We are prone to think of them as mysterious, complex, costly and delicate machines of dubious durability.

Now consider the general-purpose steam turbine as an example to make a point that a broader understanding is the keynote to a wide acceptance of gas expanders. The popularity of the steam turbine is only surpassed by the common squirrel-cage, induction electric motor. The turbine is the most reliable element in a refinery or petrochemical complex that is capable of providing 30 months or longer continuous service. The steam turbine is categorically a *gas expander*.

The operator apparently believes that the energy recovered from a gas expander is not worth the price, the effort and the distraction from the plant's main objective. It is not uncommon to find a catalytic cracker where 20,000 to 200,000 scfm of combustion gas products 20 psig and greater at approximately 1,400° F, available from pressurized furnaces, are currently being vented to the air (23). A gas expander could recover 20,000 hp from the greater of such flow. This energy is worth a $1 million per year in equivalent electric power. The payout is 40 to 50 percent on such an investment. Figure 10.1 represents such an expander.

The introduction of cryogenics to refineries and chemical plants bring a new interest in gas expanders. When the process involves condensing temperatures that are below minus 200° F, each additional degree of refrigeration can be worth several ten-thousands of dollars of production per year. It is highly probable that within the next generation (or two), cryogenic and process gas expanders will be as acceptable as steam turbines are today. This statement presumes that the technical gap between the aero-dynamist and the industrial turbo-designer will have been bridged in the interim.

Basic Types

There are two basic types of turbine designs—the impulse and the reaction concept. Complete gas expansion occurs in the nozzle of an impulse turbine. The spouting gas passes through one or more rotating buckets thereby reducing the initial velocity. The number of buckets depends upon the desired efficiency and power requirement. Turbines smaller than 300 hp are usually single-stage machines. Two

Figure 10.1. Gas expander having integral, two-stage radial, 20,000 hp air compressor. The expander is motivated by the waste gas from a pressurized furnace. Three, 50-percent reaction expander blades are shown on the left. The compressed air is used to charge the furnace. Photo courtesy the Elliott Company.

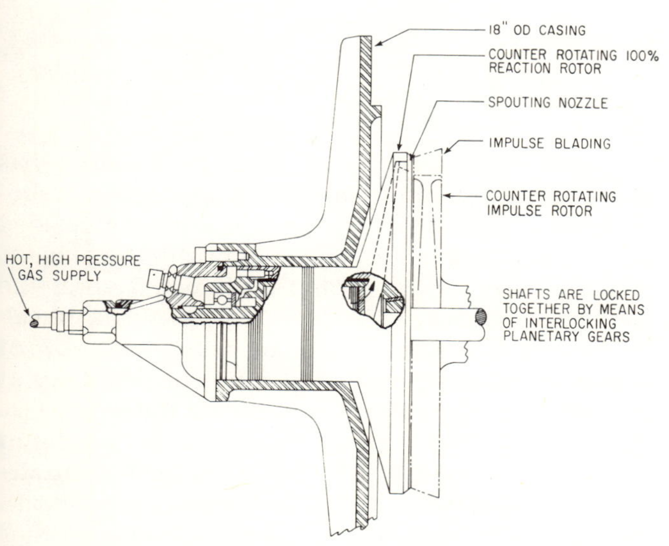

Figure 10.2. Reaction-Impulse turbine with high pressure gas entering through a central bore in the shaft.

and three rows of impellers are applied to low pressure exhaust turbines for greater power capacity and efficiency.

Large steam turbines are usually combinations of both impulse and reaction design features. The impulse design is best suited for the high pressure section of the turbine. Impulse design provides multiple nozzles to regulate load variations. At the other end of the velocity spectrum are the axial flow turbines. They involve large vapor volumes as experienced in the operation of high-vacuum section of steam turbines, gas turbines and air-borne, turbojet engines. These turbines function on a 50 percent reaction principle. Half of the enthlapy is expended through the nozzle, and half is released in passing through the rotor blading. The turbine may contain 20 or more sets of reaction blading.

There are several examples of full reaction design turbines. The simplest application is found in the whirling lawn sprinkler heads. There are two other historical examples of full-reaction turbines. One is the 2,000-year-old Hero's turbine, and the second is the early American Barker's grist mill.

U.S. Naval Ordinance developed a *reaction-impulse* turbine which was quite successful. The initial prototype developed an overall efficiency of 55 percent and had a realistic potential of 65 percent. This compares with 35 to 45 percent for the fully developed existing impulse turbines used for equivalent service. The initial *spin wheel* section is driven by velocity reaction. The high pressure gas is introduced through a central bore through the shaft and a passage drilled in the rotor disk (See Figure 10.2). The nozzles are drilled along the peripheral rim, spouting axially, instead of radially as did its historical predecessors. The spouting jets are directed into the buckets of a matching impulse rotor supported on an independent rotating shaft. The two conversely rotary assemblies are locked into a common rotation by means of planetary gearing. The Ljungstrom (Sweden) steam turbine functions on a similar design premise. They do not interlock the two shafts mechanically. They operate two counter-rotating generators which are tied together electrically. The Navy prototype was capable of 800 hp and only weighed 212 pounds. The Ljungstrom turbine is perhaps a 100-fold heavier for the same power. The average size Ljungstrom turgine is estimated to be about 5,000 hp. These turbines are of interest because they suggest a method of circumventing the condensation problems experienced with the cryogenic *radial inflow* (RIF) turbine.

The offshore satellites present several potential applications for turboexpanders. One feasibility study indicated that it would be possible to generate one megawatt of electric power by expanding 125 MMscfd of natural gas from 2,500 to 1,500 psig. This would have provided sufficient power to dehydrate and ship 8,000 barrels of oil production per day. Each satellite would be self-sustaining insofar as heating and power utilities are concerned.

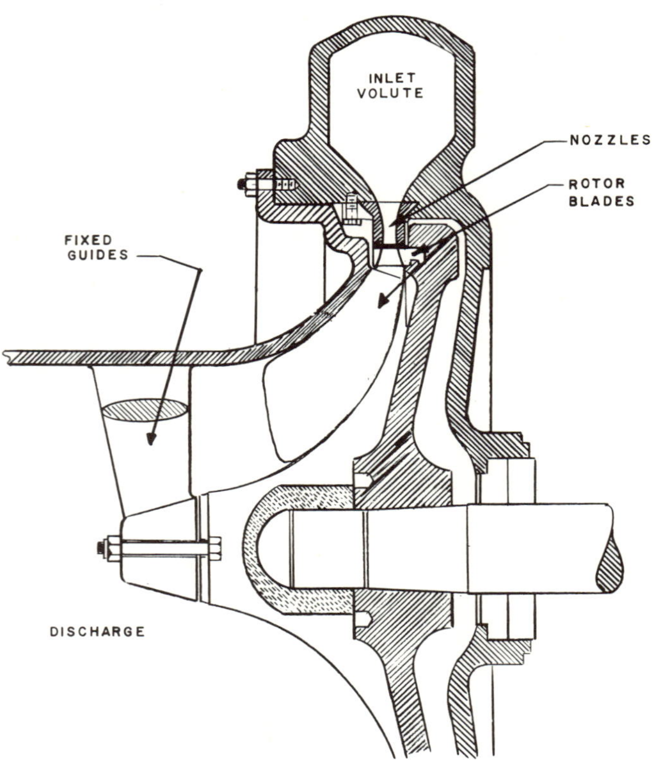

Figure 10.3. Cross-sectional view of the Linde cryogenic expander.

Figure 10.4. Rotor and nozzle wedges used in a U.S. radial inflow cryogenic expander. Courtesy of Rotoflow Corp.

The study also showed that it would have been possible to operate all of the facilities with hydraulic machinery and pneumatic controls. The amine-glycol dehydration solution was to be "stripped" by the heat of inefficient pumping. This can be accomplished with a high-head, low-capacity type pump. Another study concerned a power expander that could develop 2.7 megawatts and 850 tons of refrigeration by expanding 50 MMscfd of 3,000 psig gas expanded to 50 psig.

The main trouble with cryogenic gas expanders was that the blading which could not withstand the erosion from the condensate and abrasion from hydrate formation. The longest service was reported to be about 200 hours. The expanded, wet gas only passes through the contra-rotating impulse blading. This type of blading has a high degree of tolerance for wet steam. Exhaust steam may carry as much as 15 percent moisture; 10 percent is common practice. The centrifugal force carries the condensate to the inner walls of the casing where it is collected and trapped to the exterior.

Radial Inflow Turbines

Linde Eismachinen AG of West Germany has manufactured more than 2,000 cryogenic turbo-expanders in the past 40 years. They have avoided the erosion problem by not permitting the exhaust pressure to cross the saturation line and enter the condensing dome. The Linde expander is a radial-inflow (RIF) design. The rotor has a single row of blades which are cantilevered from the periphery of the disk rim. The radial tail vanes are stationary (see Figure 10.3). The German design includes a series of fixed nozzles, usually 10 in number. These nozzles are chambered in lots of 4, 3, 2 and a single nozzle. By selectively combining these four chambers, a partial admission system is available ranging from 10 to 100 percent in 10 equal increments. These compartments are automatically manipulated to control the flow requirement in the same manner that multiple admission valves are used on electric-generator turbines.

The U.S. version of the RIF turboexpander has an impeller resembling that of a conventional end suction process pump. The vanes and shroud are an integral component of the rotor. The droplets of condensate formed by the expansion are influenced by the centrifugal force of the absolute tangential vector. This tends to carry the condensate radially and countercurrently against the centripetal gas

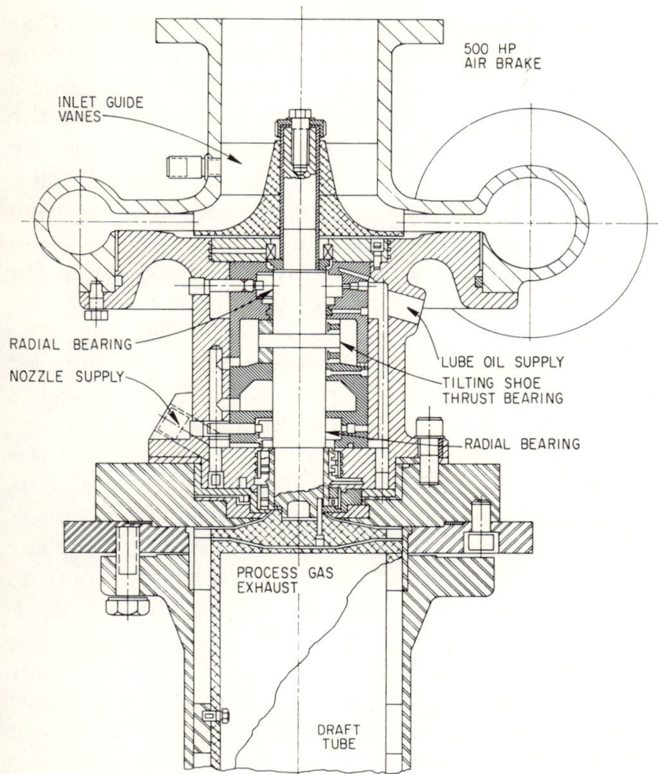

Figure 10.5. Velocity ratio versus average potential turbine efficiency for impulse, 50 and 100 percent reaction designs.

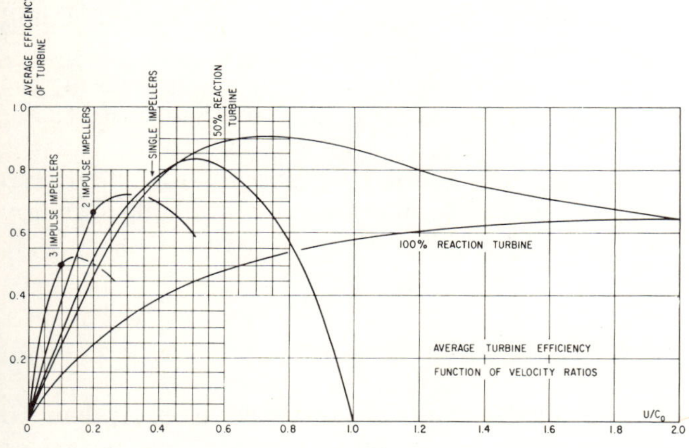

Figure 10.6. Propylene ethalpy-entropy chart showing effect of turbine efficiency on exhaust temperature. Note: Joule-Thompson effect is 25° F drop in gas temperature. A 75 percent dynamic efficient turbine shows a 105° F drop.

flow. The condensate can not exit against the spouting flow, but this interference can detract from the efficiency and be an erosion nuisance. The General Electric Co. made a study of such blade erosion and concluded that each percent of moisture carried by the vapor deducted at least one and in some instances as much as 2 percent from the dynamic efficiency.

The domestic cryogenic expanders have a radial-inflow (RIF) reaction design. The gas is admitted completely around the periphery as shown in Figure 10.4. The tip of the impeller and four nozzle wedges are shown. These wedges are symmetrically pivoted around the impeller offering a variable orifice flow control.

Another American-designed, axial-flow, partial-admission cryogenic expander is shown in Figure 10.5. The cryogenic impeller is mounted on the lower end of the vertical turbine shaft. The top impeller is the air brake to dissipate the energy absorbed by the lower wheel.

Velocity Ratios

The velocity U/C_O ratio is the first turbine checkpoint for the most suitable type of design, size, speed and likely efficiency. A scan of the velocity ratios and the average potential turbine efficiencies are given in Figure 10.6. This ratio is the tip speed divided by the nozzle spouting velocity. It is commonly known as the U/C_O ratio. The tip speed and the spouting velocities are found from

$$U = (\text{Mean diameter of impeller})\,(\text{rpm})/229. \tag{10.1}$$

$$C = 224(\Delta H)^{0.5} = 8.05(L_x)^{0.5} \tag{10.2}$$

where L_x represents the adiabatic head of expansion developed later in Equation 10.3. ΔH represents the difference in enthalpy between initial pressure and temperature and the exhaust pressure.

Figure 10.6 shows a single-stage impulse turbine having a U/C ratio of 0.47 to produce its maximum potential efficiency of 83 percent. Most general purpose steam turbines and expanders would be in this category. The efficiency stays well above 70 percent from 0.3 to 0.7 U/C ratio. Adding a second impeller would raise the efficiency of a single impeller from 50 to 68 percent at 0.2 U/C ratio. Adding a third impeller would raise a 30 percent efficient, single impeller to 50 percent at a ratio of 0.1 U/C.

The 50 percent reaction-design turbine has a potential dynamic efficiency of 90 percent where the U/C ratio holds between 0.6 and 0.9. The

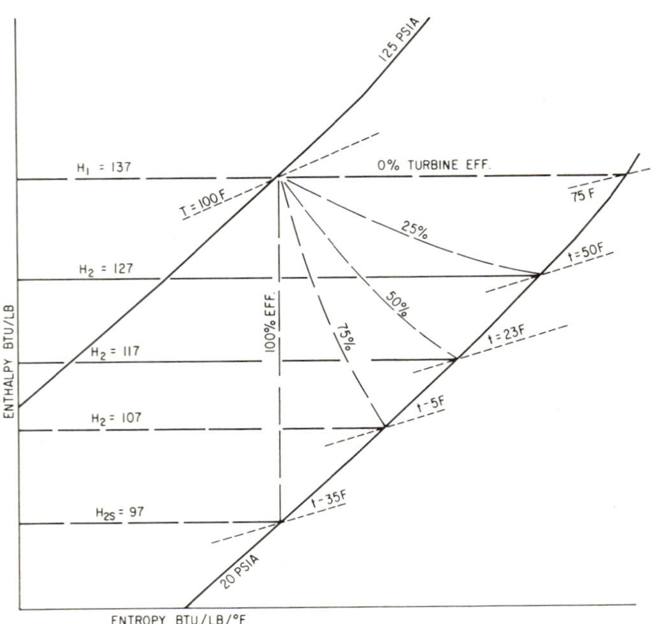

Figure 10.7. A cross-sectional view of a vertical, partial-admission impulse cryogenic expander with integral air brake on top side.

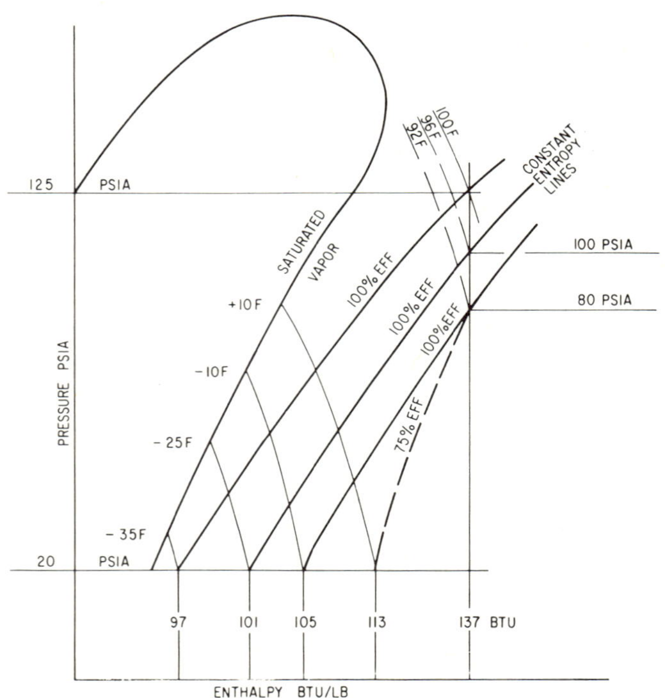

Figure 10.8. Pressure-enthalpy diagram illustrating the effect of a throttling governor on the turbine efficiency (137 – 113)/(137 – 97) = 60 percent overall efficiency.

greater ratio may give the erroneous impression that the reaction turbine involves higher tip speeds. The effect of the larger velocity ratio is diluted by the 50 percent enthalpy split between the nozzle and the blading. The spouting velocity is further diluted by the large number of reaction stages. The velocities involved in a reaction turbine are generally less than those experienced in an impulse design.

Figure 10.7 is an enthalpy-entropy diagram for propylene. The two positive-slope lines are the inlet and the exhaust pressures, 125 and 20 psia. The upper horizontal line represents the throttling of propylene gas from 125 psia to 20 psia through a turbine with a locked rotor, having zero efficiency. The enthalpy remains constant. If the *perfect gas law* held true, the exhaust temperature would be 100° F. Actually, the temperature drops from 100° F to 75° F because of the Joule-Thomson effect. The ideal path of energy flow is the vertical line from 125 psia, 100° F and 137 Btu/lb to 20 psia, —35° F and 97 Btu/lb. If the turbine efficiency is 100 percent, the exhaust would be —35° F. The dynamic imperfections are evident by the actual exhaust being —5° F. The enthalpy for —5° F and 20 psia is 107 Btu/lb. The efficiency of the turbine is (137 — 107)/(137 —97) = 75 percent.

Figure 10.8 is a pressure-enthalpy diagram depicting the same operation, except that a throttling governor valve is used. The Joule-Thomson effect is noted to be 95° F at 100 psia and 92° F at 80-psia nozzle inlet. Extending these points isentropically to 20 psia shows the exhaust to be —35° F, —25° F and —10° F for the ideal case.

Where the turbine efficiency is 75 percent, the 80 psia example has an exhaust enthalpy of (137 — 105) (1.0 — 0.75) + 105 = 113 Btu/lb, and the exhaust temperature is 10° F above zero. The overall efficiency of the turbine from the available enthalpy is (137 — 113)/(137 — 97) = 60 percent.

Adiabatic Expansion Head

The adiabatic expansion head is equivalent to ΔH, the enthalpy (or unit of energy) per pound of gas. When the enthalpy data is not available, the best substitute is to calculate the adiabatic head:

$$L_x = (1 - 1/R_x^{\sigma})\, 1{,}545\, (T_o Z_a / m\sigma) \qquad (10.3)$$

Where L_x is the ratio of expansion pressures (nozzle inlet pressure divided by the exhaust pressure, both in psia). The inverted ratio is used to avoid raising a decimal fraction by a negative exponential power,

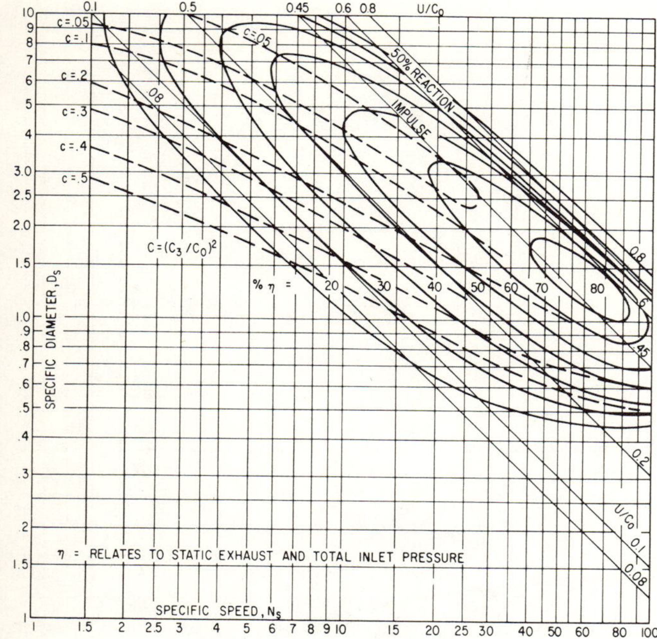

Figure 10.9. Baljé chart for performance of single-stage, full admission, axial flow, impulse turbines.

which is an unfamiliar procedure. The average compressibility factor is Z_a. It is assumed to be unity throughout this article. The nozzle inlet temperature is T_o using the Rankine scale. The mole weight of the gas is m. $\sigma = (k - 1)/k$, where k is the ratio of specific heats at the mean temperature of the expansion. Continuing the example used for Figure 10.7 and 10.8, where the energy release is 40 Btu per pound, the isentropic head is

$$L_x = (40)\ 778 = 31{,}000 \text{ ft-lb/lb or}$$
$$L_x = (1 - (1/(125/20)^{0.13}))\ 1{,}545\ (560)\ 1/42\ (0.13) = 33{,}700 \text{ ft-lb/lb.}$$

The above adiabatic head did not include a compressibility factor Z_a, which is 0.92 to make the two equations balance. The isentropic approach is more accurate than the adiabatic when the operation lies close to the saturated vapor line. The one-fourth and the three-fourths power of the 31,000-ft-lb/lb head are 13.26 and 2,333, respectively.

Specific Speeds

The specific speeds and the specific diameters that are used to establish the ordinates for the performance maps, Figures 10.9, 10.10 and 10.11, were conceived and developed by Dr. O.E. Baljé (12). They indicate the nominal obtainable efficiency and the optimum design geometry. The basic parameters are determined from these equations:

$$N_s = NQ^{0.5}/L_x^{0.75}, \quad (10.4)$$
$$D_s = DL_x^{0.25}/Q^{0.5} \quad (10.5)$$

Again continuing the reference turbine, assume that the flow rate is 4.37 MMscfd. From a Mollier diagram, the specific volume at exhaust conditions is 5.7 cubic feet per pound. The weight flow, W pounds per second (pps), is

$$W = (\text{MMscfd})(m)/32.8, \quad (10.6)$$

where the constant 32.8 is 1,000,000/380 (1,440)60.

Then $W = 4.37(42)/32.8 = 5.6$ pps. The exhaust volume is $Q = 5.6(5.7) = 32$ acfs; $Q^{0.5} = 5.65$.

Referring to Figure 10.10, assuming that a 75 percent dynamic efficiency is acceptable, the impulse design velocity-ratio is intercepted at $34N_s$ and $2.0D_s$. The speed to satisfy this condition is $N = N_sL_x^{0.75}/Q^{0.5} = 34(2{,}333)/5.65 = 14{,}000$ rpm.

The mean diameter of the impulse turbine rotor is $D = 2.0(5.65)/13.26 = 0.855$ feet or 10.25 inches.

The peripheral speed is $U = 10.25\ (14{,}000)/229 = 626$ fps. The spouting velocity is $C = 224(40)^{0.5} = 1{,}415$ fps. $U/C = 626/1{,}415 = 0.443$ which is an adequate check.

Any turbomachine having a $D_s - N_s$ design point above and to the left of the maximum efficiency island on the Baljé charts is a *low specific speed* turbomachine (refer to Figure 10.11). The characteristics that constitute *low specific speeds* are relatively small volumes, small nozzle apertures and high heads. This also means that the flow must be increased to enhance the efficiency with all other factors constant. This is demonstrated by increasing the flow 225 percent; the specific speed is increased from 33 to 50, where the efficiency is 80 percent. The flow must be increased ninefold before it exceeds the parameter of the 80 percent island. Beyond this range, the turbine becomes a candidate for multiple impellers and series flow. Beyond N_s = 200, reaction design features should be considered.

Full Admission Turbines

In *full admission* turbines, the steam or process gas is admitted completely around the periphery. The Rotoflow design shown in Figure 10.4 is a *full admission* RIF type expander. The Linde cryogenic expander is also a *full admission* design when all

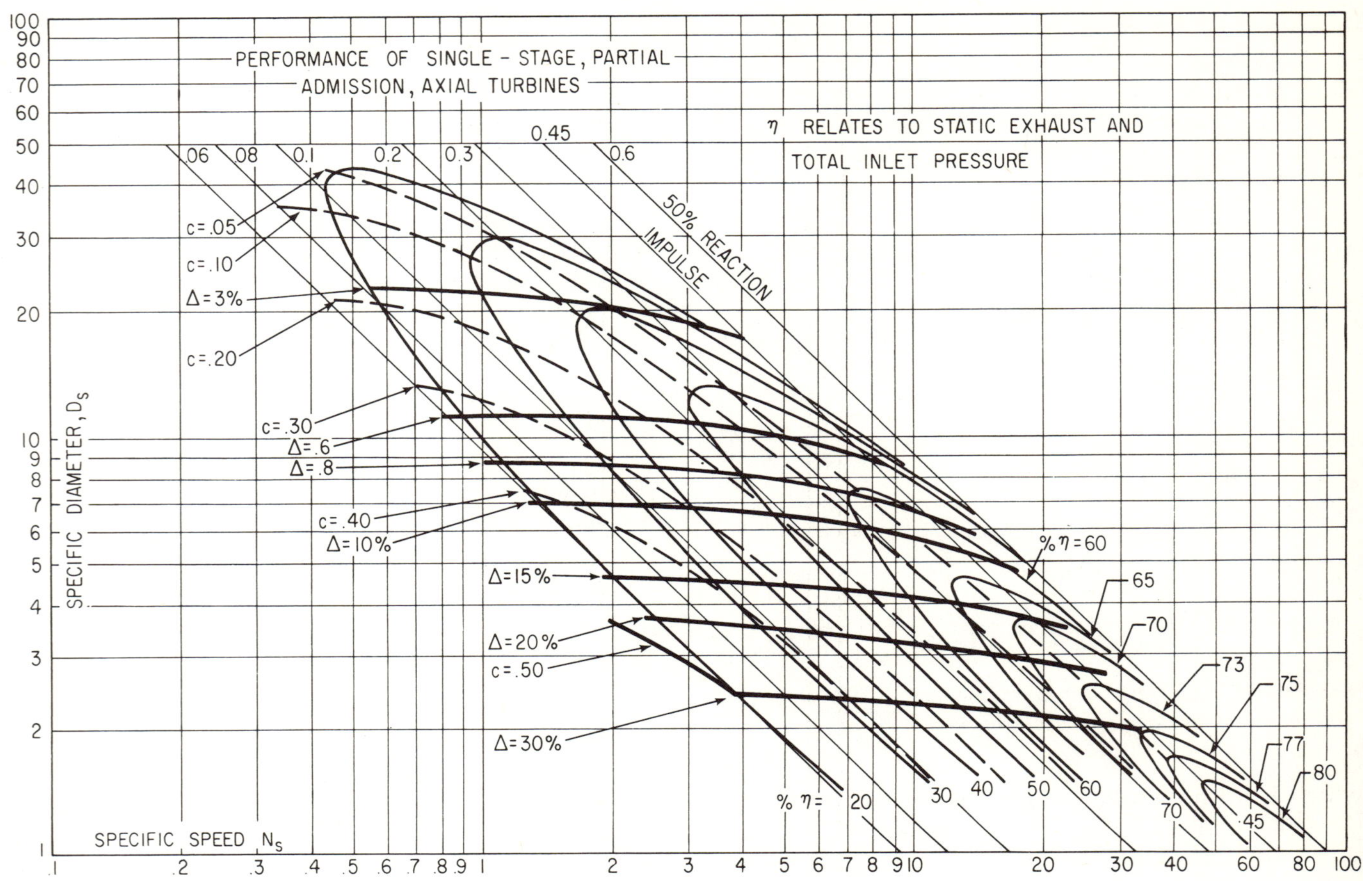

Figure 10.10. Baljé chart for performance of partial admission, single-stage, axial flow turbines.

nozzles are active. The closing of one nozzle makes it a *partial admission design.* The vertical expander shown in Figure 10.5 and most general purpose steam turbines are *partial admission* turbines.

Partial Admission Turbines

When the height of the blade is less than one-half inch, the production cost is excessive. The rotor clearances are also exaggerated and unrealistic. The size of the nozzles are affected by the same logic.

The losses associated with such miniaturized components make the *partial admission* design more practical to produce and efficient to operate. This design confines the required nozzles to a restricted arc span of the rotor, which carries conventional sized blades. For example, each nozzle applied to the expander shown in Figure 10.5 occupies 24° of peripheral arc. The five-nozzle cryogenic expander is rated 250 hp at 31,300 rpm. The addition of 10 more nozzles would make the expander a *full admission* machine and increase the power rating to 750 hp. Continuing the propylene expander example from the specific speed paragraph, the specific volume at nozzle entrance conditions is 1.02 cu ft/lb. The flow is (5.6 lb/sec)(1.02) = 5.7 cfs. The spouting velocity through the nozzle bore is 1,415 (0.93 nozzle coefficient) = 1,300 fps.

A single nozzle requires a bore area of 5.7(144)/1,300 = 0.633 square inches or 0.90 inch in diameter. The nozzle exhaust at the face of the impeller blades is about 1.75 inch at the minor axis. This would require a nominal 2-inch blade height. The ratio of the blade height to the mean diameter of the rotor (h/D) is 0.17. The normal range of the h/D ratios are from 0.05 to 0.11, for a 17° to 22° nozzle angle approach to the impeller. An expander having <0.17 ratio would experience excessive windage and other frictional losses. The ratio of the peripheral arc span to the nozzle bore is approximately 6/1. This makes the single nozzle, *percent admission*, 6(0.90)/11.25 (3.1416) = 15 percent.

Figure 10.10 outlines the performance of *partial admission axial impulse expanders*. The Δ numbers on Figure 10.10 represent the *percent admission*. The potential efficiency for this example is 70 percent, the locus of Δ = 15 percent and 0.45(U/C) on Figure 10.10. The effect of a greater *percent admission* is determined by trying three nozzles. The new nozzle bores are $(0.90 \times 0.67)^{0.5}$ = 0.52 inch.

The percent admission is (3)(6)(0.52)/11.25(3.1416) = 30 percent. The potential efficiency for a locus of Δ = 30 percent and 0.45(U/C) is 75 percent, which is the same efficiency expected from the *full admission* expander, providing it has reasonable geometry.

A continuous circumferential aperture is 0.633/(11.25 + 1.25)3.1416 = 0.016 inch. The blade height would be absurdly small. The h/D factor would be about 0.003, which further emphasizes its impracticability.

A four-nozzle array would require bores of $(0.633/3.1416)^{0.5}$ = 0.448 inch. The percent admission is (4)(6)(0.448)/11.25 (3.1416) = 35 percent. Reference to Figure 10.10 indicates an efficiency of 80 percent is expected.

The small letter *c* on Figures 10.9 and 10.10 represent the percent of kinetic energy in the exhaust with reference to the total energy entering the expander. The factor is the square of the ratio of the exhaust gas and the spouting velocity. This *c* value indicates the total energy that may be recovered with an exhaust diffuser or extra velocity staging. The loci for the three points considered are all on the c = 5 percent line. This means that the exhaust velocities are approximately 300 fps $(300/1{,}300)^2$ = 0.05. A minimal velocity of 200 fps only represents 2.5 percent of the total energy. The exhaust energy lines of 10 percent (400 fps) to 5 percent (300 fps) follow the peninsular projections of peak efficiencies on Figure 10.10. This emphasizes the fact that there is little or no benefit to be realized from series impellers or volute refinements on partial admission turbines.

Radial-Inflow Expanders

The optimized efficiencies for radial-inflow expanders having 90° blades are shown in Figure 10.11. The line of optimum efficiency follows the 0.5 to 0.7 U/C ratio lines from N_S = 25 to 100 and D_S = 3 to 1. The degree of reaction follows the same 0.6 velocity-ratio line. The centripetal-flow pattern experiences a natural restriction at the shortened radii approaching the exhaust eye. The wheel disk friction is also increased by the greater exhaust density. The 1.5 and 1.7 ϵ curves give the approximate ratio of the impeller outside diameter to the outlet-eye diameter. The $(C_3/C_o)^2$ kinetic energy loss to the exhaust is about 4 percent at the lower cup of the 80 percent island. The exhaust/spouting-velocity loss is about 8 percent and follows the lower cup of the 55 percent efficiency profile. Figures 10.3 and 10.4 illustrate RIF expanders.

Total Efficiency

The optimized efficiencies shown on the three Baljé charts assume that all of the kinetic energy leaving the impeller is lost. The square of the ratio of the exhaust velocity to the spouting velocity represents the power lost in the exhaust. Referring to the RIF turbine, the exhaust velocity at the 80 percent efficiency must be on the order of 260 fps to have an exhaust loss of 4 percent; $(260/1{,}300)^2$ = 0.04. The lower profile having an 8 percent exhaust loss implies that the exhaust velocity is on the order of 368 fps; $(368/1{,}300)^2$ = 0.08. It may be possible to reduce the latter exhaust velocity from 368 to 260 and thereby raise the turbine efficiency from 55 to 59 percent.

Dr. Baljé states that the efficiencies given on Figures 10.9, 10.10 and 10.11 are realistic for commercial manufacturing practice. He further states that it is possible to improve the expander efficiencies by as much as 8 percent with manufacturing refinements.

The minimum exhaust velocity loss of 2.5 percent occurs when the exhaust flow is without a whirl, and in the RIF expander, where blade height over the impeller diameter (h/D) tends to equal 0.05, assuming the hub radius is negligible. When this occurs, D_S must equal 1.0 and the effect of changing the specific diameter is shown in the following equation

$$(C_3/C_o)^2 = 0.025/D_S^4. \qquad (10.7)$$

Should D_S = 0.5, the exhaust loss would consume $0.025/(0.5)^4$ or 0.025/0.0625 = 40 percent of the total energy. Other pertinent expander equations are

$$U/C = 0.00654\, N_S D_S \qquad (10.8)$$

$$N_S = N/L_{ad}^{1.25}(550 \text{ hp } RT_3 \eta P_3)^{0.5}. \qquad (10.9)$$

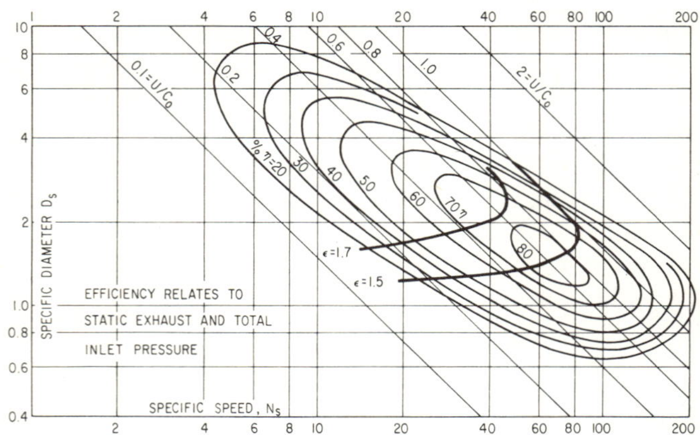

Figure 10.11. Baljé performance chart for radial inflow, single-stage turbine.

Figure 10.12. Open view of general purpose, partial-admission, inpulse turbine having two impellers in series. Courtesy of Elliott Company.

Turbine Steam Rates

The potential efficiencies given in Figures 10.7, 10.9, 10.10 and 10.11 are suitable for estimating the steam rates for turbines as well as estimating the performance of gas expanders. The mechanical inefficiencies are equal to the 0.4 root of the nominal rating; i.e., a 500 hp turbine has $(500)^{0.4}$ = 24 hp of mechanical losses or 95.2 mechanical efficiency, (η_m).

The Theoretical Steam Rate Table (24) gives the enthalpy differential per Kwh, based on isentropic expansion for the entire gamut of steam inlet and exhaust conditions. The turbine steam rate is readily obtained by using this enthalpy in the following equation:

$$\text{lb/Bhp hr} = 0.7455\ (\text{lb/Kwh})\ \eta_{dy}\eta_m \qquad (10.10)$$

The impeller diameters for popular, general service steam turbines are 14, 18, 22 and 28 inches.

The tip speeds, U, for these diameters at the most common operating speed (two-pole motor) of 3,550 rpm are 217, 279, 341 and 434 fps, respectively. A common steam pressure range for hydrocarbon processing operations is 175 psig and 500°F to 15 psig. The isentropic change in enthalpy is 21.95 Btu/lb, which gives a spouting velocity of C_o = $224(21.95)^{0.5}$ = 1,050 fps.

Dividing the tip speeds by the spouting velocities gives U/C ratios of 0.207, 0.266, 0.325 and 0.413. Typical efficiencies for a single-stage, impulse turbine as given on Figure 10.9 are 52, 63, 72 and 80 percent. Using the previous developed 95.2 percent mechanical efficiency, the respective steam rates are 21.95(0.7455)/0.952(0.52) = 33.3; 26.8; 23.4; 21.1 lb/hp-hour.

The economy of the larger impeller turbine is quite impressive if the turbine is used continuously. The margin between the 28 and the 14-inch impeller represents about 20,000,000 pounds of steam per year for a 200-hp turbine. A nominal value of steam is 50 cents per 1,000 pounds which makes the earnings $10,000 per year or a payout every four months. The difference in the cost of the large and the small impeller turbine is about $3,000. A typical large wheel general purpose steam turbine is shown in Figure 10.12.

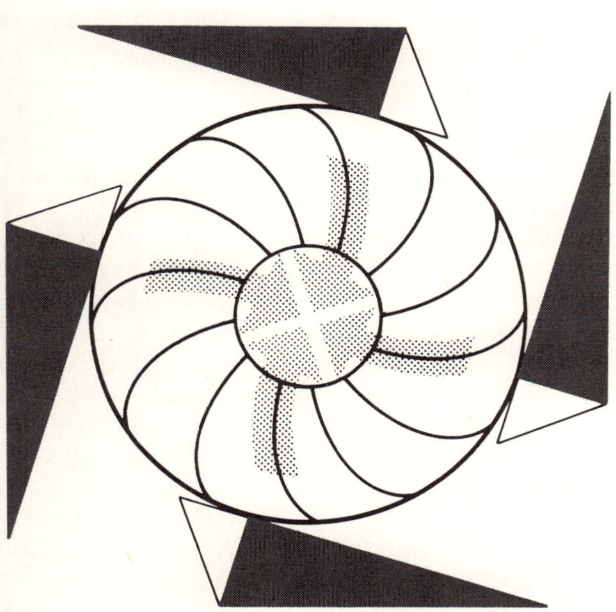

11 Compressor Computer Programs

This chapter presents computer programs for the piston and centrifugal gas compressor. These programs are designed to produce hardware dimensions and pertinent operating data to accommodate a given quantity of gas for stated terminal conditions.

A complementary series of programs can report the quantity of gas, the required power and other pertinent data that describe the capabilities of a given machine or series of machines.

Introduction

Someone has described a computer as "a moron that follows precise instructions." Many engineers are prone to accept or issue a computer analysis as an indisputable truth. There are few occasions to question the mathematical accuracy. But the quality of the program can only reflect the substance of the equations and instructions, be they excellent or invalid. The program can be designed to adjust, size and evaluate the variables which form the dimensions of an application. A certain problem may require a definite volumetric clearance for a given cylinder displacement to produce the necessary volumetric efficiency to accommodate the desired capacity. The cylinder bore is also limited by the piston rod thrust capabilities and the permissive valve area. The computer can resolve these complex circumstances and propose an equitable size which satisfies all stipulations. Such eventualities must first be conceived and then resolved by the programer.

The analysis of a series of machines proposed for a given function is one of the most valuable services that a computer can render. Each of such proposals is prepared by the respective manufacturers' proprietary methods, which are most likely to produce results of appreciable variance. Unfortunately, these variations do not reflect the real aerodynamic differences in the mechanical details and refinements.

The differences are largely the result of discrepancies in the empirical reference data used. For an example, the NGSMA Engineering Data Book gives the following overall efficiencies for the category of small compressors (200 hp and smaller) operating on air with 20 psia suction and 40 psia discharge (see Table 11.1). The last column gives the *comprehensive* dynamic and mechanical efficiency as described in References 2 and 25. This study presumes that all pertinent data are equal, except the piston-speeds which values are shown as fpm. The comprehensive equations are unique because they provide consistent and realistic evaluations. The alternative empirical method is not applicable to computer computation. Similarly, there is little or no differ-

Table 11.1
Overall Efficiencies of Small Compressors

Stroke	rpm	fpm	Ov'l Eff.	Comp. Eff.
13	400	870	83	75
11	475	870	78	75
6	685	685	67	87

ence in equipping a piston compressor for a 120 molecular weight refrigerant service than is given to the compression of a 6 mol weight gas. The lift of the valve element in light gas service is usually reduced to half normal to minimize the valve breakage. The conservative concession that is usually made to favor a refrigeration compressor using a heavy halocarbon refrigerant is to apply a 1,150 rpm motor driver in lieu of the standard 1,750 rpm motor. There is a surprising degree of unawareness concerning these intrinsic losses.

The computer operator need not be an expert to apply the intrinsic compressor analysis. It requires only a fraction of the time required for the slide rule calculations. There is no particular skill required to produce a comprehensive analysis of the machine's capabilities. The analyst has the assurance that the included methods have produced consistent results for the past several decades.

The computer can render a unique service in digesting a series of hardware dimensions submitted for a given assignment. The sizing of each unit was developed by the respective manufacturers' procedure. The computer makes an independent unilateral evaluation of each machine's capabilities. The idiosyncracies of the various methods are avoided by using a common gauge.

Scope

This chapter contains computer programs for both piston and centrifugal compressors. The programs for each type of compressor are divided into two groups. The primary purpose of the first group is to determine the capacity and power requirement for a given series of hardware dimensions. This group includes the piston computer programs UOHP, P6S and RKCPC. It includes the centrifugal compressor capacity program NSQ.

The second group is concerned with sizing the hardware for a given quantity of gas for fixed terminal conditions. This includes the piston computer programs PBS and RKBCS. The centrifugal computer programs are CCA, CKNP and NSG. There are two other preparatory programs, RKZ and MUCS. Program RKZ resolves the compressibility of the gas for each stage of piston compression. It also develops the necessary physical properties of the gas from a fundamental analysis. All programs containing the prefix RK, such as RKCPC and RKBCS, include the Redlich-Kwong equations which automatically develop all of the gas properties.

The MUCS program concerns the sizing of centrifugal compressors when the pressure range of the operation is beyond the means of a single casing. It resolves the number of stages for each casing and the optimum terminal pressures for each casing.

The programs are written in the BASIC language of the Service Bureau Corporation, a division of IBM (26). This text is not an exposition on computer language, nor is it a paragon of form and simplicity.

The computer language consists of a large number of highly organized abbreviations. For example, the molecular weight of gas is designated M2. The ratio of specific heats is K3. These are essential physical properties of the gas that are used in the computer equations. The two derivatives of K3 are K9 = 1 / K3 and S = (K3 − 1) / K3. The necessary mechanical features include the average piston speed, U, in fps, which relates to the RPM (R4) and the piston stroke (R6); U = R4*R6 / 360. These programs include about 150 such abbreviations which are cataloged in the Glossary of Symbols, Table 11.31. The plus (+) and minus (—) signs have the common arithmetical significance in the BASIC computer language. The asterisk (*) means to multiply. The slant bar (/) means to divide. The vertical arrow (↑) signifies an exponential power is to be performed. The operation indicated by the innermost parenthesis has the first priority in the calculation procedure.

Piston Compressor Performance

Program UOHP is designed to evaluate the *capacity* of an existing compressor, a series of cylinders which constitute a compressor station, a single or an array of units from a bid proposal. The program (Table 11.2, Figure 11.1) also produces the Bhp, the piston rod thrust and stress, the volumetric and the compression efficiency. Tables 11.3, 11.4, 11.5 and 11.6 show an application of program UOHP. It is only applicable to single-stage and two-stage operations. Program P6S is designed for the more complex operations. It is capable of handling six single stages, three two-stage operations, two three-stage operations, a four, a five or six-stage operation.

Program RKCPC includes the Redlich-Kwong (R-K) equation for resolving the compressibility factors for each terminal. It produces the pseudo critical pressure and temperature. It also calculates the average molecular weight, the molal heat capacity and the ratio of specific heats. Program RKCPC also balances the power load per cylinder by selecting the proper intermediate pressures for multistage operations. This program also includes an optional "long form" print-out of cylinder displacement, specific volume, the total head and the valve losses.

Each program includes an *input data sheet* where the given material is prepared for the computer entry on the respective line number. These input data sheets are illustrated in Tables 11.4, 11.7, 11.9, etc. Table 11.4a gives the necessary input data for a 150-ton propylene refrigeration compressor. The expanded flow-sheet for Program UOHP is illustrated on Figure 11.2. The detailed flow-sheets which designate the action are Figure 11.2. The program instructions and equations are given in Table 11.5. The program *print-out* in Table 11.6 shows that 195 Bhp is required to develop 150 tons of −5° F refrigeration.

Program P6S in Table 11.8 shows the performance of a two-stage natural gas compressor having three 26-inch and three 15-inch cylinders in parallel operation. The performance of each cylinder includes the original minimal clearance of 8.8 and 9.7 percent, respectively. The other two cases reflect the unloading capabilities of extraneous clearances. Program P6S also has the feature of producing an array of data for five performances having a suction pressure variation of 10 percent increments. The abbreviated flow-sheet on Figure 11.3 shows how the computer program engages the array operation by stating I5 is equal to 2. This data is useful in charting a performance curve as shown on Figure 11.4.

Program CPC produces the weight flow per stage and per cylinder, various efficiencies, etc. The flow diagram is given in Figure 11.5. The program is well suited for evaluating a bid analysis. The input data Table 11.9 and the print-out Table 11.10 show a two-stage propylene operation with maximum suction and minimal clearance. The second condition shows the effect of one psi reduction in suction pressure and the effect of extraneous clearance on the machine load. The load is reduced from 605 to 455 Bhp.

Program CPC has a short form and a long form of print-out. The short form includes 10 pertinent performance factors that are developed from the given input data. The long form adds five more performance items, which are identified as:

V (J,1) Suction specific volume
Q (J,1) Displacement, cfm
H (J,1) Adiabatic head ft-lb/lb
B (J,2) Suction valve loss, decimal fraction
B (J,3) Discharge valve loss, decimal fraction.

Program RKCPC is a combination of programs RKZ (developed later) and CPC. It offers the advantage of resolving the physical properties from the gas analysis and fixing the interstage pressures. The input data is posted in Table 11.11 and the production is printed on Table 11.12. Figure 11.6 are the program flow diagrams.

Cylinder Sizing

The next most prevalent problem in a gas compressor application concerns the cylinder sizing. This sizing, the power requirement and other performance data are developed in Program PBS. These code letters refer to the *piston bore sizing*. This program is limited to single or two-stage operations. The usual mechanical details, gas properties and conditions at the cylinder terminals are prepared as the necessary input on the data sheet, identified as Table 11.13. Figure 11.7 illustrates the data flow pattern; Table 11.14 is the program.

Program RKBCS is prepared for the more complex cylinder sizing procedures. The code letters refer to the BASIC (language) *cylinder sizing*. The cylinder size is adjusted to the nearest nominal cylinder bore by adding to the minimal clearance. The nominal sizing increment is assumed to be one-half inch. If the apparent compression efficiency is less than 85 percent, the computer will nominate the necessary P/V area ratio to correct the compression efficiency to 90 percent. The extra items on the *long form* include:

B (J,10) Adjusted intrinsic valve loss factor
B (J, 8) Adjusted suction valve loss
F (J, 1) Volumetric efficiency, original conditions
B (J, 1) Preliminary cylinder bore; no size restrictions
B (J, 5) Adjusted cylinder bore; size and thrust qualified

B (J,12) Tentative cylinder bore; without integer corrections

E (J, 1) Compression efficiency, original conditions.

If the cylinder bore exceeds 35 inches (or any assigned limit) or if the piston rod thrust stress exceeds 10,000 psi, the cylinder displacement is reduced to half and, if necessary, to one-third of the initial sizing. This sizing produces two and three parallel cylinder arrangements. The program is suitable for one to four stages. The RK equation is applied in the prescribed manner. The input data is posted on Table 11.15. The data flow diagram is shown on Figure 11.8. The product is shown on Table 11.16. The print-out describes a three-stage, natural gas operation. The compressor station involves 27,000 hp in pumping 100 MMscfd from 23 to 1,500 psia. Where the required mechanical features are not specifically known, it is possible to use realistic generalized values. A rundown on these generalized values follows.

Piston Speeds

The average piston speed "U" is a common denominator that can be applied to all categories of compressors. The simple, straight-line crosshead guided compressors are rated at piston speeds of 10 to 13.3 fps. The average piston speed for all large compressors is 13.3 fps.

One of the principal manufacturers recently announced that they are marketing compressor designs capable of 18.3 fps piston speeds. This appears to be a technological breakthrough. Actually, it is only an admission of the validity of the intrinsic compression concept. This manufacturer has been most generous in providing valve areas. They are applying this surplus valve area to increase their piston speed. This is a perfectly valid manuever. The valve losses are affected by the square of the change in piston speed and P/V area ratio.

The piston speed U is obtained by the equation:

$$U = \text{RPM* STROKE} / 360, \text{ fps}$$

$$U = R4 * R6 / 360 \text{ (see Table 11.3).} \qquad (11.1)$$

P/V Area Ratio

Not only is the piston velocity significant per se, but it also establishes the velocity at any point in the flow system. For an example, the velocity through the lifted valve area is a product of the piston speed and the ratio of the piston/valve area. The effective or the valve lift area is the product of the valve lift, usually about 0.080 inch, and the sum of the flowing peripheries or edges. The area reported for most disk, plate and poppet valves is consistent with this definition. The passage area through the valve seat is sometimes offered as the effective area for strip-type valves. The effective area is about 30 percent of the seat passage area. The effective valve area can be estimated from the size of the valve port by the following:

$$\text{For 0.100 lift disk valves:} \quad A_v = D_o^2 / 17 \qquad (11.2)$$

$$\text{For 0.25 lift poppet valves:} \quad A_v = D_o^2 / 8 \qquad (11.3)$$

$$\text{For 0.100 lift strip valves:} \quad A_v = D_o^2 / 25, \qquad (11.4)$$

where A_v is the effective valve area and D_o is the valve port diameter in inches. The maximum port diameter can be estimated by the following equation:

$$D_o = (4\,D / \text{No. } V) - 3 \text{ inches,} \qquad (11.5)$$

where D is the cylinder diameter in inches and No. V is the number of valves per end of the cylinder.

The average P/V area ratio is 10 for most industrial and process type compressors. Commercial machines restricted to air and lighter gases have P/V ratios of 12 to 15. A ratio of 20 will provoke exorbitant losses, unless the machine is working with gas having a mol weight less than 10. Where the gas mol weight is 40 or heavier, the P/V ratio should be 8 or smaller. These ratios times the average piston speed gives the average valve velocity. The maximum surge velocity through the central zone of piston travel is 50 percent greater than the average piston speed (2).

Temperatures

The temperature of natural gas from buried pipelines in the Temperate Zone is 60°F, with perhaps a 5° plus and minus. Gas from water-cooled process heat exchangers and intercoolers is generally taken to be 95°F for design purposes. It is 110°F for an 85°F ambient when using air cooled equipment. These temperatures are raised 15°F for the warm gulf and desert countries. The reference temperatures are reduced 10°F or more in the Frigid Zone.

The most equitable discharge temperature is a product of the Rankine suction temperature and the apparent ratio of compression. This would be 3.38 *R* for the forthcoming illustration, raised to the exponential power of (1.244 − 1.0)/1.244 or 0.196. The discharge temperatures are 520* 1.269 = 660 R or 200°F and 555* 1.269 = 705 *R* or 245°F.

Balanced Interstage Pressures

A 5 percent pressure drop is allowed for the aggregate sum of two pulse dampers, an intercooler, a separator and the piping between each stage. The pulse damper consumes about one-half percent and as much as one percent pressure drop by the API Standards. The intercooler consumes one to 2 percent. Piping for each of the last two items consumes one-half to one percent. The following example illustrates the slide-rule approach and the subsequent computer analysis.

Determine the pressure array for an operation having 100 MMscfd of 20 *m* gas, where $k = 1.244$, the intake is 25 psia and 60°F and the discharge is 2,500 psia:

Total System $R' = 2{,}500 / 25 = 100$.

The suction system includes a pulse damper, a separator and piping which are satisfied with a 2 percent pressure drop. The discharge includes a pulse damper, an aftercooler, a separator, a check or control valve and piping which total 5 percent. The selection of a suitable ratio of compression follows:

Total $R = 2{,}500\,(1.05) / 25\,(0.98) = 107$.
The two-stage $R_2 = 107^{0.5} = 10.33$, which is excessive.
The three-stage $R_3 = 107^{0.33} = 4.64$, 3.8 *R* and still too large.
The four-stage $R_4 = 107^{0.25} = 3.18$, which is satisfactory.

The interstage ratio of compression is 3.35, the product of 3.18 and 1.05, the interstage loss. The expanded pressure, temperature and compressibility factors for the four-stage operation are given in Table 11.17.

The compressibility factors were determined from the charts in the NGPSA Data Book. The computer program RKZ automatically resolves the interstage pressures and the compressibility factors.

Program RKZ

This program is named for the Redlich-Kwong equation. It is an algebraic equation for establishing gas compressibility correction factors. These corrections are normally taken from charts prepared by Brown, Katz, etc. The procedure is tedious. The objective of this program is to resolve the compressibility factors for both terminals of each stage. In addition, the interstage pressures are selected to balance an optimized load for each stage, in the manner previously described.

The cubic form of the Redlich-Kwong equation is

$$Z^3 = Z^2 - BP\,((A^2 \,/ -B) - BP - 1)\,Z + (A^2 / B)\,(BP)^2 \qquad (11.6)$$

A solution for *Z* is developed from the following arrangement:

$$\begin{aligned}
\text{Let } p &= -1 \\
q &= BP\,(A^2 / B - BP - 1) \\
r &= -(A^2 / B)\,(BP)^2 \\
a &= 6.289\,R^2\,T_c^{2.5} / P_c \\
b &= 1.273\,R\,T_c / P_c \\
A^2 / B &= (a / b)\,R\,T^{1.5} \\
B &= b / RT \\
DN &= (DM/3 + r - 0.074) \\
DM &= (q - 0.333).
\end{aligned}$$

These expressions reduce Equation 11.6 to proper form for the BASIC language:

$$\begin{aligned}
DM = {} & 6.289^*\,(P / P_c)^*\,(T_c / T) \uparrow 2.5 \\
& - 1.625^*\,(T_c\,P / T\,P_c) \uparrow 2.0 \\
& - 1.273^*\,(T_c / T)\,(P / P_c) - 0.333 \qquad (11.7) \\
DN = {} & DM / 3 - 8^*\,(T_c / T) \uparrow 3.5^*\,(P / P_c) \uparrow 2 \\
& + 0.037 \qquad (11.8)
\end{aligned}$$

Values of *q* and *r* are determined for the given (A^2 / B) and (BP) conditions. From these values the term

$$(DN^2 / 4) + (DM^3 / 27) \qquad (11.9)$$

can be calculated.

When this term is greater than zero, the compressibility factor is in the monophase. This involves the gas in the dense phase above saturation or above the critical temperature area. The other condition

Table 11.17

Balance Stage Pressures

Stage	R_c	Suction				Discharge		
		R	psia	Z	v_s	psia	°F	Z
1	3.35	520	24.4	1.00	11.5	82	200	1.000
2	3.35	555	78	0.99	3.78	261	245	0.970
3	3.35	555	250	0.96	1.14	835	245	0.944
4	3.35	555	795	0.86	0.32	2680	245	0.894

where the term is equal to or less than zero concerns the two-phase conditions which fortunately are not experienced in a gas compressor. This information was abstracted from "Applied Hydrocarbon Thermodynamics," (4).

Concerning the compressibility factors, it is well to remember that they are less than unity as they approach the dew point. The minimum factors are attained when the gas temperature is slightly greater than the pseudo critical temperature and when the pseudo critical pressure (P_r) is in the range of 1 to 4. Correction factors greater than unity are experienced at P_r values of 11 and greater, with temperatures above the pseudo critical.

The input data sheet for this program is given in Table 11.18. The physical properties for 10 common gases are given for a mean compression temperature of 140°F. The data for other temperatures or other gases may be substituted by typing in the program line number, followed by the new data. The flow chart for RKZ is given in Figure 11.9. A production print-out is shown on Table 11.19.

Computer Programs for Centrifugal Compressors

These programs are designed to resolve an industrial compressor operation into the optimum number of stages and sizes of centrifugal compressor casings. The first detail to establish is the number of casings required. Eight to 10 stages or impellers are the usual limitations per casings. Ten stages are used in small casings where the shaft diameter is small and the speed is relatively high. Conversely, where the shaft and impellers are large and the speed is slow, the staging is usually limited to six, and in some instances, five stages are the limit. There is a sensitive balance between the impeller count, the shaft dimensions and the critical speeds. The head capacity for each stage also has a significant function in sizing series casings. The following equation is a generalization of common practice:

$$H(s) = 15{,}000 - 1500^{*}\ ((M2) \uparrow 0.35)$$

Programs CCA, CKNP and NSG are concerned with the sizing of the hardware for given quantity of gas and terminal conditions. The feature of program CCA is the ability to select the optimum number of stages and rationalize the dynamic efficiency. Program CKNP accomplishes the same results, plus selects the optimum impeller diameter for a given nominal speed. Program NSG develops the production data without a given nominal diameter or speed. It develops an array of hardware dimensions for a spread of specific speeds, from 150 to 25. Program NSQ produces the same performance array, except that it develops the optimum quantity for a given impeller diameter.

Program MUCS

This program code (Table 11.20 and Figure 11.10) refers to its "Multi Unit Casing Sizing" capabilities. The required input data includes the quantity, the terminal pressure, the initial pressure and temperature as shown on Table 11.21. The cooled interstage temperature, the ratio of specific heats, the mol weight and the approximate compressibility factors are the usual gas information necessary to resolve the hardware description.

This program is designed to break down the head capacity per stage, the number of impellers per casing and determine the number of casings for the service. Two and three casings in series are common; four and sometimes five casings are used for synthetic ammonia compression and hydrogenation operations. Allis-Chalmers has built a supersonic wind tunnel that involves eight casings in series (2).

After this program determines the number of casings and the number of stages per case, the performance data are developed by an appropriate program. It is presumed that the gas is intercooled to 95°F by water cooling. An intercasing loss of 3 percent of the system pressure should be allowed between each casing. The flow diagram for MUCS is given in Figure 11.11. The production print-out is shown in Table 11.22.

Program CKNP

This program is designed for *centrifugal* compressors at constant speed and high *NS performance* (CKNP). The optimum diameter is selected and adjusted for the interstage frictional loss of 5 percent plus a 17 percent dynamic "slip" loss. The latter is also corrected for the molecular weight of the gas. The head per stage is limited by the sonic velocity and by the industrial trade practices. This program is useful in resolving the impeller diameter for a given speed and specific operating conditions. It gives the optimum efficiency, power requirement, discharge temperature and other performance factors. The input data is presented on Table 11.23. The flow diagram is shown on Figure 11.12. The production is printed on Table 11.24.

Program NSG

The usual procedure after determining the adiabatic compressor head is to select the optimum speed, the impeller diameter, the size of the casing, etc. The program initials refer to the specific speed (NS) used to develop a series of centrifugal compressor geometrical dimensions (NSG). The input data is collected on Table 11.25. This includes quantity of gas, its molecular weight, the "k" value and the terminal conditions. The flow of data through the equations is shown in Figure 11.13.

The production print-out on Table 11.26 gives the hardware details for an array of likely specific speeds from 150 to 25 Ns. The two examples for 20 mol weight natural gas and R-12 refrigerant show that the 100 Ns is the most efficient position. The 150 Ns value is only off 2 percent and 75 Ns is off 5 percent from the peak efficiency attained at 100 Ns performance. See Figure 6.8 and Reference 6 for more details.

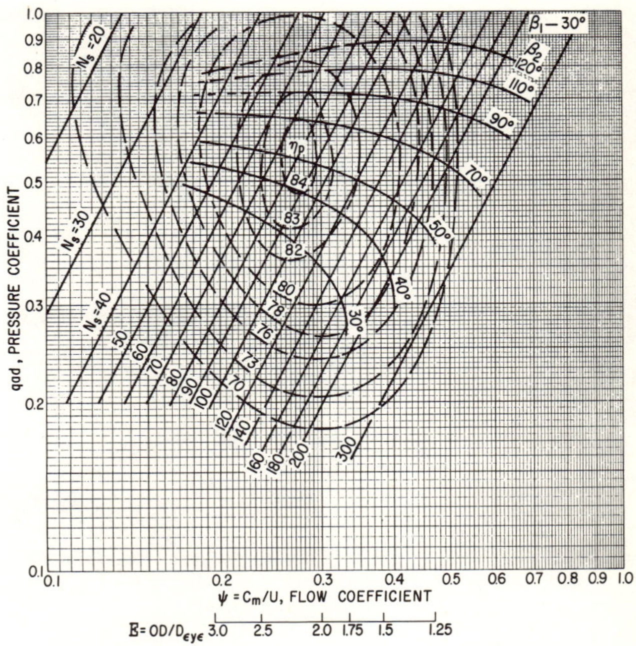

Figure 11.16. Eckert chart giving pertinent adiabatic performance data for impellers having various vane angles.

A good index for regulating the optimum flow capacity of an impeller is the ratio of the outside diameter to the suction eye diameter. The practical minimum value of this ratio, ϵ (epsilon), is 1.4. The maximum ϵ ratio is about 3.0. The "GULP" factor is another useful index in sizing impellers. This ψ (psi) factor is the ratio of the meridional eye velocity divided by the tip speed. The optimum ratio is about 0.30. There is a 10 percent tolerance in applying this factor when the specific speed is maintained in the 90 to 115 range (see Figure 11.16).

An interesting application of the Production Table 11.26 indicates that it is possible to establish a power correction factor for the two uncontrolled factors which influence the refrigeration compressor power load. The temperature of the return brine establishes the suction pressure for the compressor. The temperature of the condensing water determines the compressor discharge pressure. Conceding that the return brine temperature can vary plus and minus 5°F, an incremental power correction of 13 hp per degree F is credited for each degree below the design point. A similar debit of 9 hp is charged for each degree that brine leaves the chiller above the design point.

Every degree that the condensing water varies from the design point of 88°F influences the motor

Figure 11.17. Giant sized, 12,000-hp, Model CLBA-10, balanced-opposed compressor for liquid hydrogen plant. (Courtesy of Clark Bros. Division of Dresser Industries.)

load by 7.5 hp. An adverse temperature above 88°F creates a credit deductible from the observed motor load, which makes a more favorable check against the guaranteed power. Conversely, a more favorable condenser water temperature, lower than 88°F, creates a debit which adds to the observed motor load. Whereas it is highly improbable that the plant process load and the condensing water temperature would ever coincide with the specified conditions in order to perform an acceptance test, the incremental power correction offers a likely method for making an equitable acceptance test under any full capacity process load conditions.

Program NSQ

This program (Table 11.27) concerns the capacity (Q) capabilities for a centrifugal compressor having a specific series of impellers and operating speed. The optimum capacity is printed in Table 11.28 for the full array of likely specific speeds. The identifying program initials, NSQ, refer to the specific speed (NS) development of the capacity (Q) capabilities. The input data is entered on Table 11.27. The program is developed according to the flow diagram given on Figure 11.14.

Program CCA

The code letters for this program refer to the *Centrifugal Compressor Applications.* This program is most useful in evaluating a series of bid proposals for a given operation. It compares the proposed impeller diameters and speed for a given gas and set of terminal conditions with the optimized blade geometry as described by Dr. Baljé (12). This analysis describes the specific diameter versus the specific speed. The plotting of this data with reference to the optimized dynamic efficiency and the pressure coefficients produces the topographical-like map and a series of 45° slope lines. The contours represent lines of equal efficiencies. The sloping lines represent the D_sN_s profile for each category of turbo-compressors. The lines at the right hand side of Figure 6.8 are applicable to axial compressors where the peak pressure coefficient, q_{ad} is about 0.3 and D_sN_s is between 200 and 250. Centrifugal compressors with back-lay vanes have a peak q_{ad} of about 0.53 and a D_sN_s product of 150. Radial bladed machines have a peak q_{ad} of about 0.63 and D_sN_s of 165. Rotary displacement machines behave like q_{ad} is equal to 4 and D_sN_s = 26. Regenerative compressors occupy the extreme left hand side of Figure 6.8 and experience pressure coefficients as high as 10. The relationship of the pressure coefficient q_{ad} to the adiabatic head is

$$H_{ad} = qad\ U^2\ /\ g.$$

Program CCA presents the performance data for a centrifugal compressor. The input data is entered in Table 11.29. The simplified flow diagram (Figure 11.15) describes the program operation. The computer program is given in Table 11.3. This program is suited for developing the performance data for a single design point. This data would be most useful for making a unilateral study of the machinery offered for a proposal bid. The service rendered by program CCA is identical to that of program CKNP, except CKNP selects an optimum size impeller. The impeller size is given for program CCA. Program NSG renders the same service as program CKNP, except that the former presents a full array of likely specific speed values.

Closure

The computer programs in this text offer a unique, unilateral and proven method of evaluating the performance and capabilities of all forms for piston and centrifugal compressors (one type is illustrated in Figure 11.17).

Operational Comments for Computer Programers

The following is a series of notes which may prove helpful to a computer programer applying this data.

1. Programs all have data tables at the very beginning. The data shown with the program and "saved" with it can be used for check-out when needed.
2. New data may be entered on any line by entering a complete line containing the new data. Use the same line number as the line being replaced (see Paragraph 8 for input format).
3. When input is complete, adjust paper to start a new sheet and give "run" command.
4. Some programs contain a sequence for entering additional data for another set of calculations while program is compiled. In such cases, follow instructions. Other programs must start each time from "run" command after modifying data table.
5. Programs RKZ, RKBCS, RCKPC have gas properties stored in first part of data table, and these values must not be altered unless different gases with appropriate property values are substituted.
6. Initially, enter programs exactly as shown, line by line. Be sure of "NAME." Save and lock.
7. To use a program on time share:
   a. Sign on line with terminal—use same code as for entering program initially.
   b. Load program—(NAME).
   c. Enter new data—prepared in advance.
   d. Run.
8. Input data sheets show the values required and the format for compatibility with the program. Always fill in data sheet first. Be sure each entry is in the correct units as indicated. Each line must be complete even when repeating most of a line. However, if existing line needs no change, make no entry at all for that line number and the existing data will be used in calculations.
9. For each program, the following information is furnished:
   a. Input data sheet—blank spaces.
   b. Input data sheet—filled in spaces corresponding to data shown in program data table.
   c. Flow chart— one of the following:
      1. Simplified flow diagram—this shows briefly the program organization.
      2. Flow details—this chart shows the complete organization of the program and will aid in de-bugging. (By inference, the chart for one program may aid in understanding the details of other programs for which detailed charts are not shown.)
   d. Program listing.
   e. Sample calculation.

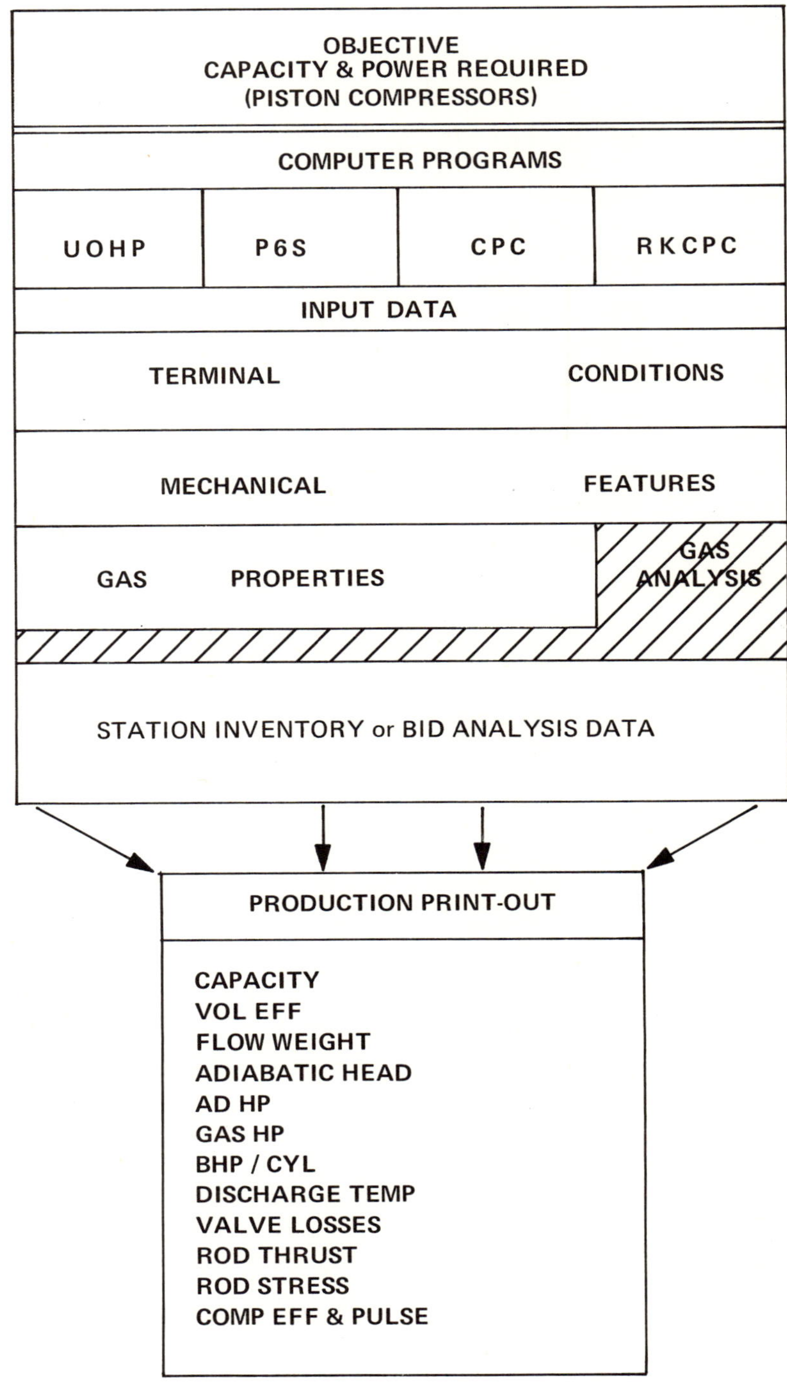

Figure 11.1a. General procedure of developing the capacity and power from programs UOHP, P6S, etc.

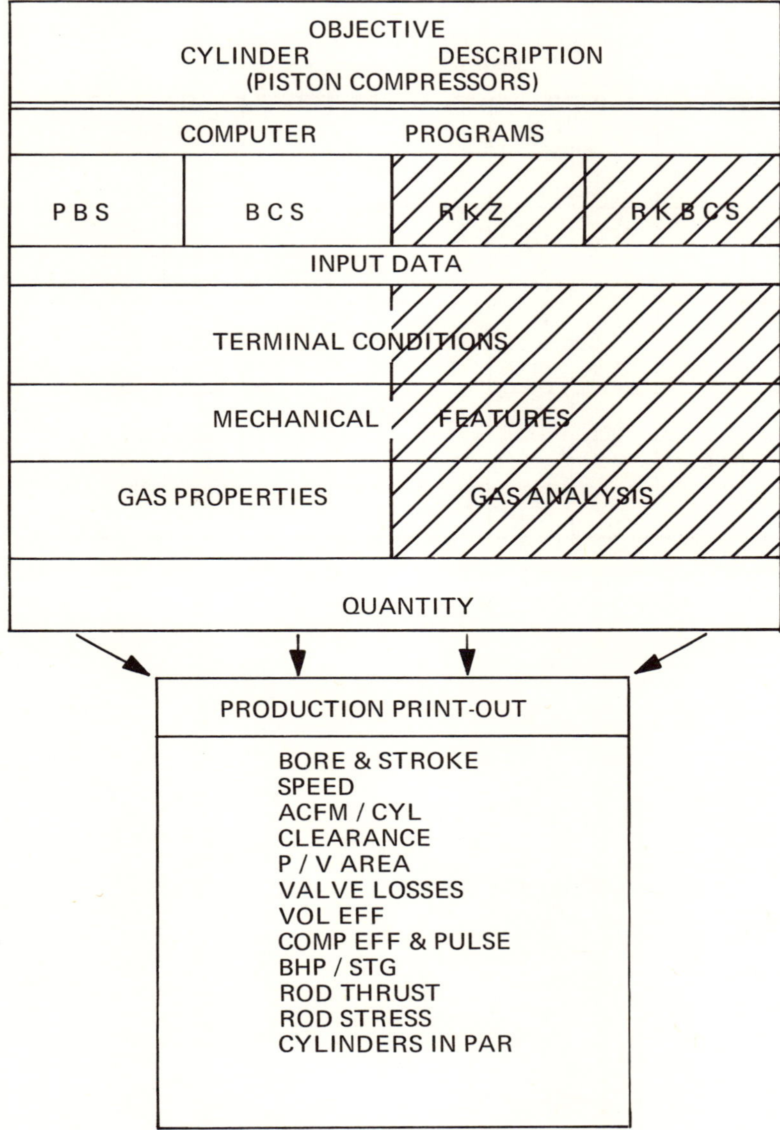

Figure 11.1b . Procedure to size the hardware necessary to handle a stated gas flow, by means of programs PBS, BCS, etc.

Table 11.2
Computer Program Input and Production for Piston Gas Compressors

PROGRAM	Q (MMscfd)	K3, M2, Z	P, P1 T1	Bore	CL	P/V	E_V	η	Piston Rod		5-Suct. Var.	No. of Stages	Bhp	Disch Temp.
									Thrust	Stress				
UOHP	C	G	G	G	G	G	C	C	C	C		2	C	C
P6S	C	G	G	G	G	G	C	C	C	—	C	6	C	C
CPC	C	C	G	G	G	G	C	C	C	—	C	6	C	C
RKCPC	C	C	G′	G	G	G	C	C	C	—	C	6	C	C
PBS	G	C	G	C	G	G	C	C	C	C	—	2	C	C
RKBCS	G	C	G′	A	A	A	A	A	A	A	—	6	A	C
RKZ		C	G′									6		

Notes: G = given data; C = calculated data; A = adjusted calculation.
G′ prime indicates that stage loads are balanced.

Table 11.3
Program UOHP Input Data Sheet (Blank)

PROGRAM LINE		SYSTEM CONSTANTS					
NO.	NAME	GAS MOL WEIGHT	1/K3	(K3−1)/K3	PISTON SPEED (FPS)	ROD DIAM (INCHES)	AMBIENT PR (PSIA)
		M2	K9	S	U	R5	P1
10	DATA						
		FUNCTION NUMBER (OR STAGE NUMBER)					
		ONE	TWO				
		SUCTION PRESSURE–(PSIA)–P (J, 1)					
20	DATA						
		DISCHARGE PRESSURE–(PSIA)–P (J, 2)					
30	DATA						
		SUCTION TEMPERATURE–(°R)–T (J, 1)					
40	DATA						
		CYLINDER BORE–(INCHES)–B (J, 1)					
50	DATA						
		CLEARANCE–(DECIMAL–E.G.–0.200)–C (J, 1)					
60	DATA						
		P/V–A RATIO A (J, 1)					
70	DATA						
		COMPRESSIBILITY–'Z'–SUCTION–Z (J, 1)					
80	DATA						
		COMPRESSIBILITY–'Z'–DISCHARGE–Z (J, 2)					
90	DATA						

NOTES: 1. SEE GAS COMPRESSION–GLOSSARY OF SYMBOLS USED IN COMPUTER PROGRAMS

Table 11.4
Program UOHP Input Data Sheet (Sample Data)

PROGRAM LINE		SYSTEM CONSTANTS					
NO.	NAME	GAS MOL WEIGHT	1/K3	(K3−1)/K3	PISTON SPEED (FPS)	ROD DIAM (INCHES)	AMBIENT PR (PSIA)
		M2	K9	S	U	R5	P1
10	DATA	42	0.87	0.1305	9.75	2.0	14.5
		FUNCTION NUMBER (OR STAGE NUMBER)					
		ONE	TWO				
		SUCTION PRESSURE–(PSIA)–P(J,1)					
20	DATA	15.5	69.7				
		DISCHARGE PRESSURE–(PSIA)–P(J,2)					
30	DATA	73.	260.				
		SUCTION TEMPERATURE–(°R)–T(J,1)					
40	DATA	455	480				
		CYLINDER BORE–(INCHES)–B(J,1)					
50	DATA	21.0	12.5				
		CLEARANCE–(DECIMAL–E.G.–0.200)–C(J,1)					
60	DATA	0.1632	0.2138				
		P/V–A RATIO A(J,1)					
70	DATA	8.6	9.2				
		COMPRESSIBILITY–'Z'–SUCTION–Z(J,1)					
80	DATA	1.00	1.00				
		COMPRESSIBILITY–'Z'–DISCHARGE–Z(J,2)					
90	DATA	1.00	0.915				

NOTES: 1. SEE GAS COMPRESSION – GLOSSARY OF SYMBOLS USED IN COMPUTER PROGRAMS

Figure 11.2. Flow diagram for Program UOHP.

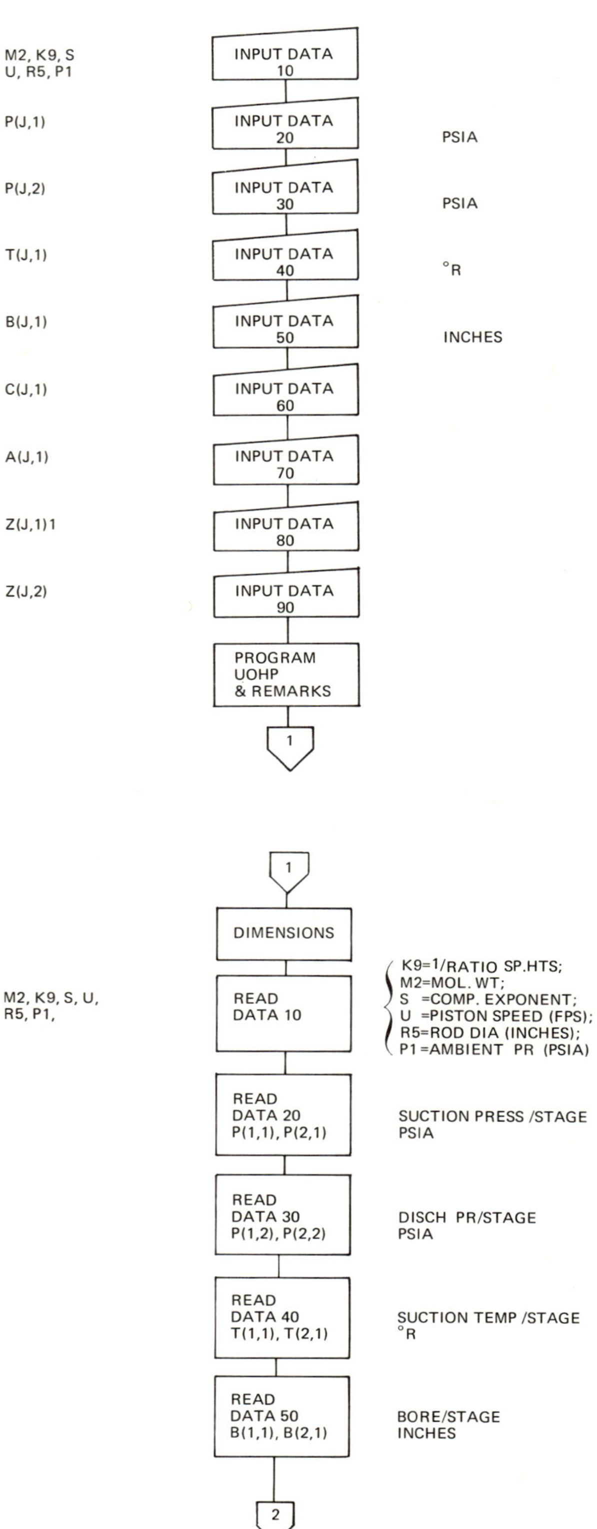

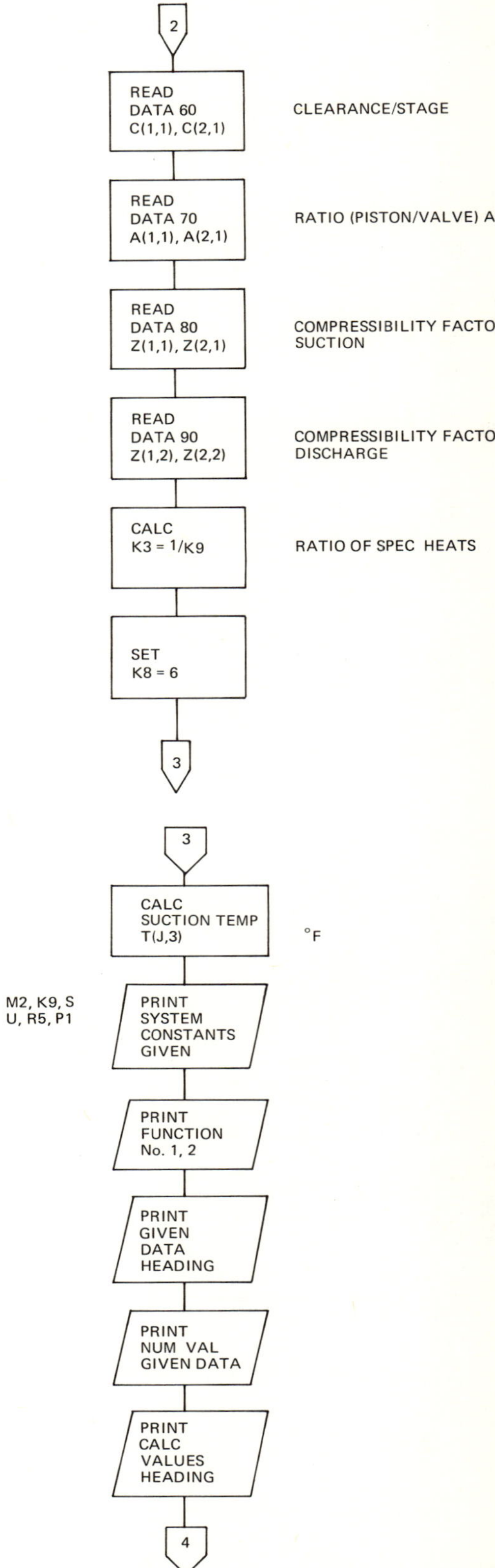

Figure 11.2 (*continued*)

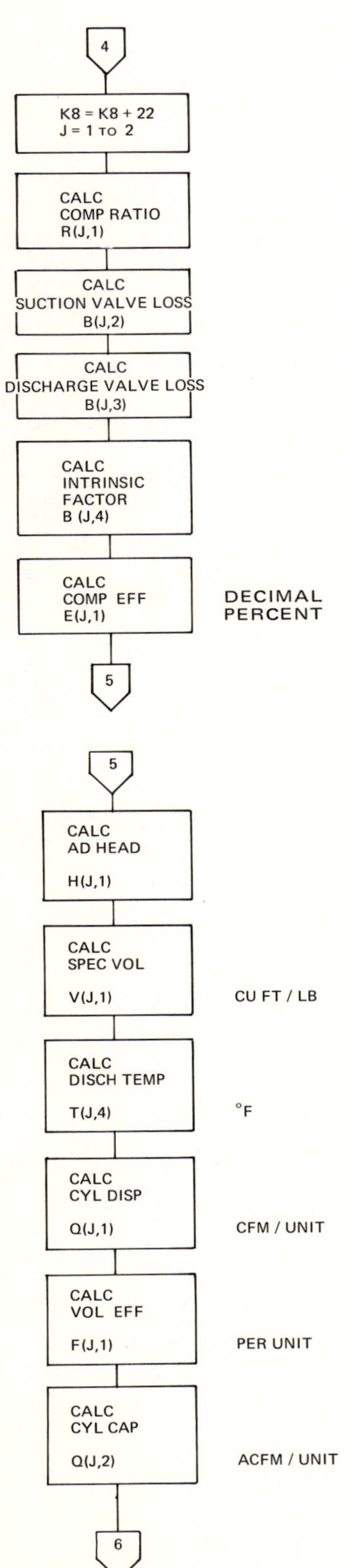

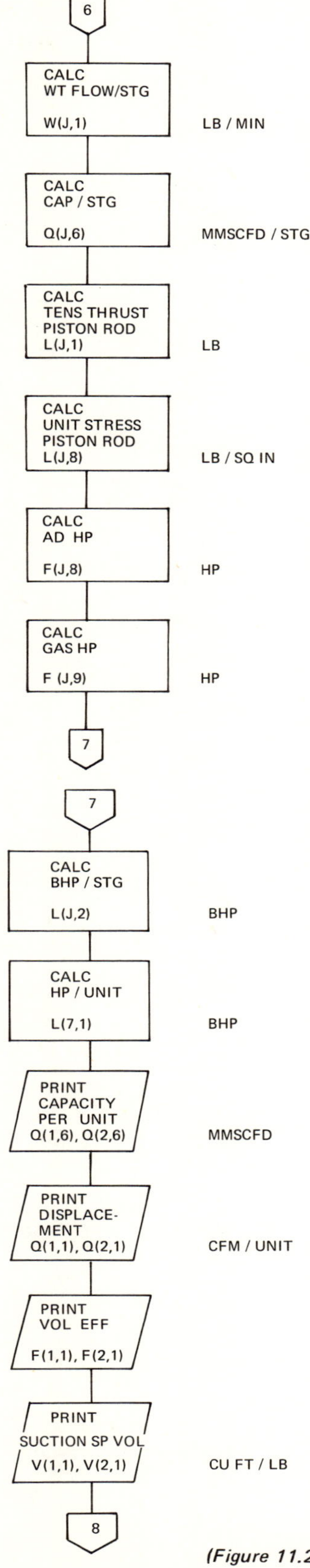

(Figure 11.2 continued on p. 114)

Figure 11.2 (*concluded*)

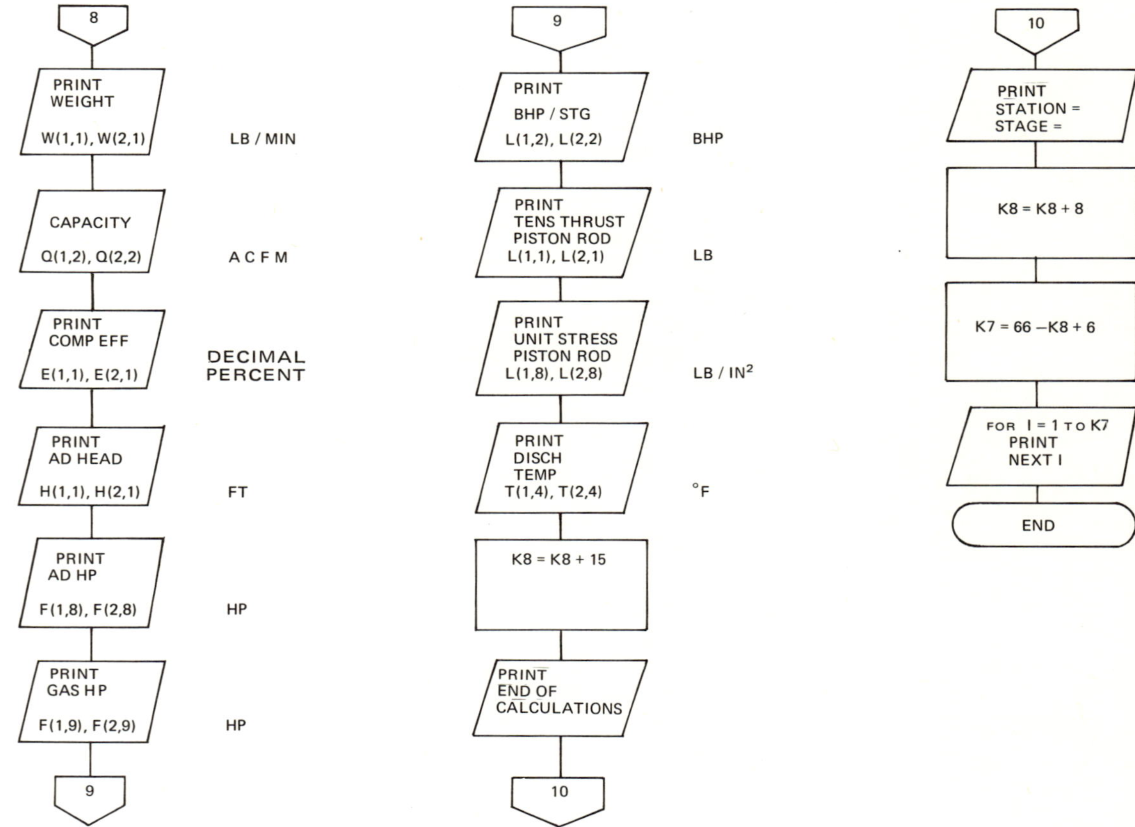

Table 11.5
Program UOHP Listing

```
10 DATA 42,0.87,0.1305,9.75,2.0,14.5,'LFS-1'
20 DATA 15.5,69.7
30 DATA 73,260
40 DATA 455,480
50 DATA 21.0,12.5
60 DATA 0.1632,0.2138
70 DATA 8.6,9.2
80 DATA 1.00,1.00
90 DATA 1.00,0.915
200  REM 01   PROGRAM UOHP
210  REM 02   THIS PROGRAM USES GIVEN GAS PROPERTIES, INCLUDING
220  REM 02   COMPRESSIBILITY FACTOR 'Z', AND GIVEN PHYSICAL
230  REM 02   DATA SUCH AS 'BORE','CLEARANCE', 'P/V-A RATIO, ETC
240  REM 02   AND CALCULATES H P AND OTHER OPERATING PARAMETERS
250  REM 02   FOR ONE TWO-STAGE UNIT OR TWO SINGLE STAGE UNITS.
260  REM 02   CALCULATIONS ARE MADE AT THE DESIGN POINT ONLY.
270 DIM B(2,4), F(2,9), L(7,8), T(2,4),Q(2,6)
300 READ M2,K9,S,U,R5,P1,K$
310 READ P(1,1),P(2,1)
320 READ P(1,2), P(2,2)
330 READ T(1,1), T(2,1)
340 READ B(1,1), B(2,1)
350 READ C(1,1),C(2,1)
360 READ A(1,1), A(2,1)
370 READ Z(1,1), Z(2,1)
380 READ Z(1,2), Z(2,2)
390 K3=1/K9
400 K8=6
410 FOR J=1 TO 2
420 T(J,3)=T(J,1)-460
430 NEXT J
440 PRINT
450 PRINT
460 PRINT USING 1090 ,'SYSTEM CONSTANTS',K$
470 PRINT
480 PRINT USING 1100 ,M2,K9,S
490 PRINT USING 1110 , U,R5,P1
500 PRINT
510 PRINT USING 1150 ,'FUNCTION NUMBER','ONE','TWO'
520 PRINT
530 PRINT USING 1090 ,'GIVEN DATA'
540 PRINT
550 PRINT USING 1120 ,'SUCTION PR (PSIA)',P(1,1),P(2,1)
560 PRINT USING 1120 ,'DISCH PR (PSIA)',P(1,2),P(2,2)
570 PRINT USING 1140 ,'SUCT TEMP (°F)', T(1,3),T(2,3)
580 PRINT USING 1130 , 'BORE (IN.)',B(1,1),B(2,1)
590 PRINT USING 1130 ,'CLEARANCE',C(1,1),C(2,1)
600 PRINT USING 1160 ,'P/V-A RATIO',A(1,1),A(2,1)
610 PRINT USING 1130 ,'Z-SUCTION',Z(1,1),Z(2,1)
```

(*Table 11.5 continued on p. 116*)

Table 11.5 (*continued*)

```
620 PRINT USING 1130 ,'Z-DISCHARGE',Z(1,2),Z(2,2)
630 PRINT
640 PRINT USING 1090 ,'CALCULATED VALUES'
650 PRINT
660 K8=K8+22
670 FOR J=1 TO 2
680 R(J,1)=P(J,2)/P(J,1)
690 B(J,2)=(A(J,1)*U)↑2*M2/(T(J,1)*10000)
700 B(J,3)=B(J,2)/R(J,1)↑S
710 B(J,4)=(1+B(J,3))/(1-B(J,2))
720 E(J,1)=(R(J,1)↑S-1)/((B(J,4)*R(J,1))↑S-1)
730 H(J,1)=(R(J,1)↑S-1)*772.5*(Z(J,1)+Z(J,2))*T(J,1)/(M2*S)
740 V(J,1)=10.73*T(J,1)*Z(J,1)/(P(J,1)*M2)
750 T(J,4)=T(J,1)*R(J,1)↑S-460
760 Q(J,1)=0.3276*(B(J,1)↑2-0.50*R5↑2)*U
770 F(J,1)=1.0+C(J,1)-(1.10*C(J,1)*R(J,1)↑(1/K3))*(Z(J,1)/Z(J,2))
780 Q(J,2)=Q(J,1)*F(J,1)
790 W(J,1)=Q(J,2)/V(J,1)
800 Q(J,6)=0.546*W(J,1)/M2
810 L(J,1)=B(J,1)↑2*0.785*(P(J,2)-P(J,1))-(R5↑2*0.785*P(J,2)-15)
820 L(J,8)=L(J,1)/(R5↑2*0.785)
830 F(J,8)=W(J,1)*H(J,1)/33000
840 F(J,9)=F(J,8)/E(J,1)
850 L(J,2)=F(J,9)+SQR(F(J,9))
860 L(7,1)=L(7,1)+L(J,2)
870 NEXT J
880 PRINT USING 1130 ,'CAPACITY/UNIT MMSCFD',Q(1,6),Q(2,6)
890 PRINT USING 1140 ,'DISPLACEMENT CFM/UNIT',Q(1,1),Q(2,1)
900 PRINT USING 1130 ,'VOL EFF',F(1,1),F(2,1)
910 PRINT USING 1130 ,'SUCT VOL (CU FT/LB)'V(1,1),V(2,1)
920 PRINT USING 1130 ,'WEIGHT (LB/MIN)',W(1,1),W(2,1)
930 PRINT USING 1140 ,'CAPACITY (ACFM)',Q(1,2),Q(2,2)
940 PRINT USING 1130 ,'COMP EFF',E(1,1),E(2,1)
950 PRINT USING 1140 ,'AD HEAD (FT)',H(1,1),H(2,1)
960 PRINT USING 1120 ,'ADIABATIC H P',F(1,8),F(2,8)
970 PRINT USING 1120 ,'GAS H P',F(1,9),F(2,9)
980 PRINT USING 1120 ,'BHP/STAGE',L(1,2),L(2,2)
990 PRINT USING 1140 ,'ROD TENS THRUST (LB)',L(1,1),L(2,1)
1000 PRINT USING 1120 ,'ROD STRESS (LB/SQ IN.)',L(1,8),L(2,8)
1010 PRINT USING 1120 ,'DISCH TEMP (°F)', T(1,4),T(2,4)
1020 PRINT
1030 PRINT
1040 PRINT
1050 PRINT
1060 PRINT USING 1180 ,'-- END OF CALCULATIONS --'
1070 K8=K8+19
1080 GO TO 1190
1090:   ########################    ##########
1100:  MOL WT=##.## , (1/K3)=#.### , S=##.###
1110:  U=##.## (FPS), ROD =##.# (IN.), AMBIENT (PSIA) =##.##
1120:   ###########################   #####.#     #####.#
1130:   ###########################     ###.###     ###.###
```

Table 11.5 (*concluded*)

```
1140:    ###########################   ######.      ######.
1150:    ###########################      ###          ###
1160:    ###########################    ####.##      ####.##
1170:    ###########################       ######.
1180:                 ###########################
1190 PRINT
1200 PRINT
1210 PRINT
1220 PRINT   'STA=STATION,  STG=STAGE'
1230 PRINT
1240 K8=K8+5
1250 K7=66-K8+6
1260 FOR K1=1 TO K7
1270 PRINT
1280 NEXT K1
1290 END
```

Table 11.6
Program UOHP Sample Calculation Print-Out

MOL WT=42.00 , (1/K3)= .870 , S= 0.131
U= 9.75 (FPS), ROD = 2.0 (IN.), AMBIENT (PSIA) =14.50

FUNCTION NUMBER	ONE	TWO
GIVEN DATA		
SUCTION PR (PSIA)	15.5	69.7
DISCH PR (PSIA)	73.0	260.0
SUCT TEMP (°F)	-5.	20.
BORE (IN.)	21.000	12.500
CLEARANCE	0.163	0.214
P/V-A RATIO	8.60	9.20
Z-SUCTION	1.000	1.000
Z-DISCHARGE	1.000	0.915
CALCULATED VALUES		
CAPACITY/UNIT MMSCFD	1.147	1.477
DISPLACEMENT CFM/UNIT	1402.	493.
VOL EFF	0.472	0.406
SUCT VOL (CU FT/LB)	7.499	1.759
WEIGHT (LB/MIN)	88.249	113.648
CAPACITY (ACFM)	662.	200.
COMP EFF	0.921	0.902
AD HEAD (FT)	28745.	24284.
ADIABATIC H P	76.9	83.6
GAS H P	83.4	92.7
BHP/STAGE	92.6	102.4
ROD TENS THRUST (LB)	19691.	22540.
ROD STRESS (LB/SQ IN.)	6271.1	7178.4
DISCH TEMP (°F)	97.0	110.0

Table 11.7a
Program P6S Input Data Sheet (Blank)

PROGRAM LINE NO.	NAME	SYSTEM CONSTANTS: GAS MOL WEIGHT	INV. RATIO SPEC. HEATS 1/K3	AMBIENT PRESS (PSIA)	DESIGN POINT FACTOR INDICATOR	DATA IDENTIFICATION CODES BY FUNCTION COLUMNS (NOTE 4): 1-2	3-4	5-6
		M2	K9	P1	I5 *	I$	J$	K$
10	DATA							

PROGRAM LINE NO.	NAME	FUNCTION NUMBER (OR STAGE NUMBER): ONE	TWO	THREE	FOUR	FIVE	SIX
		SUCTION PRESSURE–(PSIA)–P (J,1)					
20	DATA						
		DISCHARGE PRESSURE–(PSIA)–P (J,2)					
30	DATA						
		SUCTION TEMPERATURE–(°R)–T (J,1)					
40	DATA						
		CYLINDER BORE–(INCHES)–B (J,1)					
50	DATA						
		CLEARANCE–(DECIMAL–E.G.–0.200)–C (J,1)					
60	DATA						
		P/V–AREA RATIO A(J,1)					
70	DATA						
		AVG PISTON SPEED–(FPS)–U (J,1)					
80	DATA						
		PISTON ROD DIA–(INCHES)–R (J,5)					
90	DATA						
		COMPRESSIBILITY–'Z'–SUCTION–Z (J,1)					
100	DATA						
		COMPRESSIBILITY–'Z'–DISCHARGE–Z (J,2)					
110	DATA						

NOTES:
1. SEE GAS COMPRESSION–GLOSSARY OF SYMBOLS COMP. PROG
2. FOR I5 = 1 CALC'S AT DESIGN POINT ONLY
   FOR I5 = 2 CALC'S AT 0.8, 0.9, 1.0, 1.1, & 1.2 X D. PT.
3. K9 REQUIRED FOR INPUT K9 = 1/K3
4. CODES UP TO 10 CHARACTERS–ENCLOSE IN ITALICS.

Table 11.7b
Program P6S Input Data Sheet (Sample Data)

PROGRAM LINE NO.	NAME	SYSTEM CONSTANTS: GAS MOL WEIGHT	INV. RATIO SPEC. HEATS 1/K3	AMBIENT PRESS (PSIA)	DESIGN POINT FACTOR RANGE INDICATOR	DATA IDENTIFICATION CODES BY FUNCTION COLUMNS (NOTE 4): 1-2	3-4	5-6
		M2	K9	P1	I5 *	I$	J$	K$
10	DATA	41	0.885	14.7	2	'CPT'	'WON'	'JOY'

PROGRAM LINE NO.	NAME	FUNCTION NUMBER (OR STAGE NUMBER): ONE	TWO	THREE	FOUR	FIVE	SIX
		SUCTION PRESSURE–(PSIA)–P (J,1)					
20	DATA	15.7	57.0	15.7	57.0	15.7	57.0
		DISCHARGE PRESSURE–(PSIA)–P (J,2)					
30	DATA	60.0	218.	60.0	218.	60.0	218.
		SUCTION TEMPERATURE–(°R)–T (J,1)					
40	DATA	590	560	590	560	590	560
		CYLINDER BORE–(INCHES)–B (J,1)					
50	DATA	13	6	13.5	6.0	13.	6.0
		CLEARANCE–(DECIMAL–E.G.–0.200)–C (J,1)					
60	DATA	0.12	0.15	0.095	0.116	0.104	0.115
		P/V–AREA RATIO A(J,1)					
70	DATA	12	10	15	15	11.75	9.55
		AVG. PISTON SPEED–(FPS)–U (J,1)					
80	DATA	9.6	9.6	10.0	10.0	9.16	9.16
		PISTON ROD DIA–(INCHES)–R (J.5)					
90	DATA	2.0	2.0	2.0	2.0	2.0	2.0
		COMPRESSIBILITY–'Z'–SUCTION–Z (J,1)					
100	DATA	1.00	0.92	1.00	0.92	1.00	0.94
		COMPRESSIBILITY–'Z'–DISCHARGE–Z (J,2)					
110	DATA	0.96	0.88	0.96	0.88	0.97	0.88

NOTES:
1. SEE GAS COMPRESSION–GLOSSARY OF SYMBOLS COMP PROG
2. FOR I5 = 1 CALC'S AT DESIGN POINT ONLY
   FOR I5 = 2 CALC'S AT 0.8, 0.9, 1.0, 1.1, & 1.2 X D.PT.
3. K9 REQUIRED FOR INPUT K9 = 1/K3
4. CODES UP TO 10 CHARACTERS–ENCLOSE IN ITALICS

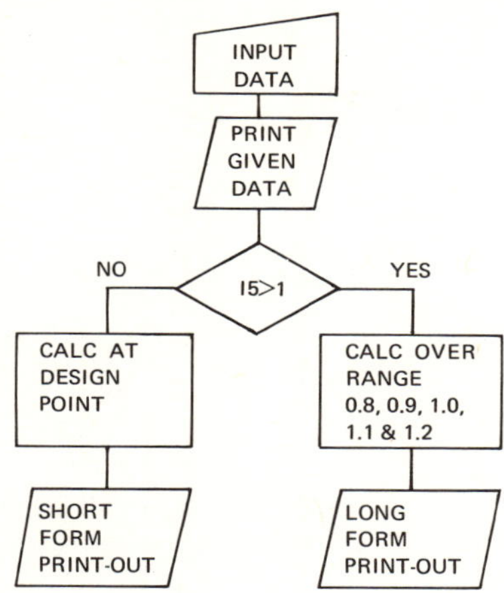

Figure 11.3. Simplified flow diagram for Program P6S.

Table 11.8
Program P6S Sample Calculation Print-Out

MOL WT =41.00 , (1/K3)= .885 , S= 0.115
AMBIENT (PSIA) = 14.70

FUNCTION NUMBER	ONE	TWO	THREE	FOUR	FIVE	SIX
GIVEN DATA	CPT		WON		JOY	
SUCT PR (PSIA)	15.7	57.0	15.7	57.0	15.7	57.0
DISCH PR (PSIA)	60.0	218.0	60.0	218.0	60.0	218.0
SUCT TEMP (°R)	590.	560.	590.	560.	590.	560.
BORE (IN)	13.000	6.000	13.500	6.000	13.000	6.000
CLEARANCE	0.120	0.150	0.095	0.116	0.104	0.115
P/V-A RATIO	12.0	10.0	15.0	15.0	11.7	9.6
PISTON (FPS)	9.60	9.60	10.00	10.00	9.16	9.16
ROD DIA (IN)	2.00	2.00	2.00	2.00	2.00	2.00
Z-SUCTION	1.000	0.920	1.000	0.920	1.000	0.940
Z-DISCHARGE	0.960	0.880	0.960	0.880	0.970	0.880

CALCULATED VALUES

DESIGN POINT FACTOR 1.00

SUCT PR (PSIA)	15.70	57.00	15.70	57.00	15.70	57.00
COMP. RATIO,R(C)	3.822	3.825	3.822	3.825	3.822	3.825
COMPRESSION EFF	0.877	0.907	0.805	0.796	0.891	0.922
VOLUMETRIC EFF	0.668	0.585	0.737	0.679	0.712	0.682
TENS.THRUST (M-LB)	5.877	4.550	6.338	4.550	5.877	4.550
ROD STRESS (PSI)	1872.	1449.	2018.	1449.	1872.	1449.
DISCH. TEMP (°F)	228.	193.	228.	193.	228.	193.
BRAKE H.P./STG	45.	29.	60.	39.	45.	31.
MMSCFD CAPACITY	0.475	0.352	0.589	0.426	0.483	0.383

(Table 11.8 continued on p. 120)

Table 11.8 *(continued)*

SUCT PR (PSIA)	15.7	57.0	15.7	57.0	15.7	57.0
DISCH PR (PSIA)	60.0	218.0	60.0	218.0	60.0	218.0
SUCT TEMP (°R)	590.	560.	590.	560.	590.	560.
BORE (IN)	13.000	6.000	13.500	6.000	13.000	6.000
CLEARANCE	0.120	0.150	0.095	0.116	0.104	0.115
P/V-A RATIO	12.0	10.0	15.0	15.0	11.7	9.6
PISTON (FPS)	9.60	9.60	10.00	10.00	9.16	9.16
ROD DIA (IN)	2.00	2.00	2.00	2.00	2.00	2.00
Z-SUCTION	1.000	0.920	1.000	0.920	1.000	0.940
Z-DISCHARGE	0.960	0.880	0.960	0.880	0.970	0.880

CALCULATED VALUES DESIGN POINT FACTOR .80

SUCT PR (PSIA)	12.56	45.60	12.56	45.60	12.56	45.60
COMP. RATIO,R(C)	4.777	4.781	4.777	4.781	4.777	4.781
COMPRESSION EFF	0.892	0.919	0.828	0.820	0.905	0.932
VOLUMETRIC EFF	0.569	0.461	0.659	0.583	0.627	0.587
TENS.THRUST (M-LB)	6.294	4.872	6.787	4.872	6.294	4.872
ROD STRESS (PSI)	2004.	1552.	2161.	1552.	2004.	1552.
DISCH. TEMP (°F)	246.	210.	246.	210.	246.	210.
BRAKE H.P./STG	36.	22.	50.	32.	38.	26.
MMSCFD CAPACITY	0.324	0.222	0.422	0.293	0.340	0.264

DESIGN POINT FACTOR .90

SUCT PR (PSIA)	14.13	51.30	14.13	51.30	14.13	51.30
COMP. RATIO,R(C)	4.246	4.250	4.246	4.250	4.246	4.250
COMPRESSION EFF	0.885	0.913	0.816	0.808	0.898	0.927
VOLUMETRIC EFF	0.624	0.529	0.702	0.636	0.674	0.639
TENS.THRUST (M-LB)	6.085	4.711	6.562	4.711	6.085	4.711
ROD STRESS (PSI)	1938.	1500.	2090.	1500.	1938.	1500.
DISCH. TEMP (°F)	237.	201.	237.	201.	237.	201.
BRAKE H.P./STG	41.	26.	55.	36.	42.	29.
MMSCFD CAPACITY	0.399	0.287	0.505	0.359	0.412	0.323

DESIGN POINT FACTOR 1.00

SUCT PR (PSIA)	15.70	57.00	15.70	57.00	15.70	57.00
COMP. RATIO,R(C)	3.822	3.825	3.822	3.825	3.822	3.825
COMPRESSION EFF	0.877	0.907	0.805	0.796	0.891	0.922
VOLUMETRIC EFF	0.668	0.585	0.737	0.679	0.712	0.682
TENS.THRUST (M-LB)	5.877	4.550	6.338	4.550	5.877	4.550
ROD STRESS (PSI)	1872.	1449.	2018.	1449.	1872.	1449.
DISCH. TEMP (°F)	228.	193.	228.	193.	228.	193.
BRAKE H.P./STG	45.	29.	60.	39.	45.	31.
MMSCFD CAPACITY	0.475	0.352	0.589	0.426	0.483	0.383

DESIGN POINT FACTOR 1.10

SUCT PR (PSIA)	17.27	62.70	17.27	62.70	17.27	62.70
COMP. RATIO,R(C)	3.474	3.477	3.474	3.477	3.474	3.477
COMPRESSION EFF	0.869	0.901	0.793	0.784	0.884	0.917

Table 11.8 *(concluded)*

VOLUMETRIC EFF	0.705	0.630	0.766	0.714	0.744	0.717
TENS.THRUST (M-LB)	5.669	4.389	6.113	4.389	5.669	4.389
ROD STRESS (PSI)	1805.	1398.	1947.	1398.	1805.	1398.
DISCH. TEMP (°F)	221.	186.	221.	186.	221.	186.
BRAKE H.P./STG	49.	32.	64.	42.	48.	34.
MMSCFD CAPACITY	0.551	0.417	0.674	0.493	0.555	0.443
DESIGN POINT FACTOR			1.20			
SUCT PR (PSIA)	18.84	68.40	18.84	68.40	18.84	68.40
COMP. RATIO,R(C)	3.185	3.187	3.185	3.187	3.185	3.187
COMPRESSION EFF	0.860	0.894	0.781	0.772	0.876	0.911
VOLUMETRIC EFF	0.735	0.669	0.790	0.744	0.771	0.746
TENS.THRUST (M-LB)	5.460	4.228	5.889	4.228	5.460	4.228
ROD STRESS (PSI)	1739.	1346.	1875.	1346.	1739.	1346.
DISCH. TEMP (°F)	214.	180.	214.	180.	214.	180.
BRAKE H.P./STG	52.	34.	67.	45.	51.	35.
MMSCFD CAPACITY	0.628	0.483	0.758	0.560	0.627	0.503

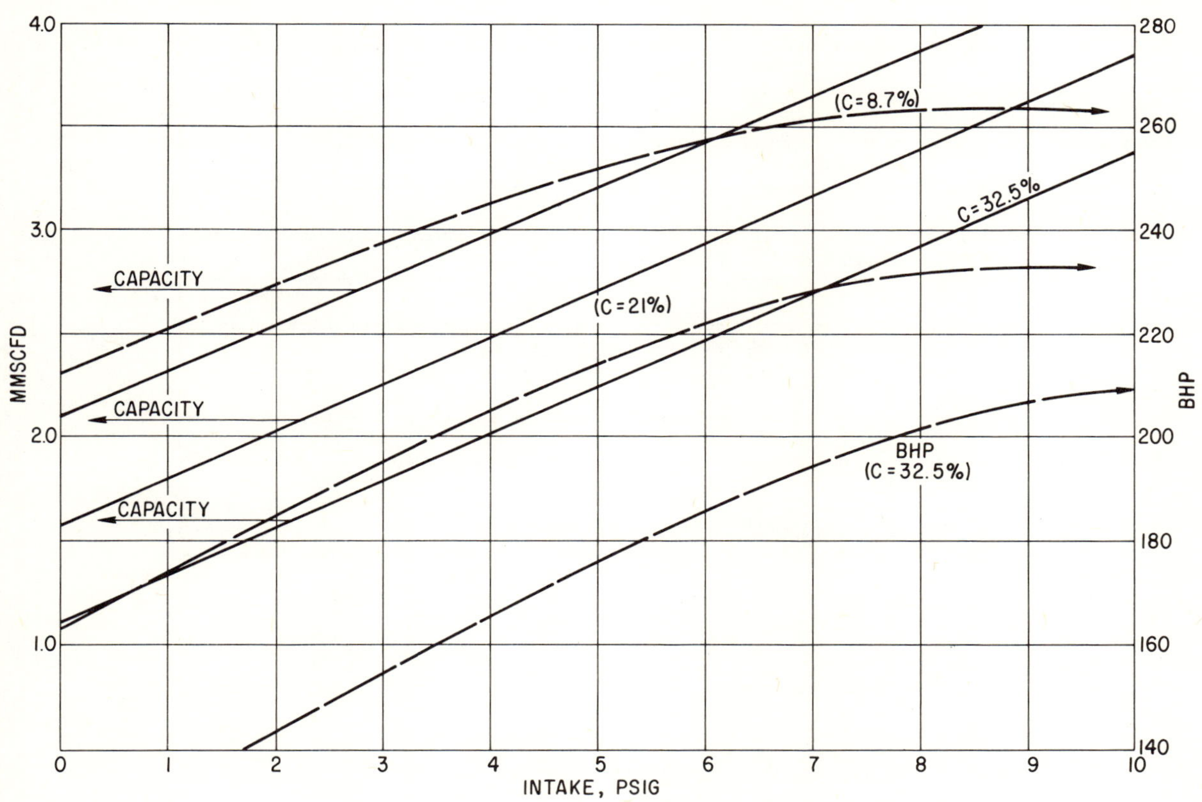

Figure 11.4. Performance chart for 22.5 x 17 — 300 rpm compressor cylinder with discharge = 62.1 psia, suction = 95F and A(J.1) = 9.3. This chart was plotted from the array of data developed in Table 11.8.

Table 11.9a
Program CPC Input Data Sheet (Blank)

PROGRAM LINE		SYSTEM CONSTANTS							DATA IDENT CODE (NOTE 3)
		BHP/ UNIT	NO OF UNITS/ STA	NO CYL /STG/ UNIT	RATED HP PER CYL	UNIT SPEED RPM	PISTON		
NO.	NAME						ROD DIA INCHES	STROKE INCH	
		U1	U3	U4	U5	R4	R5	R6	K$
10	DATA								

NO.	NAME	GAS MOL WT LB/MOL	RATIO OF SPEC HEATS	NO. OF STAGES	CALCULATION INDICATOR (NOTE 2)
		M2	K3	S2	I5
20	DATA				

NO.	NAME	FUNCTION NUMBER (OR STAGE NO.)				
		ONE	TWO	THREE	FOUR	
		SUCT PRESS–PSIA–P (J,1)				
30	DATA					
		DISCH PRESS–PSIA–P (J,2)				
40	DATA					
		SUCT TEMP–°R–T (J,1)				
50	DATA					
		CYLINDER BORE–INCHES–B (J,1)				
60	DATA					
		CLEARANCE–DECIMAL–E.G. 0.20) C (J,1)				
70	DATA					
		P/V–AREA RATIO AREA (J,1)				
80	DATA					
		SUCT COMPRESSIBILITY–Z (J,1)				
90	DATA					
		DISCH COMPRESSIBILITY–Z (J,2)				
100	DATA					

NOTES:
1. SEE GLOSSARY OF SYMBOLS FOR GAS COMPRESSION COMPUTER PROGRAMS.
2. SHORT FORM CALC I5 =1; LONG FORM I5 = 2
3. DATA IDENTIFICATION – UP TO 10 CHARACTERS - ENCLOSED IN ITALICS. 'TEST–1'

Table 11.9b
Program CPC Input Data Sheet (Sample Data)

PROGRAM LINE		SYSTEM CONSTANTS							DATA IDENT CODE (NOTE 3)
							PISTON		
NO.	NAME	BHP/ UNIT	NO. OF UNITS/ STA	NO. CYL /STG/ UNIT	RATED HP PER CYL	UNIT SPEED RPM	ROD DIA INCHES	STROKE INCH	
		U1	U3	U4	U5	R4	R5	R6	K$
10	DATA	600	1	3	100	585	2.0	6	'LFS–2'

NO.	NAME	GAS MOLWT LB/MOL	RATIO OF SPEC HEATS	NO. OF STAGES	CALCULATION INDICATOR (NOTE 2)
		M2	K3	S2	I5
20	DATA	42	1.15	4	2

NO.	NAME	FUNCTION NUMBER (OR STAGE NO.)				
		ONE	TWO	THREE	FOUR	
		SUCT PRESS–PSIA–P (J,1)				
30	DATA	15.6	68.0	14.6	67.0	
		DISCH PRESS–PSIA–P (J,2)				
40	DATA	72.0	215	71.0	215	
		SUCT TEMP–°R–T (J,1)				
50	DATA	460	515	460	515	
		CYLINDER BORE–INCHES–B (J,1)				
60	DATA	21	12.5	21	12.5	
		CLEARANCE (DECIMAL–E.G. 0.20) C (J,1)				
70	DATA	0.1632	0.2138	0.2002	0.2595	
		P/V–AREA RATIO A(J,1)				
80	DATA	7.7	5.4	7.7	5.4	
		SUCT COMPRESSIBILITY–Z (J,1)				
90	DATA	1.00	1.00	1.00	1.00	
		DISCH COMPRESSIBILITY–Z (J,2)				
100	DATA	1.00	1.00	1.00	1.00	

NOTES:
1. SEE GLOSSARY OF SYMBOLS FOR GAS COMPRESSION COMPUTER PROGRAMS.
2. SHORT FORM CALC I5 = 1; LONG FORM I5 = 2
3. DATA IDENTIFICATION–UP TO 10 CHARACTERS–ENCLOSED IN ITALICS: 'TEST–1'

Figure 11.5. Simplified flow diagram for Program CPC.

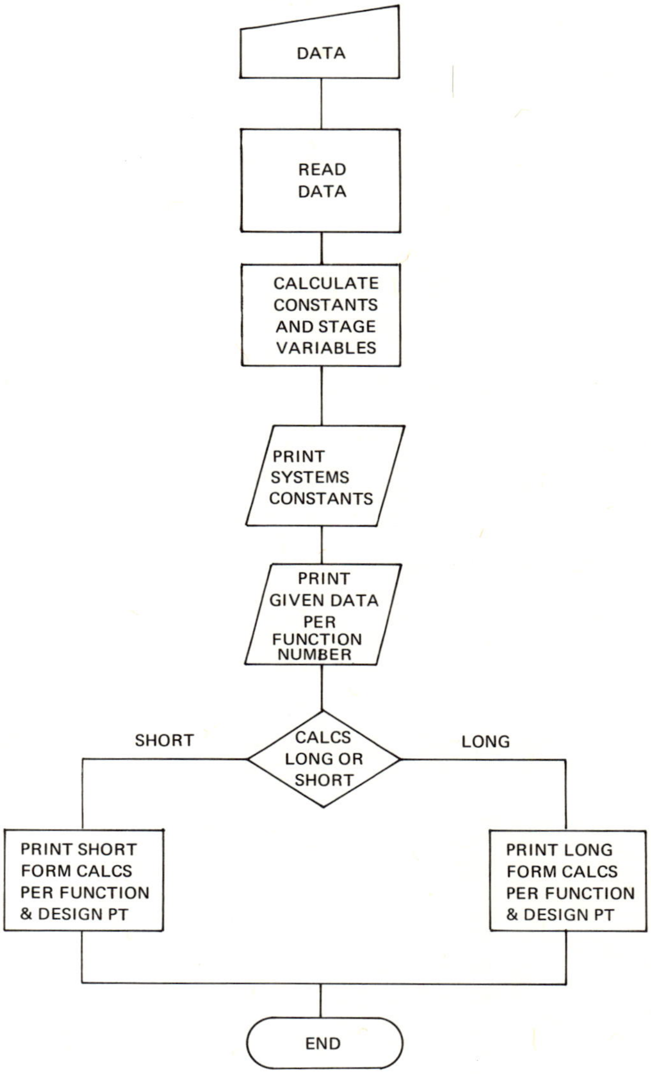

DATA

READ DATA

CALCULATE CONSTANTS AND STAGE VARIABLES

PRINT SYSTEMS CONSTANTS

PRINT GIVEN DATA PER FUNCTION NUMBER

CALCS LONG OR SHORT

SHORT

LONG

PRINT SHORT FORM CALCS PER FUNCTION & DESIGN PT

PRINT LONG FORM CALCS PER FUNCTION & DESIGN PT

END

Table 11.10
Program CPC Sample Calculation Print-Out

```
BHP/UNIT (U1)= 600., NO. UNITS (U3)= 1. ,NO.CYL/STG/UNIT (U4)=3
RATED HP/CYL (U5)=100. ,UNIT RPM (R4)= 585. ,ROD DIA (R5),INCHES = 2.0
PISTON STROKE (R6),INCHES= 6., NO. OF STAGES (S2)= 4.
MOL WT OF GASES HANDLED (M2)=42.0, RATIO OF SPEC HEATS (K3)=1.150

SYSTEM CONSTANTS - CALCULATED

COMPRESSION EXPONENT S= .130 ,PISTON SPEED (U)=  9.7 (FPS)
MECHANICAL EFFICIENCY (K5)= .950
```

STAGE	1	2	3	4
GIVEN DATA				
SUCTION PR (PSIA)	15.6	68.0	14.6	67.0
DISCH PR (PSIA)	72.0	215.0	71.0	215.0
SUCTION TEMP (°R)	460.	515.	460.	515.
CYL BORE (INCH)	21.00	12.50	21.00	12.50
CLEARANCE	0.1632	0.2138	0.2002	0.2595
P/V-A RATIO	7.70	5.40	7.70	5.40
Z-SUCTION	1.000	1.000	1.000	1.000
Z-DISCHARGE	1.000	1.000	1.000	1.000
CALCULATED RESULTS				
W/CYL W(J,3),LB/MIN-	90.02	145.87	57.19	118.39
W/STG W(J,1),LB/MIN-	270.05	437.60	171.57	355.17
ACFM/CYL,Q(J,3) -	678.11	282.23	460.35	232.48
ROD TENS L(J,1),M-LB-	19.35	17.40	19.35	17.53
ROD STRESS, LB/SQ.IN-	6161.1	5542.2	6162.1	5581.2
DISCH TEMP T(J,4)-°F-	101.56	138.43	105.40	139.59
BHP/STG L(J,4)	263.74	341.65	173.54	281.00
BHP/CYL L(J,2) -	87.91	113.88	57.85	93.67
VOL EFF F(J,1) DEC-	0.4845	0.5739	0.3289	0.4727
COMP EFF E(J,1) DEC-	0.9361	0.9620	0.9380	0.9624
INTRINSIC FACT B(J,4)	1.0987	1.0430	1.0984	1.0430
SUCT VOL V(J,1) CF/LB-	7.5333	1.9349	8.0492	1.9637
DISPL Q(J,1) CFM-	1399.64	491.79	1399.64	491.79
ADIAB HEAD H(J,1) FT-	28659.90	23545.32	29743.91	23871.87
SUCT VALVE LOSS (DEC)-	0.0515	0.0226	0.0515	0.0226
DISCH VALVE LOSS (DEC)-	0.0422	0.0195	0.0419	0.0194

Table 11.11a
Program RKCPC Input Data Sheet (Blank)

PROGRAM LINE		GAS	GAS PROPERTIES				GAS	GAS PROPERTIES			
NO.	NAME	NO.	MOL WT LB/MOL	CRIT TEMP °R	CRIT PR PSIA	MOL SPEC HEAT CPM	NO.	MOL WT LB/MOL	CRIT TEMP °R	CRIT PR PSIA	MOL SPEC HEAT CPM
10	DATA	1					2				
20	DATA	3					4				
30	DATA	5					6				
40	DATA	7					8				
50	DATA	9					10				

PROGRAM LINE		GAS COMPOSITION – (PER UNIT – %) – NOTE 2									
NO.	NAME	NO.	%	NO.	%	NO.	%	NO.	%	NO.	%
60	DATA	1		2		3		4		5	
70	DATA	6		7		8		9		10	

GAS NO.	GAS NAME
1	CH_4
2	C_2H_6
3	C_3H_8
4	C_4H_{10}
5	C_5H_{12}
6	AIR
7	H_2
8	CO
9	CO_2
10	H_2S
NOTE 1	

PROGRAM LINE		SYSTEM CONDITIONS			DATA IDENTITY
		INITIAL SUCT PR – PSIA	FINAL SYSTEM PR – PSIA	INITIAL SUCTION TEMP °F	CODE – (10 CHAR IN ITALICS)
		P (1,1)	P	T1	K$
80	DATA				

PROGRAM LINE		SYSTEM CONSTANTS						CALC INDICATOR
		HP/ UNIT	NO. OF UNITS	CYL/ UNIT	SPEED RPM	ROD DIA INCHES	STROKE INCHES	STD=1 LONG=2 5 STAGES ONLY
		U1	U3	U4	R4	R5	R6	I5
90	DATA							

PROGRAM LINE		FUNCTION NUMBER (OR STAGE NO.)					
		ONE	TWO	THREE	FOUR	FIVE	SIX
		BORE – INCHES – B (J,1)					
100	DATA						
		NUMBER OF CYLINDERS PER/UNIT/STAGE U(J,1)					
110	DATA						
		CLEARANCE (DECIMAL – E.G. 0.20) C (J,6)					
120	DATA						
		P/V – AREA RATIO A (J,1)					
130	DATA						

NOTES: 1. GAS PROPERTIES STORED. LEAVE 10 — 50 BLANK
2. NO. GASES HAVING COMPOSITION = 1.00 WILL BE USED

Table 11.11b
Program RKCPC Input Data Sheet (Sample Data)

PROGRAM LINE NO.	PROGRAM LINE NAME	GAS NO.	GAS PROPERTIES MOL WT LB/MOL	CRIT TEMP °R	CRIT PR PSIA	MOL SPEC HEAT CPM	GAS NO.	GAS PROPERTIES MOL WT LB/MOL	CRIT TEMP °R	CRIT PR PSIA	MOL SPEC HEAT CPM
10	DATA	1	16.04	673	344	8.88	2	30.07	710	550	13.6
20	DATA	3	44.09	617	666	19.2	4	58.12	551	766	25.5
30	DATA	5	72.15	489	846	31.2	6	28.97	547	239	6.98
40	DATA	7	2.02	188	60	6.93	8	28.01	507	240	6.97
50	DATA	9	44.01	1071	548	9.24	10	34.08	1306	673	8.30

GAS NO.	GAS NAME
1	CH_4
2	C_2H_6
3	C_3H_8
4	C_4H_{10}
5	C_5H_{12}
6	AIR
7	H_2
8	CO
9	CO_2
10	H_2S
NOTE 1	

		GAS COMPOSITION — (PER UNIT — %) — NOTE 2									
		NO.	%	NO.	%	NO.	%	NO.	%	NO.	%
60	DATA	1	0.750	2	0.060	3	0.050	4	0.020	5	0.020
70	DATA	6	0.020	7	0.020	8	0.020	9	0.020	10	0.020

		SYSTEM CONDITIONS			
		INITIAL SUCT PR — PSIA	FINAL SYST PR — PSIA	INITIAL SUCT TEMP °F	DATA IDENT CODE — (10 CHAR IN ITALICS)
		P (1,1)	P	T1	K$
80	DATA	14.50	6000	95	'LFS — 10'

		SYSTEM CONSTANTS						CALC INDICATOR
		HP PER UNIT	NO. OF UNITS	CYL PER UNIT	SPEED RPM	ROD DIA INCHES	STROKE INCHES	STD=1 LONG=2 5 STAGES ONLY
		U1	U3	U4	R4	R5	R6	I5
90	DATA	2500	12	5	300	4	18	2

		FUNCTION NUMBER (OR STAGE NO.)					
		ONE	TWO	THREE	FOUR	FIVE	SIX
		BORE — INCHES — B (J,1)					
100	DATA	30.5	20.5	16.5	9.5	5.8	4
		NO. CYL / UNIT / STAGE U (J,1)					
110	DATA	3	2	1	1	1	1
		CLEARANCE (DECIMAL — E. G. 0.20) C (J,6)					
120	DATA	0.08	0.10	0.12	0.16	0.20	0.25
		P/V — AREA RATIO A (J,1)					
130	DATA	12	11	10	9	8	8

NOTES 1. GAS PROPERTIES STORED. LEAVE 10 — 50 BLANK

2. NO. GASES HAVING COMPOSITION = 1.00 WILL BE USED

Figure 11.6. Simplified flow diagram for Program RKCPC.

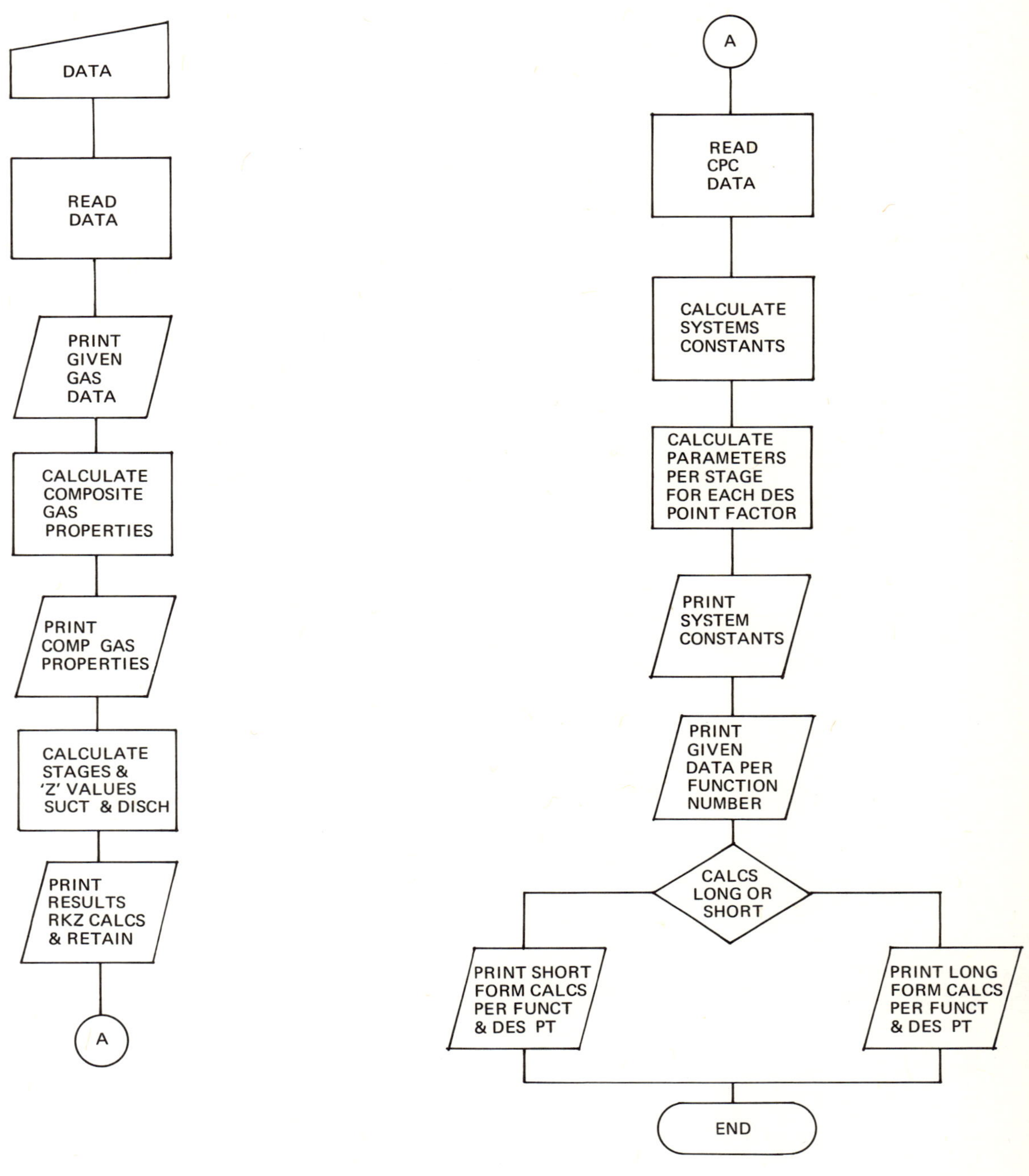

Table 11.12
Program RKCPC Sample Calculation Print-Out

```
     SPECIFIC COMPOSITION            LFS-10
       GAS                 MOL PERCENT

      CH4                    75.00
      C2H6                    6.00
      C3H8                    5.00
      C4H10                   2.00
      C5H12                   2.00
      AIR                     2.00
      H2                      2.00
      CO                      2.00
      CO2                     2.00
      H2S                     2.00

   TOTAL     10                    100.00


     COMPOSITE GAS PROPERTIES

      MOL WT        CPM        K         T(CRIT)    P(CRIT)
                                           °R        PSIA

       21.39       10.34     1.238       392.       671.


     OPERATING CONDITIONS
  STAGE     SUCTION              DISCHARGE                  COMPRESSIBILITY
   NO.     P         T          P            T                    'Z'
          PSIA       °F        PSIA          °F

   1       14.50    95.                                          0.9975
   1                           50.81        246.                 0.9963
   2       48.39    95.                                          0.9915
   2                          169.54        246.                 0.9877
   3      161.47    95.                                          0.9718
   3                          565.75        246.                 0.9617
   4      538.81    95.                                          0.9073
   4                         1887.91        246.                 0.9115
   5     1798.01    95.                                          0.7573
   5                         6300.00        246.                 1.1034
 FINAL                       6000.00         95.

SYSTEM CONSTANTS

BHP/UNIT (U1)=2500 , NO. OF UNITS (U3)=12 , CYL/UNIT (U4)= 5
SPEED - RPM (R4)= 300 , ROD DIA - INCH- (R5)=4 , STROKE - INCH (R6)=18
```

Table 11.13a
Program PBS Input Data Sheet (Blank)

PROGRAM LINE		SYSTEM CONSTANTS						
NO.	NAME	GAS MOL WEIGHT	RATIO OF SPECIFIC HEATS	DATA IDENT CODE (2)				
		M2	K3	K$				
10	DATA							
		BHP PER UNIT	NO. UNITS PER STATION	CYL PER UNIT	RATED HP PER CYL	RPM	ROD DIAM INCHES	STROKE INCHES
		U1	U3	U4		R4	R5	R6
20	DATA							

NO.	NAME	FUNCTION NUMBER (OR STAGE NO.)	
		ONE	TWO
		MMSCFD CAPACITY – Q (J,6)	
30	DATA		
		SUCTION PRESSURE – (PSIA) – P (J,1)	
40	DATA		
		DISCH PRESSURE – (PSIA) – P (J,2)	
50	DATA		
		SUCTION TEMPERATURE – (°R) – T (J,1)	
60	DATA		
		CLEARANCE – (DECIMAL) – C (J,1)	
70	DATA		
		P/V–AREA RATIO–A (J,1)	
80	DATA		
		COMPRESSIBILITY 'Z' – SUCT Z (J,1)	
90	DATA		
		COMPRESSIBILITY 'Z' – DISCH – Z (J,2)	
100	DATA		

NOTES:
1. SEE GAS COMPRESSION – GLOSSARY OF SYMBOLS USED FOR COMPUTER PROGRAMS
2. DATA IDENTIFICATION CODE – 10 CHARACTERS OR LESS – ENTER IN ITALICS – E.G.–'TEST–1'.

Table 11.13b
Program PBS Input Data Sheet (Sample Data)

PROGRAM LINE NO.	PROGRAM LINE NAME	SYSTEM CONSTANTS: GAS MOL WEIGHT	SYSTEM CONSTANTS: RATIO OF SPECIFIC HEATS	SYSTEM CONSTANTS: DATA IDENTITY CODE (2)
		M2	K3	K$
10	DATA	44	1.123	'LF S–5'

NO.	NAME	BHP PER UNIT	NO. UNITS PER STATION	CYL PER UNIT	RATED HP PER CYL	RPM	ROD DIAM INCHES	STROKE INCHES
		U1	U3	U4		R4	R5	R6
20	DATA	600	1.0	4	150	720	2.0	6

NO.	NAME	FUNCTION NUMBER (OR STAGE NO.): ONE	FUNCTION NUMBER (OR STAGE NO.): TWO
		MMSCFD CAPACITY – Q (J,6)	
30	DATA	3.45	3.45
		SUCTION PRESSURE–(PSIA)– P (J,1)	
40	DATA	14.3	55
		DISCHARGE PRESSURE–(PSIA)– P (J,1)	
50	DATA	58	225
		SUCTION TEMPERATURE–(°R) – T (J,1)	
60	DATA	538	550
		CLEARANCE–(DECIMAL) – C (J,1)	
70	DATA	0.12	0.15
		P/V–AREA RATIO–A (J,1)	
80	DATA	11	11
		COMPRESSIBILITY 'Z' – SUCT Z (J,1)	
90	DATA	1.00	1.00
		COMPRESSIBILITY 'Z' – DISCH–Z (J,2)	
100	DATA	1.00	0.975

NOTES: 1. SEE GAS COMPRESSION–GLOSSARY OF SYMBOLS USED FOR COMPUTER PROGRAMS.

2. DATA IDENTIFICATION CODE–10 CHARACTERS OR LESS–ENTER IN ITALICS – E.G. 'TEST–1'

Figure 11.7. Simplified flow diagram for Program PBS.

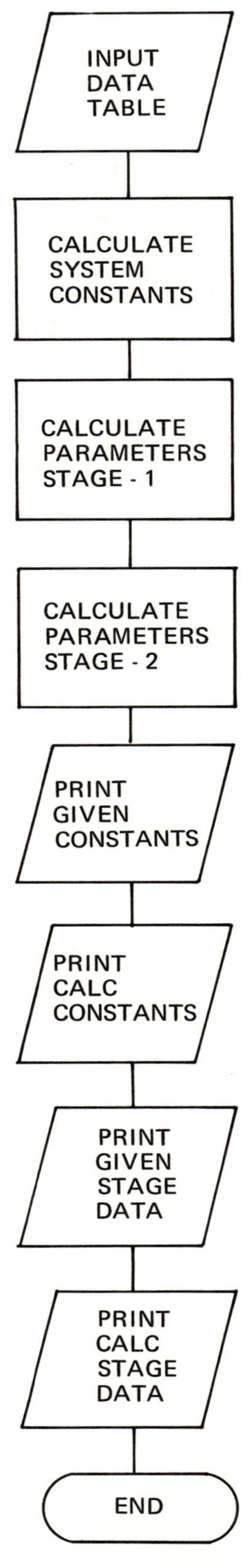

Table 11.14
Program PBS Sample Calculation Print-Out

```
    SYSTEM CONSTANTS - GIVEN              LFS-5

M2,GAS MOL WT =44.0,    K3=1.123,    STAGES = 2
U1,BHP/UNIT= 600,    U3,UNITS/STA= 1,    U4,CYL/UNIT= 4
U5,RATED HP/CYL= 150,      R4,RPM = 720
R5,ROD DIAM (IN.) = 2.0,    R6,STROKE (IN.) =  6.0

    SYSTEM CONSTANTS - CALCULATED

 S,COMP EXPONENT= .110,   U,PISTON SPEED (FPS)=  12.0
 K5, MECH. EFF =  .918


  STAGE NUMBER                              1.          2.

    GIVEN DATA

  MMSCFD CAP/STG                           3.450       3.450
  SUCT PR (PSIA)                          14.30       55.00
  DISCH PR (PSIA)                         58.00      225.00
  SUCT TEMP (°F)                          78.0        90.0
  CLEARANCE (DEC)                          0.120       0.150
  P/V-AREA RATIO                          11.00       11.00
  COMP Z-SUCTION                           1.000       1.000
  COMP Z-DISCHARGE                         1.000       0.975

    CALCULATED VALUES

  CYL DIA (IN)                            31.39       17.67
  ROD TENS THRUST (LB)                  33676.2     41018.9
  ROD UNIT STRESS (LB/SQ.IN.)           10724.9     13063.3
  ACFM/CYL                               2548.68      677.44
  DISCH TEMP (°F)                        167.2       181.8
  VOL EFFICIENCY                           0.661       0.557
  BHP/CYL                                317.2       320.7
  COMP EFF                                 0.826       0.830


                  -- END OF CALCULATIONS --


STA= STATION,   STG= STAGE.
```

Table 11.15a
Program RKBCS Input Data Sheet (Blank)

PROGRAM LINE	GAS	GAS PROPERTIES				GAS	GAS PROPERTIES			
NO. NAME	NO.	MOL WT	CRIT TEMP	CRIT PR	MOL SPEC HEAT	NO.	MOL WT	CRIT TEMP	CRIT PR	MOL SPEC HEAT
		LB/MOL	°R	PSIA	CPM		LB/MOL	°R	PSIA	CPM
10 DATA	1					2				
20 DATA	3					4				
30 DATA	5					6				
40 DATA	7					8				
50 DATA	9					10				

GAS NO.	GAS NAME
1	CH_4
2	C_2H_6
3	C_3H_8
4	C_4H_{10}
5	C_5H_{12}
6	AIR
7	H_2
8	CO
9	CO_2
10	H_2S

(NOTE 1)

	GAS COMPOSITION – (PER UNIT – %) – NOTE 2									
	NO.	%	NO.	%	NO.	%	NO.	%	NO.	%
60 DATA	1		2		3		4		5	
70 DATA	6		7		8		9		10	

	SYSTEM CONDITIONS			
	INITIAL SUCT PR	FINAL SYSTEM PR	INITIAL SUCT TEMP	DATA IDENTITY CODE – (UP TO 10 CHARACTERS)
	PSIA	PSIA	°F	
	P(1,1)	P	T1	K$
80 DATA				

	SYSTEM CONSTANTS					
	STATION CAPACITY	HP/UNIT	NO. CYL/ UNIT	UNIT SPEED	PISTON ROD DIA	PISTON STROKE
	MMSCFD	BHP		RPM	INCHES	INCHES
	Q	U1	U4	R4	R5	R6
90 DATA						

	FUNCTION NUMBER (OR STAGE NUMBER)					
	ONE	TWO	THREE	FOUR	FIVE	SIX
	CLEARANCE – (DECIMAL – E.G. 0.20) – C(J,1)					
100 DATA						
	P/V–AREA RATIO–A (J,1)					
110 DATA						
120 DATA		I5 – CALCULATION INDICATOR STD=1; LONG=2				

NOTES 1. GAS PROPERTIES STORED. LEAVE 10 – 50 BLANK

2. NUMBER OF GASES HAVING TOTAL COMPOSITION = 1.00 WILL BE USED.

Table 11.15b
Program RKBCS Input Data Sheet (Sample Data)

PROGRAM LINE NO.	PROGRAM LINE NAME	GAS NO.	GAS PROPERTIES: MOL WT (LB/MOL)	CRIT TEMP (°R)	CRIT PR (PSIA)	MOL SPEC HEAT (CPM)	GAS NO.	GAS PROPERTIES: MOL WT (LB/MOL)	CRIT TEMP (°R)	CRIT PR (PSIA)	MOL SPEC HEAT (CPM)
10	DATA	1	16.04	673	344	8.88	2	30.07	710	550	13.6
20	DATA	3	44.09	617	666	19.2	4	58.12	551	766	25.5
30	DATA	5	72.15	489	846	31.2	6	28.97	547	239	6.98
40	DATA	7	2.02	188	60	6.93	8	28.01	507	240	6.97
50	DATA	9	44.01	1071	548	9.24	10	34.08	1306	673	8.30

GAS NO.	GAS NAME
1	CH_4
2	C_2H_6
3	C_3H_8
4	C_4H_{10}
5	C_5H_{12}
6	AIR
7	H_2
8	CO
9	CO_2
10	H_2S
(NOTE 1)	

GAS COMPOSITION–(PER UNIT–%)–NOTE 2

LINE NO.	LINE NAME	NO.	%	NO.	%	NO.	%	NO.	%	NO.	%
60	DATA	1	0.76	2	0.08	3	0.02	4	0.01	5	0.01
70	DATA	6	0.01	7	0.05	8	0.02	9	0.03	10	0.01

SYSTEMS CONDITIONS

LINE NO.	LINE NAME	INITIAL SUCT PR	FINAL SYSTEM PR	INITIAL SUCT TEMP	DATA IDENTITY CODE–(UP TO 10 CHARACTERS)
		PSIA	PSIA	°F	
		P (1,1)	P	T1	K$
80	DATA	23	1500	90	'LFS–3'

SYSTEM CONSTANTS

LINE NO.	LINE NAME	STATION CAPACITY	HP/UNIT	NO.CYL/ UNIT	UNIT SPEED	PISTON ROD DIA	PISTON STROKE
		MMSCFD	BHP		RPM	INCHES	INCHES
		Q	U1	U4	R4	R5	R6
90	DATA	100	2500	5	300	4	18

FUNCTION NUMBER (OR STAGE NUMBER)

LINE NO.	LINE NAME	ONE	TWO	THREE	FOUR	FIVE	SIX
		CLEARANCE–(DECIMAL–E.G. 0.20) – C (J,1)					
100	DATA	0.12	0.15	0.18	0.18	0.21	0.24
		P/V–AREA RATIO A (J,1)					
110	DATA	10	10	10	10	10	10
120	DATA	1	I5 – CALCULATION INDICATOR STD = 1; LONG = 2				

NOTES:
1. GAS PROPERTIES STORED. LEAVE 10 – 50 BLANK
2. NO. OF GASES HAVING TOTAL COMPOSITION = 1.00 WILL BE USED.

Figure 11.8. Simplified flow diagram for Program RKBCS.

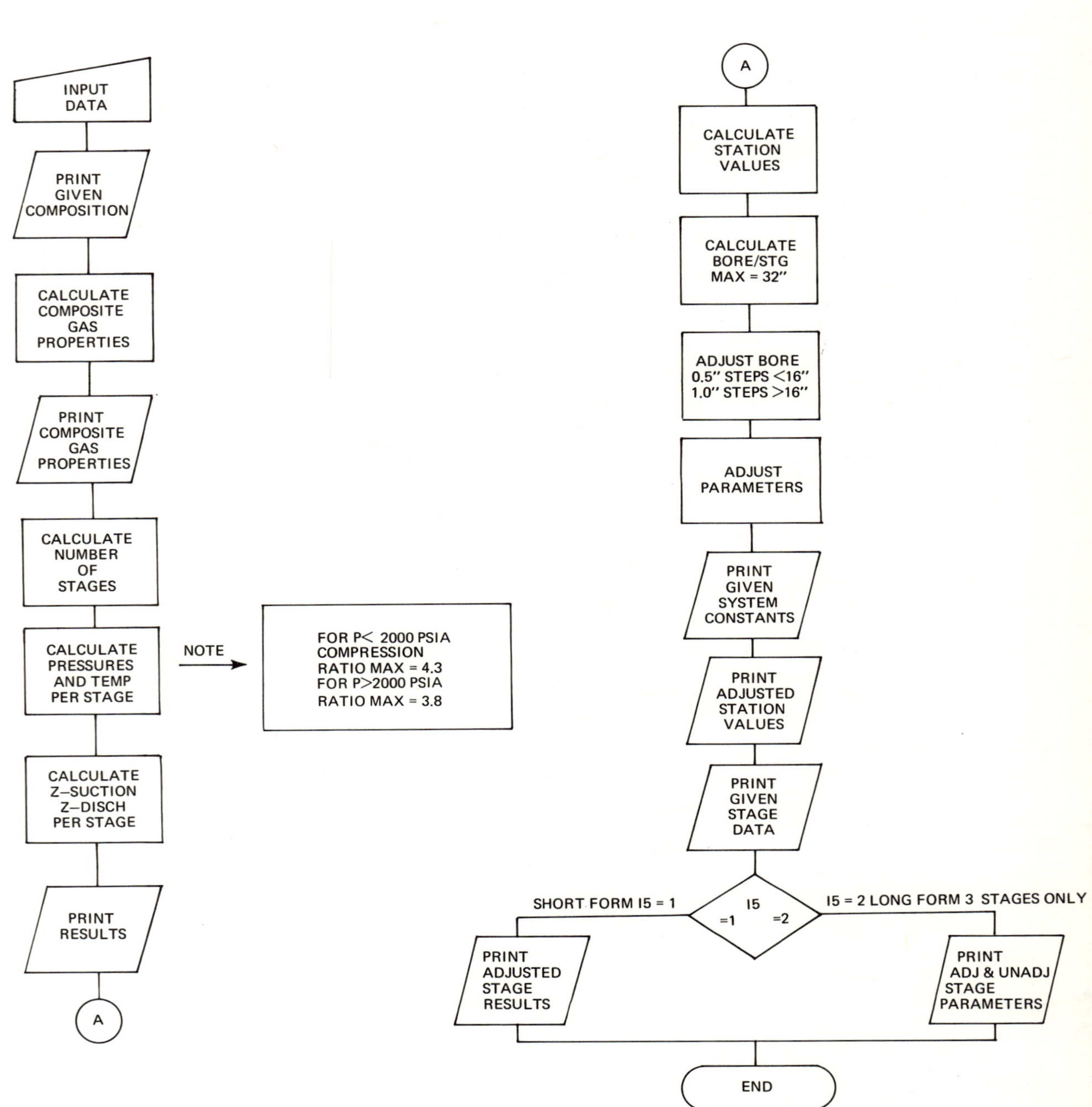

Table 11.16
Program RKBCS Sample Calculation Print-Out

```
     SPECIFIC COMPOSITION                 LFS-3
       GAS                 MOL PERCENT

       CH4                  76.00
       C2H6                  8.00
       C3H8                  2.00
       C4H10                 1.00
       C5H12                 1.00
       AIR                   1.00
       H2                    5.00
       CO                    2.00
       CO2                   3.00
       H2S                   1.00

   TOTAL     10                   100.00


     COMPOSITE GAS PROPERTIES
       MOL WT       CPM        K       T(CRIT)    P(CRIT)
                                         °R        PSIA

       19.39       9.70     1.257      368.       661.


     OPERATING CONDITIONS
  STAGE     SUCTION             DISCHARGE                COMPRESSIBILITY
   NO.      P         T          P            T                'Z'
          PSIA       °F        PSIA          °F

   1       23.00     90.                                     0.9966
   1                           97.21        279.             0.9954
   2       92.58     90.                                     0.9863
   2                          391.28        279.             0.9828
   3      372.65     90.                                     0.9456
   3                         1575.00        279.             0.9522
 FINAL                       1500.00         90.


     SYSTEM CONSTANTS - GIVEN

STATION CAP -MMSCFD- Q= 100 , BHP/UNIT - U1=2500. ,CYL/UNIT U4= 5
RPM - R4=300 , ROD DIA -INCHES -R5= 4 , STROKE -INCHES- R6=18


ADJUSTED VALUES

NUMBER OF UNITS (U8)-             11.
STATION BHP (L2)-              26984.
```

(*Table 11.16 continued on p. 138*)

Table 11.16 (*concluded*)

STAGE	1	2	3
GIVEN DATA			
CLEARANCE	0.120	0.150	0.180
P/V-A RATIO	10.	10.	10.
CALCULATED RESULTS - SHORT FORM			
CYL.IN PAR	2	1	1
CYL.DIA.	25.000	19.000	10.000
CLEARANCE	0.121	0.160	0.182
P/VA RATIO	10.000	10.000	10.000
ROD TENS M-LB	35.376	79.922	74.791
ROD STR PSI	2527.	5709.	5342.
ACFM/CYL	2127.	1046.	249.
HP/STAGE	9174.	9069.	8741.
H.P./CYL.	417.	824.	795.
VOL.EFF	0.703	0.605	0.553
COMP EFF	0.898	0.898	0.898

Table 11.18a
Program RKZ Input Data Sheet (Blank)

PROGRAM LINE NO.	PROGRAM LINE NAME	GAS NO.	GAS PROPERTIES MOL WT LB/MOL M(J,11)	CRIT TEMP °R D(J,1)	CRIT PR PSIA D(J,2)	MOL SPEC HEAT CPM H (J,2)	GAS NO.	GAS PROPERTIES MOL WT LB/MOL M(J,11)	CRIT TEMP °R D(J,1)	CRIT PR PSIA D(J,2)	MOL SPEC HT CPM H(J,2)	
10	DATA	1					2					SEE NOTE 1
20	DATA	3					4					
30	DATA	5					6					
40	DATA	7					8					
50	DATA	9					10					

GAS COMPOSITION – (PER UNIT – %) – NOTE 2

PROGRAM LINE NO.	NAME	NO.	%	NO.	%	NO.	%	NO.	%	NO.	%
60	DATA	1		2		3		4		5	
70	DATA	6		7		8		9		10	

SYSTEM CONDITIONS

PROGRAM LINE NO.	NAME	INITIAL SUCTION PRESSURE PSIA	FINAL SYSTEM PRESSURE PSIA	INITIAL SUCTION TEMPERATURE °F	DATA IDENT CODE – (UP TO TEN CHAR. IN ITALICS)
		P(1,1)	P	T1	K$
80	DATA				

NOTES:

1. GAS PROPERTIES STORED LINES 10 – 50.
   LEAVE BLANK TO USE STORED DATA SHOWN ON SAMPLE.
   GASES ARE 1 - CH_4, 2 - C_2H_6, 3 - C_3H_8, 4 - C_4H_{10}
   5 - C_5H_{12}, 6 - AIR, 7 - H_2, 8 - CO, 9 - CO_2, 10 - H_2S

2. NO. OF GASES HAVING TOTAL COMPOSITION = 1.00
   WILL BE USED IN CALCULATIONS.

Table 11.18b
Program RKZ Input Data Sheet (Sample Data)

PROGRAM LINE NO.	PROGRAM LINE NAME	GAS NO.	GAS PROPERTIES: MOL WT	CRIT TEMP	CRIT PR	MOL SPEC HEAT	GAS NO.	GAS PROPERTIES: MOL WT	CRIT TEMP	CRIT PR	MOL SPEC HT	
			LB/MOL	°R	PSIA	CPM		LB/MOL	°R	PSIA	CPM	
			M (J,1)	D (J,1)	D (J,2)	H (J,2)		M (J,11	D (J,1)	D (J,2)	H (J,2)	
10	DATA	1	16.04	673	344	8.88	2	30.07	710	550	13.6	SEE
20	DATA	3	44.09	617	666	19.2	4	58.12	551	766	25.5	NOTE 1
30	DATA	5	72.15	489	846	31.2	6	28.97	547	239	6.98	
40	DATA	7	2.0	188	60	6.93	8	28.01	507	240	6.97	
50	DATA	9	44.01	1071	548	6.93	10	34.08	1306	673	8.30	

GAS COMPOSITION—(PER UNIT—%)—NOTE 2

LINE NO.	NAME	NO.	%	NO.	%	NO.	%	NO.	%	NO.	%
60	DATA	1	0.75	2	0.06	3	0.05	4	0.02	5	0.02
70	DATA	6	0.02	7	0.02	8	0.02	9	0.02	10	0.02

SYSTEM CONDITIONS

LINE NO.	NAME	INITIAL SUCTION PRESSURE PSIA	FINAL SYSTEM PRESSURE PSIA	INITIAL SUCTION TEMPERATURE °F	DATA IDENT CODE—(UP TO TEN CHARACT. IN ITALICS)
		P (1,1)	P	T1	K$
80	DATA	14.5	6000	95	'LFS—10'

NOTES:

1. GAS PROPERTIES STORED LINES 10 THRU 50.
   LEAVE BLANK TO USE STORED DATA SHOWN ON SAMPLE.
   GASES ARE: 1-CH_4, 2-C_2H_6, 3-C_3H_8, 4-C_4H_{10}
   5-C_5H_{12}, 6-AIR, 7-H_2, 8-CO, 9-CO_2, 10-H_2S
2. NO. GASES HAVING TOTAL COMPOSITION = 1.00
   WILL BE USED IN CALCULATIONS.

Figure 11.9. Simplified flow diagram for Program RKZ.

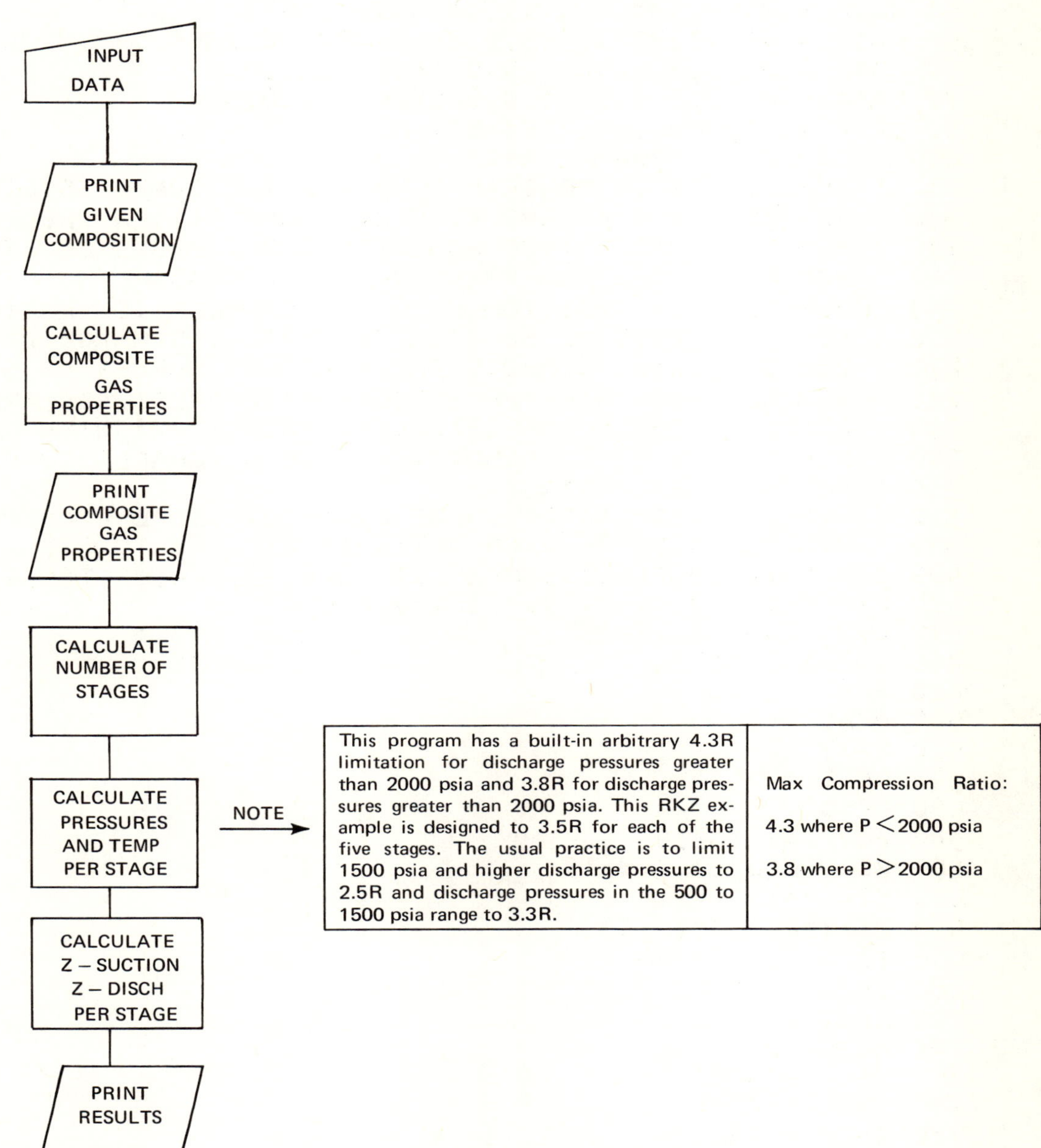

Table 11.19
Program RKZ Listing and Sample Calculation Print-Out

```
10 DATA 1,16.04,673,344,8.88,2,30.07,710,550,13.6
20 DATA 3,44.09,617,666,19.2,4,58.12,551,766,25.5
30 DATA 5,72.15,489,846,31.2,6,28.97,547,239,6.98
40 DATA 7,2.02,188,60,6.93,8,28.01,507,240,6.97
50 DATA 9,44.01,1071,548,9.24,10,34.08,1306,673,8.30
60 DATA 1,0.75,2,0.06,3,0.05,4,0.02,5,0.02
70 DATA 6,0.02,7,0.02,8,0.02,9,0.02,10,0.02
80 DATA 14.5,6000,95,'LFS-4'
200    REM 01    PROGRAM RKZ
210    REM 02    THIS PROGRAM COMPUTES THE COMPRESSIBILITY FACTOR 'Z'
220    REM 02    FOR A COMPOSITE GAS COMPOSITION WHEN GIVEN THE MOL
230    REM 02    PERCENT COMPOSITION AND THE SYSTEM TEMPERATURE 'T1'
240    REM 02    (°F) AND PRESSURE 'P' (PSIA).
250    REM 02    IN ADDITION WHEN THE COMPRESSION RATIO EXCEEDS 4.3 FROM
260    REM 02    INITIAL SUCTION TO FINAL DISCHARGE PRESSURE THE PROGRAM
270    REM 02    WILL COMPUTE THE SMALLEST NUMBER OF STAGES REQUIRED AND
280    REM 02    COMPUTE THE INTERMEDIATE PRESSURES AND TEMPERATURES.
290    REM 02    THE COMPRESSIBILITY FACTOR 'Z' WILL BE CALCULATED FOR
300    REM 02    SUCTION CONDITIONS AND DISCHARGE CONDITIONS FOR
310    REM 02    EACH STAGE.
320    REM 03    FOR CASES WHERE FINAL PRESSURE EXCEEDS 2000 PSIA THE
330    REM 03    COMPRESSION RATIO IS LIMITED TO 3.8
400 DIM O(10,5),P(6,10),M(10,11),N(6,4),X(6,2),T(6,13),H(6,13)
420 DIM D(10,2),G(10,2),Y(10,2),R(6,6)
440 DIM S(6),Z(6,2)
460 FOR I=1 TO 10
480 FOR J=1 TO 5
500 READ O(I,J)
520 NEXT J
540 NEXT I
560 K2=0
580 FOR J=1 TO 10
600 READ G(J,1),Y(J,1)
620 K2=K2+1
640 Y1=Y1+100*Y(J,1)
660 IF Y1≥100 THEN 720
680 IF J>10 THEN 3620
700 NEXT J
720 K1=K2
740 O$(1)='CH4'
760 O$(2)='C2H6'
780 O$(3)='C3H8'
800 O$(4)='C4H10'
820 O$(5)='C5H12'
840 O$(6)='AIR'
860 O$(7)='H2'
880 O$(8)='CO'
900 O$(9)='CO2'
920 O$(10)='H2S'
940 READ P(1,1),P,T1,K$
```

Table 11.19 (*continued*)

```
960 K8=6
980 PRINT
1000 PRINT
1020 PRINT USING 2680 ,'SPECIFIC COMPOSITION',K$
1040 PRINT USING 2720
1060  PRINT
1080 K8=K8+5

1100 FOR J=1 TO K1
1120 Y(J,2)=100*Y(J,1)
1140 PRINT USING 2740 O$(G(J,1)),Y(J,2)
1160 K8=K8+1
1180 NEXT J
1200 PRINT
1220 PRINT USING 2760 K1,Y1
1240 K8=K8+2
1260 M2=0
1280 FOR J=1 TO K1
1300 M(J,11)=O(G(J,1),2)*Y(J,1)
1320 M2=M2+M(J,11)
1340 NEXT J
1360 C2=0
1380 FOR J=1 TO K1
1400 D(J,1)=Y(J,1)*O(G(J,1),3)
1420 C2=C2+D(J,1)
1440 NEXT J
1460 D2=0
1480 FOR J=1 TO K1
1500 D(J,2)=Y(J,1)*O(G(J,1),4)
1520 D2=D2+D(J,2)
1540 NEXT J
1560 H2=0
1580 FOR J=1 TO K1
1600 H(J,2)=Y(J,1)*O(G(J,1),5)
1620 H2=H2+H(J,2)
1640 NEXT J
1660 PRINT
1680 PRINT
1700 PRINT USING 2680 ,'COMPOSITE GAS PROPERTIES'
1720 PRINT
1740 PRINT USING 2780
1760 PRINT USING 2800
1780 PRINT
1800 K8=K8+7
1820 K3=H2/(H2-1.986)
1840 S=(K3-1)/K3
1860 PRINT USING 2820 M2,H2,K3,D2,C2
1880 PRINT
1900 PRINT
1920 PRINT USING 2680 ,'OPERATING CONDITIONS'
1940 PRINT USING 2840
1960 PRINT USING 2860
```

(*Table 11.19 continued on p. 144*)

Table 11.19 (*continued*)

```
1980 PRINT USING 2880
2000 PRINT
2020 K8=K8+8
2040 K4=0
2060 FOR J=1 TO 6
2080 P2=((P/P(1,1))↑(1/J))*P(1,1)*1.05
2100 K4=K4+1
2120 IF P>2000 THEN 2180
2140 IF P2≤P(1,1)*4.3 THEN 2300
2160 GO TO 2200
2180 IF P2≤P(1,1)*3.8 THEN 2300
2200 IF K4=6 THEN 2240
2220 NEXT J
2240 PRINT ' SIX STAGES NOT SUFFICIENT . PLEASE CHECK INITIAL DATA '
2260 PRINT ' AND LIMIT OF SIX STAGES FOR THIS PROGRAM.'
2280 GO TO 3820
2300 S2=K4
2320 FOR J=1 TO S2
2340 T(J,1)=T1+460
2360 K=1
2380 R(J,1)=P2/P(1,1)
2400 M(J,1)=0.427*(P(J,K)/C2)*(D2/T(J,1))↑2.5
2420 M(J,2)=-0.00752*(P(J,K)/C2)↑2*(D2/T(J,1))↑2
2440 M(J,3)=-0.0867*(D2/T(J,1))*(P(J,K)/C2)-0.3333
2460 M(J,4)=M(J,1)+M(J,2)+M(J,3)
2480 N(J,1)=M(J,4)/3-0.0372*(D2/T(J,1))↑3.5*(P(J,K)/C2)↑2+0.037
2500 X(J,1)=((N(J,1))↑2)/4+((M(J,4))↑3)/27
2520 M(J,9)=(-(N(J,1))/2+(X(J,1))↑(1/2))↑(1/3)
2540 N(J,3)=(-(N(J,1))/2-(X(J,1))↑(1/2))↑(1/3)
2560 Z(J,1)=M(J,9)+N(J,3)+1/3
2580 T(J,3)=T(1,1)-460
2600 PRINT USING 2900 J,P(J,1),T(J,3),Z(J,1)
2620 K8=K8+1
2640 K=2
2660 IF S2≥2 GO TO 3040
2680:       #########################      ##########
2700:                  #########################
2720:         GAS                   MOL PERCENT
2740:        #####                ###.##
2760:     TOTAL     ##                  ###.##
2780:        MOL WT       CPM         K        T(CRIT)    P(CRIT)
2800:                                            °R         PSIA
2820:        ###.##     ###.##    ###.###     ####.       ####.
2840:   STAGE     SUCTION             DISCHARGE                COMPRESSIBILITY
2860:    NO.       P         T          P           T                   'Z'
2880:            PSIA       °F        PSIA         °F
2900:   ##   #####.##   ###.                                       ####.####
2920:   ##                      #####.##      ###.                 ####.####
2940:  FINAL                    #####.##      ###.
2960:                                                  ###.######
```

Table 11.19 *(continued)*

```
2980 P(J,K)=P*1.05
3000 R(J,1)=P(J,K)/P(1,1)
3020 GO TO 3140
3040 IF J=S2 THEN 3100
3060 P(J,2)=P(J,1)*R(J,1)
3080 GO TO 3140
3100 P(J,2)=P*1.05
3120 R(J,1)=P(J,2)/P(J,1)
3140 T(J,6)=(R(J,1))↑S*T(J,1)-460
3160 T(J,2)=T(J,6)+460
3180 M(J,5)=0.427*(P(J,2)/C2)*(D2/T(J,2))↑2.5
3200 M(J,6)=-0.00752*(P(J,2)/C2)↑2*(D2/T(J,2))↑2
3220 M(J,7)=-0.0867*(D2/T(J,2))*(P(J,2)/C2)-0.3333
3240 M(J,8)=M(J,5)+M(J,6)+M(J,7)
3260 N(J,2)=M(J,8)/3-0.0372*(D2/T(J,2))↑3.5*(P(J,2)/C2)↑2+0.037
3280 X(J,2)=((N(J,2))↑2)/4+((M(J,8))↑3)/27
3300 M(J,10)=(-(N(J,2))/2+(X(J,2))↑(1/2))↑(1/3)
3320 N(J,4)=(-(N(J,2))/2-(X(J,2))↑(1/2))↑(1/3)
3340 Z(J,2)=M(J,10)+N(J,4)+1/3
3360 T(J,4)=T(J,2)-460
3380 PRINT USING 2920 J,P(J,2),T(J,4),Z(J,2)
3400 K8=K8+1
3420 P((J+1),1)=P(J,2)/1.05
3440 NEXT J
3460 PRINT USING 2940 ,P,T(J,3)
3480 PRINT
3500 PRINT
3520 PRINT
3540 PRINT USING 2700 ,'-- END OF CALCULATIONS --'
3560 PRINT
3580 K8=K8+6
3300 GO TO 3740
3620 PRINT 'GAS COMPOSITION LESS THAN 100 PERCENT (< 1.00) FOR TEN'
3640 PRINT 'GASES.  RUN STOPPED.  PLEASE CHECK DATA.  IF NECESSARY'
3660 PRINT 'LIST ACTUAL INPUT WHICH SHOULD BE IN DECIMALS TO MAKE'
3680 PRINT 'TOTAL = 1.00 FOR TEN OR FEWER GASES.'
3700 PRINT '(E.G. 0.90+0.10 = 1.00)'
3720 PRINT
3740 K6=66-K8+6
3760 FOR J=1 TO K6
3780 PRINT
3800 NEXT J
3820 END
```

(*Table 11.19 continued on p. 146*)

Table 11.19 (*concluded*)

SPECIFIC COMPOSITION LFS-4

GAS	MOL PERCENT
CH4	75.00
C2H6	6.00
C3H8	5.00
C4H10	2.00
C5H12	2.00
AIR	2.00
H2	2.00
CO	2.00
CO2	2.00
H2S	2.00
TOTAL 10	100.00

COMPOSITE GAS PROPERTIES

MOL WT	CPM	K	T(CRIT) °R	P(CRIT) PSIA
21.39	10.34	1.238	392.	671.

OPERATING CONDITIONS

STAGE NO.	SUCTION P PSIA	SUCTION T °F	DISCHARGE P PSIA	DISCHARGE T °F	COMPRESSIBILITY 'Z'
1	14.50	95.			0.9975
1			50.81	246.	0.9963
2	48.39	95.			0.9915
2			169.54	246.	0.9877
3	161.47	95.			0.9718
3			565.75	246.	0.9617
4	538.81	95.			0.9073
4			1887.91	246.	0.9115
5	1798.01	95.			0.7573
5			6300.00	246.	1.1034
FINAL			6000.00	95.	

-- END OF CALCULATIONS --

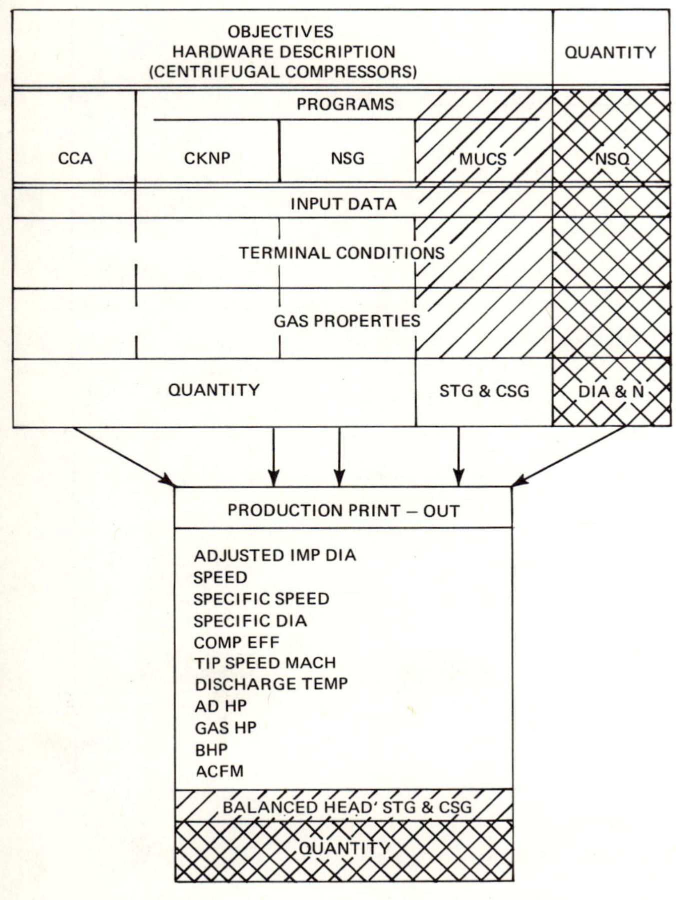

Centrifugal Compressors

Figure 11.10. Consolidated data flow diagram for all centrifugal compressors.

Table 11.20
Computer Program Input and Production for Centrifugal Compressors

PROGRAM	Q (MMscfd)	P, P1, T1	K3, M2, Z	Imp. Dia. (inches)	N (rpm)	D_s	N_s	H (Ad Head/ Stage)	No. of Stages	η_1 (Dyn. Eff. First)	η_t (Dyn. Eff. (Total)	Q/N (Cap Index)	τ (Sonic in and out)	Opt. Dia. Suct & ADFM	u (Tip Speed)	M_u* (Mach Tip-N)
CCA	G	G	G	G	G	C	C	C	C	C	A	C	C	C	C	C
CKNP	G	G	G	A	G	C	C	C	C	C	A	C	C*	C	C*	C*
NSG	G	G	G	AC	AC	A	AR	C	C	C	C	C	C	C*	C	C
NSQ	AC	G	G	G	A	C	AR	C	C	C	C	C	C	C	C	C
MUCS		G	G					C	C							

Notes: T2 = adjusted for each case; R = given in array form.
G = given data; C = calculated data; A = adjusted calculations

Table 11.21a
Program MUCS Input Data Sheet (Blank)

PROGRAM LINE		SYSTEM INPUT					
		QUANTITIES			GAS		
NO.	NAME	MOL WT	Cp/CV	IDENT CODE			
		M2	K3	K$			
10	DATA						
		SYSTEM TERMINAL CONDITIONS					
		PRESSURE (PSIA)		SUCTION TEMPERATURES °R			
		INITIAL	FINAL	INITIAL	2nd CSG	3rd CSG	
		P1	P	T1	T3	T5	
20	DATA						
		COMPRESSIBILITY FACTORS					
		Z1	Z2				
30	DATA						

Table 11.21b
Program MUCS Input Data Sheet (Sample Data)

PROGRAM LINE		SYSTEM INPUT					
		QUANTITIES			GAS		
NO.	NAME	MOL WT	Cp/Cv	IDENT CODE			
		M2	K3	KS			
10	DATA	20	1.28	'N GAS'			
		SYSTEMS TERMINAL CONDITIONS					
		PRESSURE (PSIA) SUCTION TEMPERATURES (°R)					
		INITIAL	FINAL	INITIAL	2ND CSG	3RD CSG	
		P1	P	T1	T3	T5	
20	DATA	100	2850	520	570	570	
		COMPRESSIBILITY FACTORS					
		Z1	Z2				
30	DATA	1.00	1.00				

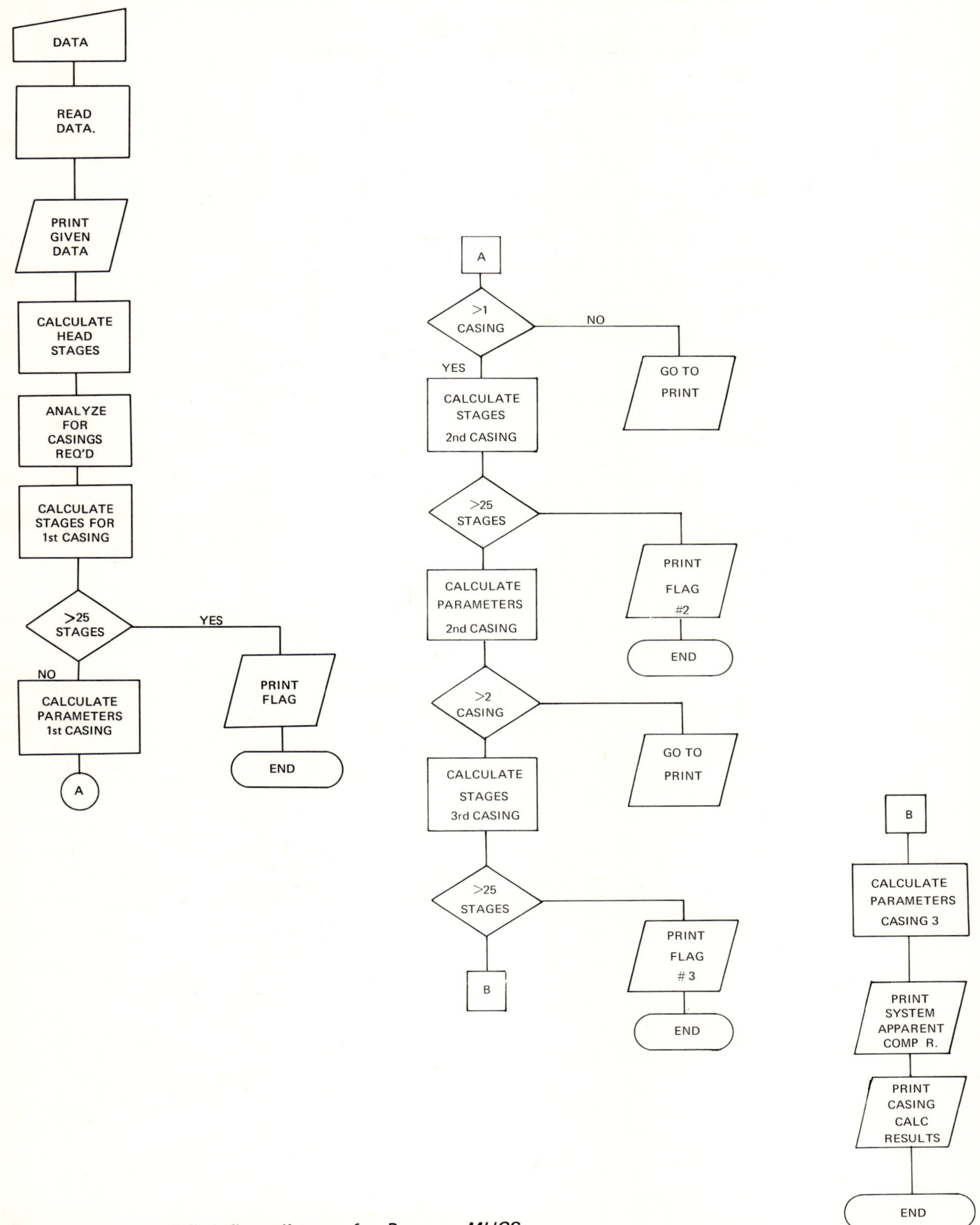

Figure 11.11. Simplified flow diagram for Program MUCS.

Table 11.22
Program MUCS Sample Calculation Print-Out

```
GIVEN DATA                              NGAS

MOL WEIGHT OF GAS                          20.
RATIO OF SPEC HEATS K3                      1.280
COMPRESSION EXPONENT S                      0.219
SUCTION PR (PSIA)                         100.00
DISCHARGE PRESS (PSIA)                   2850.00
SUCT TEMP CASING 1 (°R)                   520.0
SUCT TEMP CASING 2 (°R)                   570.0
SUCT TEMP CASING 3 (°R)                   570.0
Z-SUCT ,Z1                                  1.000
Z-DISCH, Z2                                 1.000

CALCULATED RESULTS

SYSTEM APPARENT COMP RATIO                 28.500

        CASING             SUCTION PR     DISCH PR
NO.     STAGES    COMP        PSIA          PSIA
                  RATIO

 1        9       4.888     100.00         534.1

 2        8       4.793     488.81        2485.6

 3        1       1.216    2342.96        2935.5

FINAL PR AT T1 (°F)                      2850.0          60.0

TIME      1 SECS.
```

Table 11.23a
Program CKNP Input Data Sheet (Blank)

<table>
<tr><td colspan="2">PROGRAM LINE</td><td colspan="7">SYSTEM INPUT</td></tr>
<tr><td rowspan="3">NO.</td><td rowspan="3">NAME</td><td colspan="3">QUANTITIES
CHOICE OF:</td><td colspan="2">GAS</td><td rowspan="2">RPM</td><td rowspan="2">CODE
IDENT
(NOTE 2)</td></tr>
<tr><td>MMSCFD</td><td>MOLS PER HR</td><td>LB/MIN</td><td>MOL WT</td><td>RATIO OF SP. HEATS</td></tr>
<tr><td>Q(4)</td><td>Q(5)</td><td>Q(6)</td><td>M2</td><td>K3</td><td>R4</td><td>K$</td></tr>
<tr><td>10</td><td>DATA</td><td></td><td></td><td></td><td></td><td></td><td></td><td></td></tr>
</table>

<table>
<tr><td rowspan="4"></td><td rowspan="4"></td><td colspan="3">SYSTEM TERMINAL CONDITIONS</td></tr>
<tr><td colspan="2">PRESSURE</td><td>TEMPERATURE</td></tr>
<tr><td>INITIAL
(PSIA)</td><td>FINAL
(PSIA)</td><td>INITIAL
(°R)</td></tr>
<tr><td>P1</td><td>P</td><td>T1</td></tr>
<tr><td>20</td><td>DATA</td><td></td><td></td><td></td></tr>
</table>

<table>
<tr><td rowspan="3"></td><td rowspan="3"></td><td colspan="2">SYSTEM
COMPRESSIBILITY
FACTORS</td><td rowspan="2">PRESSURE
COEFF
(NOTE 1)</td></tr>
<tr><td>INITIAL</td><td>FINAL</td></tr>
<tr><td>Z1</td><td>Z2</td><td>S3</td></tr>
<tr><td>30</td><td>DATA</td><td></td><td></td><td></td></tr>
</table>

NOTES:

1. PRESSURE COEFFICIENT FOR VANE TYPE
   1.1 RADIAL VANE S3 = 0.63
   1.2 BACKLAY VANE S3 = 0.53

2. IDENTIFY DATA WITH CODE NAME
   E.G. 'TEST' IDENTITY TO 5 CHARACTERS PLUS ITALICS

Table 11.23b
Program CKNP Input Data Sheet (Sample Data)

PROGRAM LINE		SYSTEM INPUT						
		QUANTITIES CHOICE OF:			GAS		RPM	CODE IDENT (NOTE 2)
NO.	NAME	MMSCFD	MOLS PER HR	LB/MIN	MOL WT	RATIO OF SP HEATS		
		Q(4)	Q(5)	Q(6)	M2	K3	R4	K$
10	DATA	2.26	0	0	44	1.123	20800	'TEST'

NO.	NAME	SYSTEM TERMINAL CONDITIONS		
		PRESSURE		TEMPERATURE
		INITIAL (PSIA)	FINAL (PSIA)	INITIAL (°R)
		P1	P	T1
20	DATA	14	56	520

NO.	NAME	SYSTEM COMPRESSIBILITY FACTORS		PRESSURE COEFF (NOTE 1)
		INIATIAL	FINAL	
		Z1	Z2	S3
30	DATA	1.00	1.00	0.55

NOTES:

1. PRESSURE COEFFICIENT FOR VANE TYPE
   1.1 RADIAL VANE S3 = 0.63
   1.2 BACKLAY VANE S3 = 0.53

2. IDENTIFY DATA WITH CODE NAME
   E.G. 'TEST' IDENTITY TO 5 CHARACTERS PLUS ITALICS

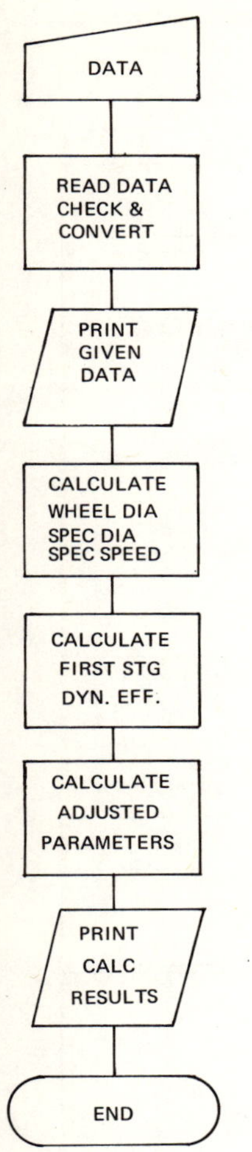

Figure 11.12 Simplified flow diagram for Program CKNP.

Table 11.24
Program CKNP Sample Calculation Print-Out

```
GIVEN DATA                                TEST

PLANT CAPACITY (MMSCFD)                              2.26
PLANT CAPACITY (MOLS/HR)                           248.35
PLANT CAPACITY (LB/MIN)                            181.98
GAS MOL WEIGHT                                      44.00
RATIO OF SPECIFIC HEATS,K3,                          1.123
COMPRESSION EXPONENT,S,                              0.110
SPEED, R4, (RPM)                                20800.
PRESSURE COEFF. S3                                   0.550
SUCTION PRESS, P1, (PSIA)                           14.0
DISCHARGE PR, P, (PSIA),                            56.0
SUCTION TEMP, T1, (°R),                            520.
COMPR,Z-SUCTION, Z1,                                 1.000
COMPR Z-DISCHARGE, Z2,                               1.000


CALCULATED RESULTS

TOTAL ADIABATIC HEAD (FT-LB/LB)                 27335.
NUMBER OF STAGES                                    3.
ADIABATIC HEAD/STG (FT-LB/LB)                    9112.
ADIABATIC H P                                     150.7
GAS H P                                           183.8
BHP                                               191.8
ADJ SPECIFIC DIA,DS,                                1.36
ADJ SPECIFIC SPEED, NS,                           110.65
X =(NS*DS)                                        150.00
ADJ IMPELLER DIA (INCHES)                           8.73
FIRST STAGE DYN EFF (PERCENT)                      82.000
UNIT DYNAMIC EFF (PERCENT)                         79.267
INLET CAPACITY,(ACFM)                            1648.3
SUCTION SONIC VEL (FPS)                           812.4
DIAMETER SUCT NOZZLE (INCHES)                       7.87
CAPACITY INDEX,Q/N, (ACFM/RPM),                     0.079
DISCHARGE TEMP, T4,(°F)                           165.78
DISCHARGE SONIC VEL (FPS)                         891.21
ROTOR TIP SPEED (FPS),                            750.30
MACH NUMBER OF TIP SPEED                            0.842


                --END OF CALCULATIONS--
```

Table 11.25a
Program NSG Input Data Sheet (Blank)

PROGRAM LINE		SYSTEM INPUT					
		CHOICE OF QUANTITIES			GAS DATA		CODE
NO.	NAME	MMSCFD	MOLS PER HR	LB/MIN	MOL WT	C_p/C_v	IDENT
		Q(6,3)	Q(6,9)	Q(6,10)	M2	K3	K$
10	DATA						

		SYSTEM TERMINAL CONDITIONS				
		PRESSURE		SUCT TEMP	COMPRESSIBILITY FACTORS	
		INITIAL	FINAL	INITIAL		
		PSIA	PSIA	°R	INITIAL	FINAL
		P1	P	T1	Z1	Z2
20	DATA					

		HUB DIA – (IN)	PRESSURE COEFF (NOTE 1)
		D8	S3
30	DATA		

NOTES:

1. PRESSURE COEFFIEIENT FOR VANE TYPE
   1.1 RADIAL VANE S3 = 0.63
   1.2 BACKLAY VANE S3 = 0.53

2. IDENTIFY DATA WITH CODE NAME, E.G. 'TEST'
   (UP TO 10 CHARACTERS ALLOWED FOR CODING.
   ADD ITALICS WHEN ENTERING DATA)

Table 11.25b
Program NSG Input Data Sheet (Sample Data)

PROGRAM LINE		SYSTEM INPUT					
		CHOICE OF QUANTITIES			GAS DATA		
NO.	NAME	MMSCFD	MOLS PER HR	LB/MIN	MOL/WT	Cp/Cv	CODE IDENT (NOTE 2)
		Q (6,8)	Q (6,9)	Q (6,10)	M2	K3	K$
10	DATA	0	0	540	29	1.40	'ASME'

		SYSTEM TERMINAL CONDITIONS				
		PRESSURE		SUCT TEMP	COMPRESSIBILITY FACTORS	
		INITIAL	FINAL	INITIAL		
		PSIA	PSIA	°R	INITIAL	FINAL
		P1	P	T1	Z1	Z2
20	DATA	14.0	56.0	520	1.00	1.00

		HUB DIA-(IN)	PRESSURE COEFF (NOTE 1)
		D8	S3
30	DATA	4	0.53

NOTES:

1. PRESSURE COEFFICIENT FOR VANE TYPE
   1.1 RADIAL VANE S3 = 0.63
   1.2 BACKLAY VANE S3=0.53

2. IDENTIFY DATA WITH CODE NAME, E.G. 'TEST'
   (UP TO 10 CHARACTERS ALLOWED FOR CODING.
   ADD ITALICS WHEN ENTERING DATA)

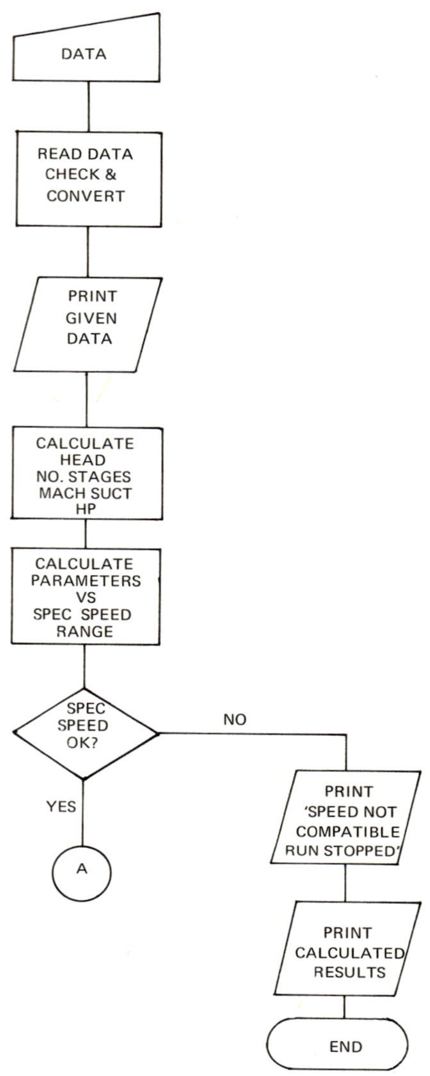

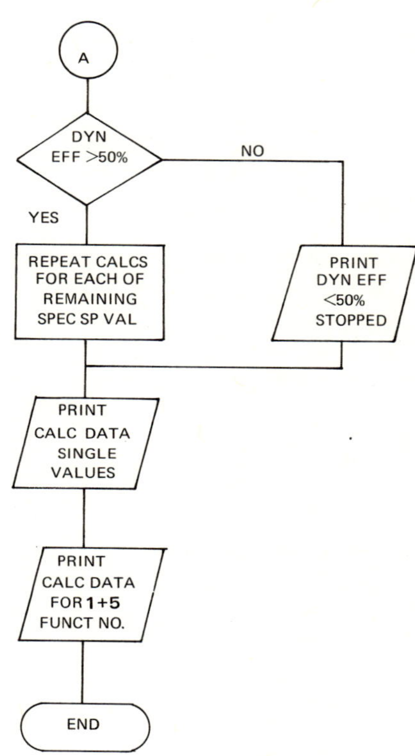

Figure 11.13. Simplified flow diagram for Program NSG.

Table 11.26
Program NSG Sample Calculation Print-Out

GIVEN DATA ASME

PLANT CAPACITY (MMSCFD)	10.18
PLANT CAPACITY (MOLS/HR)	1118.16
PLANT CAPACITY (LB/MIN)	540.00
GAS MOL WEIGHT	29.00
RATIO OF SPECIFIC HEATS, K3,	1.400
SUCTION PRESS, P1, (PSIA)	14.0
DISCHARGE PRESS, P, (PSIA)	56.0
SUCTION TEMP, T1, (°R)	520.
COMPR Z-SUCTION, Z1	1.000
COMPR Z-DISCHARGE, Z2	1.000
HUB DIA (IN)/ PRESS COEFF (S3)	4.0 / 0.530

CALCULATED RESULTS

TOTAL ADIABATIC HEAD (FT-LB/LB)	47123.
NUMBER OF STAGES	5.
INLET SONIC VEL, (FPS)	1117.3
INLET CAPACITY, (ACFM)	7421.1
CAPACITY (CFS)	123.64
CAP SURGE (CFS)	82.84
CAP CHOKE (CFS)	198.26
ADJ PRESSURE COEFFICIENT	0.458
DYN PRESS COEFF, C(J,7)	0.521
MERIDIONAL VEL, C(J,4)	245.0
ADJ TIP SPEED (FPS)	814.1
EYE DIA (IN)	9.728

FUNCTION NUMBER	1	2	3	4	5	6	7
SPEC SPEED,	150.00	100.00	75.00	60.00	40.00	30.00	25.00
ADIAB HP	770.8	770.8	770.8	770.8	770.8	770.8	770.8
GAS HP	1027.7	940.0	963.5	1027.7	1101.1	1284.7	1541.6
BHP	1043.7	955.5	979.1	1043.7	1117.6	1302.2	1560.4
SPEC DIA	1.000	1.500	2.000	2.500	3.750	5.000	6.000
ADJ NS	160.02	106.68	80.01	64.01	42.67	32.00	26.67
IMP DIA (IN)	13.54	20.31	27.08	33.86	50.78	67.71	81.25
ADJ (RPM)	13766.1	9177.4	6883.1	5506.4	3671.0	2753.2	2294.4
UNIT EFF %	72.000	78.720	76.800	72.000	67.200	57.600	48.000
FLOW Q/N	0.539	0.809	1.078	1.348	2.022	2.695	3.235
DISCH TEMP F	421.78	382.91	393.15	421.78	455.68	546.23	688.25
DISCH MACH	1455.0	1422.5	1431.1	1455.0	1482.7	1554.2	1660.3
MACH U	0.560	0.572	0.569	0.560	0.549	0.524	0.490
RATIO DIA	1.392	2.088	2.784	3.480	5.220	6.960	8.353

--END OF CALCULATIONS--

Table 11.27a
Program NSQ Input Data Sheet (Blank)

LINE		GAS & COMPRESSOR DATA					DATA IDENT CODE
		MOL WEIGHT	RATIO OF SPEC HTS	SPEED (RPM)	ROTOR DIA-(IN)	HUB DIA-(IN)	
NO.	NAME	M2	K3	R4	D	D8	KS
10	DATA						
		OPERATING CONDITIONS					
		SUCT PRESS PSIA	FINAL SYS-PR PSIA	SUCTION TEMP °R	COMPRESSIBILITY FACTORS		PRESSURE COEFF. (NOTE 1)
					SUCT	DISCH	
		P1	P	T1	Z1	Z2	S3
20	DATA						

NOTES:

1. Pressure coefficient for vane type

   1.1 Radial Vane S3 = 0.63

   1.2 Backlay Vane S3 = 0.53

*2. Identify Data with Code Name

   E.G.—'Test—1'

   (Up to 10 characters allowable for Code. Add italics when entering data)

Table 11.27b
Program NSQ Input Data Sheet (Sample Data)

LINE		GAS & COMPRESSOR DATA					DATA IDENT CODE*
		MOL WEIGHT	RATIO OF SPEC HTS	SPEED (RPM)	ROTOR DIA–(IN)	HUB DIA–(IN)	
NO.	NAME	M2	K3	R4	D	D8	K$
10	DATA	20	1.1283	10000	18	4.0	'ASME'
		OPERATING CONDITIONS					
		SUCT PRESS PSIA	FINAL SYS PR PSIA	SUCTION TEMP °R	COMPRESSIBILITY FACTORS		PRESSURE COEFF. (NOTE 1)
					SUCT	DISCH	
		P1	P	T1	Z1	Z2	S3
20	DATA	100	300	520	1.00	1.00	0.625

NOTES

1. PRESSURE COEFFICIENT FOR VANE TYPE
   - 1.1 RADIAL VANE S3 = 0.63
   - 1.2 BACKLAY VANE S3 = 0.53

*2. IDENTIFY DATA WITH CODE NAME
E.G. – 'TEST 1'
(UP TO 10 CHARACTERS ALLOWABLE
FOR CODE. ADD ITALICS WHEN ENTERING
DATA)

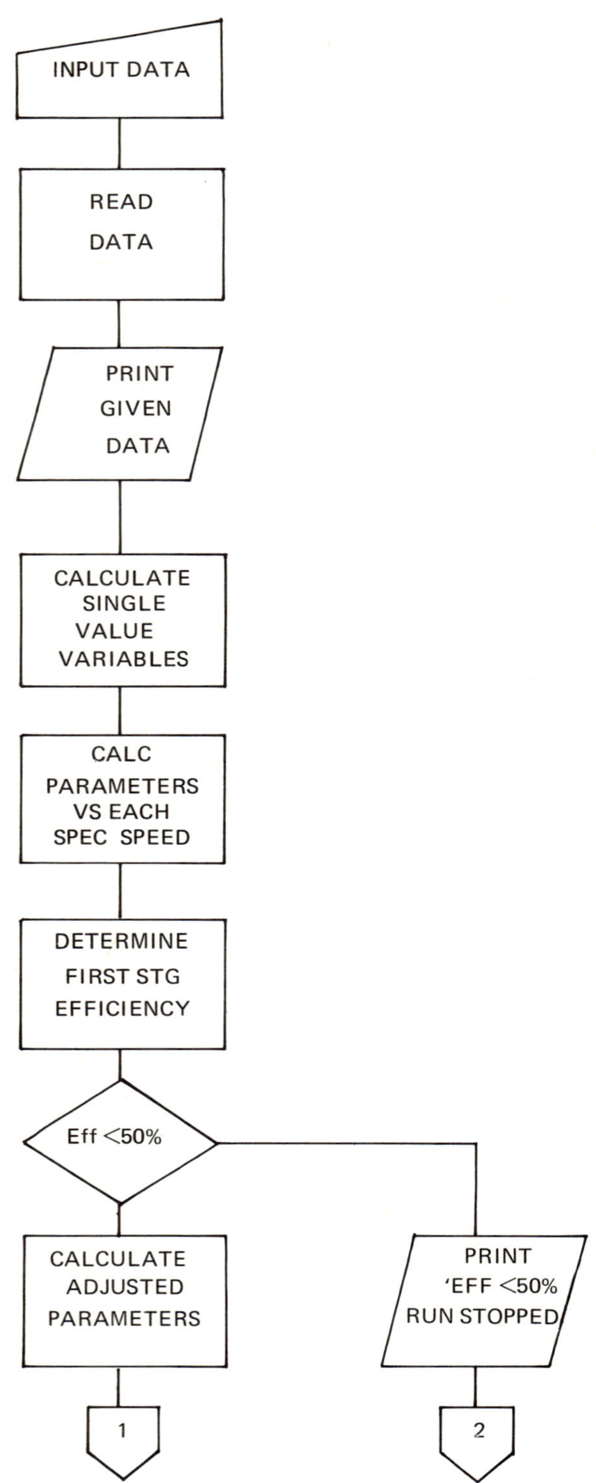

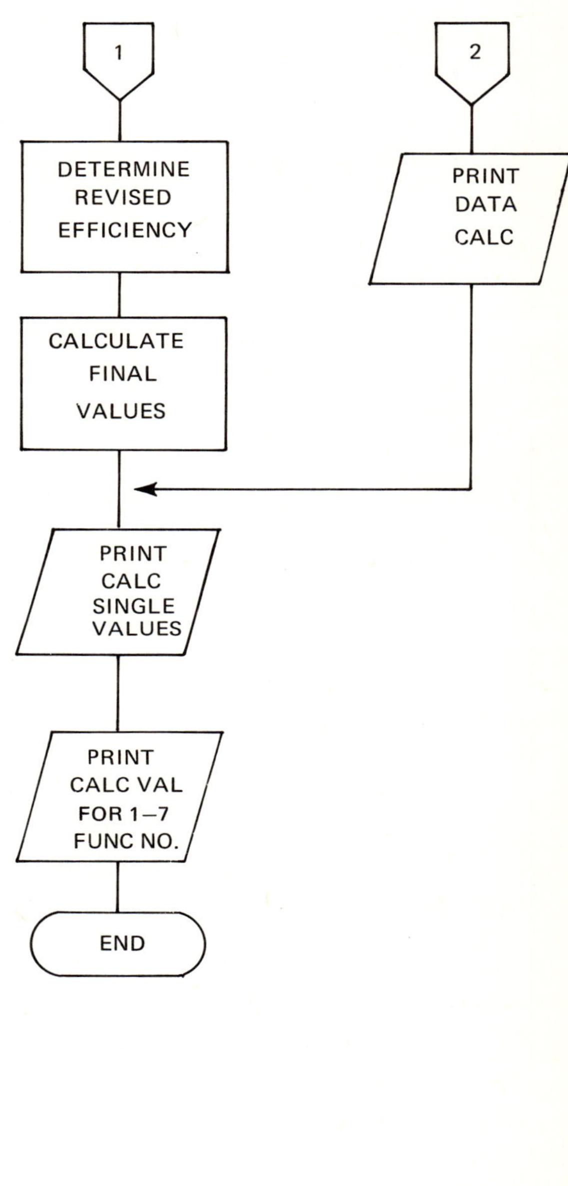

Figure 11.14. Simplified flow diagram for Program NSQ.

Table 11.28
Program NSQ Sample Calculation Print-Out

GIVEN DATA ASME

GAS MOL WEIGHT	20.00
RATIO OF SPECIFIC HEATS, K3,	1.128
NOMINAL SPEED (RPM)	10000.
IMPELLER DIA (INCHES)	18.00
HUB DIA (IN) / PRESS COEFF (S3)	4.000 / 0.625
SUCTION PRESS, P1, (PSIA)	100.0
DISCHARGE PRESS, P, (PSIA)	300.0
SUCTION TEMP, T1, (°R)	520.
COMPR Z-SUCTION, Z1	1.000
COMPR Z-DISCHARGE, Z2	1.000

CALCULATED RESULTS

TOTAL ADIABATIC HEAD (FT-LB/LB)	47006.
NUMBER OF STAGES	4.
INLET SONIC VEL, (FPS)	1207.8
ADJ SPEED (RPM)	10642.9
ADJ IMP TIP SPEED (FPS)	836.6
MERIDIONAL VEL (FPS)	251.81
SUCTION MACH	0.208
ADJ PRESS COEFF, S4,	0.541
DYN PRESS COEFF, C(J,7),	0.612

FUNCTION NUMBER	1	2	3	4	5	6	7
SPEC SPEED (GIVEN)	150.00	100.00	75.00	60.00	40.00	30.00	25.00
CAP (MMSCFD)	168.50	74.89	42.12	26.96	11.98	6.74	4.68
CAP MOLS/HR	18500.1	8222.3	4625.0	2960.0	1315.6	740.0	513.9
Q (LB/MIN)	6164.6	2739.8	1541.1	986.3	438.4	246.6	171.2
Q (ACFM)	17204.8	7646.6	4301.2	2752.8	1223.5	688.2	477.9
CAP (CFS)	286.63	127.39	71.66	45.86	20.38	11.47	7.96
CAP SURGE	192.0	85.4	48.0	30.7	13.7	7.7	5.3
CAP CHOKE	398.6	177.1	99.6	63.8	28.3	15.9	11.1
AD HP	8781.0	3902.7	2195.3	1405.0	624.4	351.2	243.9
GAS HP	11708.1	4759.4	2744.1	1873.3	892.0	585.4	487.8
BHP	11750.5	4788.9	2767.8	1893.7	907.2	598.2	499.7
SPEC DIA DS	1.000	1.500	2.000	2.500	3.750	5.000	6.000
ADJ NS	159.64	106.43	79.82	63.86	42.57	31.93	26.61
UNIT EFF %	72.19	78.92	77.00	72.19	67.37	57.75	48.12
FLOW Q/N	1.720	0.765	0.430	0.275	0.122	0.069	0.048
DISCH TEMP F	154.24	145.57	147.88	154.24	161.60	180.36	207.59
DISCH MACH	1312.7	1303.4	1305.9	1312.7	1320.6	1340.3	1368.5
ADJ MACH U	0.637	0.642	0.641	0.637	0.633	0.624	0.611
DIA EYE (IN)	13.91	9.74	7.77	6.66	5.35	4.81	4.57
DIA RATIO	1.29	1.85	2.32	2.70	3.36	3.75	3.93

--END OF CALCULATIONS--

Table 11.29a
Program CCA Input Data Sheet (Blank)

<table>
<tr><td colspan="2">PROGRAM</td><td colspan="8">SYSTEM INPUT—GAS</td></tr>
<tr><td rowspan="2">LINE
NO.</td><td rowspan="2">NAME</td><td colspan="3">CHOICE OF QUANT.</td><td colspan="4"></td><td rowspan="2">INDENT
CODE</td></tr>
<tr><td>MMSCFD</td><td>MOLS/HR</td><td>LB/MIN</td><td>MOL WT</td><td>cpCv</td><td>RPM</td><td>DIA
IN.</td></tr>
<tr><td></td><td></td><td>Q (4)</td><td>Q (5)</td><td>Q (6)</td><td>M2</td><td>K3</td><td>R4</td><td>D</td><td>KS</td></tr>
<tr><td>10</td><td>DATA</td><td></td><td></td><td></td><td></td><td></td><td></td><td></td><td>'TEST'</td></tr>
</table>

<table>
<tr><td rowspan="4"></td><td rowspan="4"></td><td colspan="4">SYSTEM TERMINAL CONDITIONS</td></tr>
<tr><td colspan="2">PRESSURE (PSIA)</td><td>TEMP</td><td rowspan="4"></td></tr>
<tr><td>INITIAL</td><td>FINAL</td><td>INITIAL</td></tr>
<tr><td>P1</td><td>P</td><td>T1 (°R)</td></tr>
<tr><td>20</td><td>DATA</td><td></td><td></td><td></td></tr>
</table>

<table>
<tr><td rowspan="3"></td><td rowspan="3"></td><td colspan="3">SYSTEM COMPRESSIBILTY FACTORS</td></tr>
<tr><td>INITIAL</td><td>FINAL</td><td rowspan="2"></td></tr>
<tr><td>Z1</td><td>Z2</td></tr>
<tr><td>30</td><td>DATA</td><td></td><td></td><td></td></tr>
</table>

Table 11.29b
Program CCA Input Data Sheet (Sample Data—"Test")

PROGRAM		SYSTEM INPUT—GAS							
		CHOICE OF QUANT.							INDENT CODE
LINE NO.	NAME	MMSCFD	MOLS/HR	LB/MIN	MOL WT	cpCv	RPM	DIA IN.	
		Q (4)	Q (5)	Q (6)	M2	K3	R4	D	KS
10	DATA	36.9	0	0	20	1.283	18000	9.5	'TEST'

		SYSTEM TERMINAL CONDITIONS		
		PRESSURE (PSIA)		TEMP
		INITIAL	FINAL	INITIAL
		P1	P	T1 (°R)
20	DATA	200	200	520

		SYSTEM COMPRESSIBILTY FACTORS	
		INITIAL	FINAL
		Z1	Z2
30	DATA	1.00	0.94

Table 11.29c
Program CCA Input Data Sheet (Sample Data—"AIE")

<table>
<tr><td colspan="2">PROGRAM</td><td colspan="8">SYSTEM INPUT—GAS</td></tr>
<tr><td rowspan="2">LINE NO.</td><td rowspan="2">NAME</td><td colspan="3">CHOICE OF QUANT.</td><td colspan="4"></td><td rowspan="2">IDENT CODE</td></tr>
<tr><td>MMSCFD</td><td>MOLS/HR</td><td>LB/MIN</td><td>MOL WT</td><td>Cp/Cv</td><td>RPM</td><td>DIA IN</td></tr>
<tr><td></td><td></td><td>Q (4)</td><td>Q (5)</td><td>Q (6)</td><td>M2</td><td>K3</td><td>R4</td><td>D</td><td>KS</td></tr>
<tr><td>10</td><td>DATA</td><td>9.40</td><td>0</td><td>0</td><td>121.</td><td>1.123</td><td>5997</td><td>19.04</td><td>'AIE'</td></tr>
</table>

<table>
<tr><td rowspan="4"></td><td rowspan="4"></td><td colspan="4">SYSTEM TERMINAL CONDITIONS</td></tr>
<tr><td colspan="2">PRESSURE (PSIA)</td><td>TEMP</td><td rowspan="4"></td></tr>
<tr><td>INITIAL</td><td>FINAL</td><td>INITIAL</td></tr>
<tr><td>P1</td><td>P</td><td>T1 (°R)</td></tr>
<tr><td>20</td><td>DATA</td><td>22.0</td><td>126.7</td><td>456</td></tr>
</table>

<table>
<tr><td rowspan="3"></td><td rowspan="3"></td><td colspan="3">SYSTEM COMPRESSIBILTY FACTORS</td></tr>
<tr><td>INITIAL</td><td>FINAL</td><td rowspan="2"></td></tr>
<tr><td>Z1</td><td>Z2</td></tr>
<tr><td>30</td><td>DATA</td><td>1.00</td><td>1.00</td><td></td></tr>
</table>

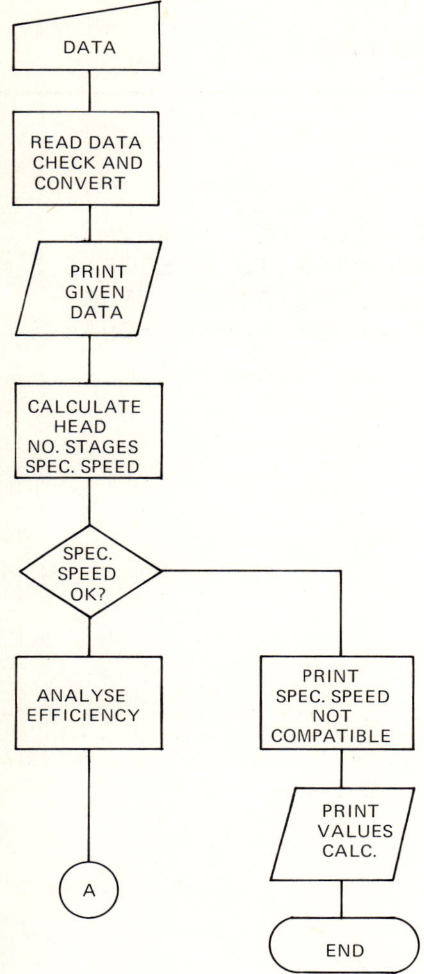

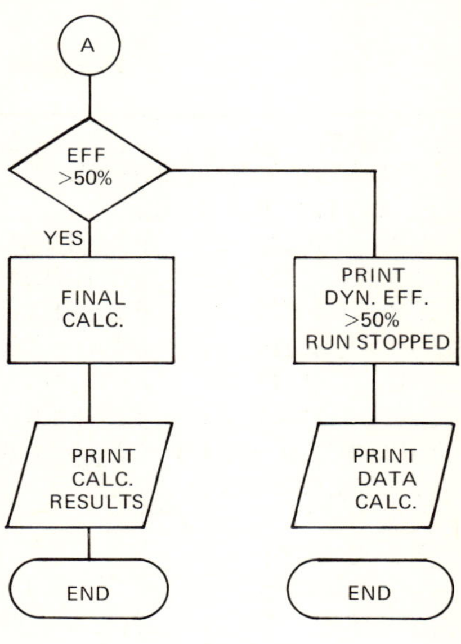

Figure 11.15. Simplified flow diagram for Program CCA.

Table 11.30
Program CCA Listing and Sample Calculation Print-Out

```
10 DATA 36.9,0,0,20,1.283,18000,9.5,'TEST'
20 DATA 200,900,520
30 DATA 1.00,0.94
100   REM 01  PROGRAM CCA
110   REM 02  PROGRAM CCA - CENTRIFUGAL COMPRESSOR BID ANALYSIS
120   REM 02  RESOLVES THE SPEED AND ROTOR DIAMETER TO SPECIFIC
130   REM 02  DIAMETER AND SPEED WHICH PRODUCES THE FIRST STAGE
140   REM 02  AND UNIT EFFICIENCY WHICH GIVES POWER AND DISCHARGE
150   REM 02  TEMPERATURE.
300 READ Q(4),Q(5),Q(6),M2,K3,R4,D,K$
310 READ P1,P,T1
320 READ Z1,Z2
330 S=(K3-1)/K3
340 K8=10
350 IF Q(4)=0 THEN 380
360 Q=Q(4)
370 GO TO 430
380 IF Q(5)=0 THEN 410
390 Q=Q(5)*0.0091
400 GO TO 430
410 IFQ(6)=0 THEN 1660
420 Q=Q(6)/(1.83*M2)
430 Q(7)=Q/0.0091
440 Q(8)=1.83*M2*Q
450 PRINT
460 PRINT USING 1600 ,'GIVEN DATA',K$
470 PRINT
480 PRINT USING 1620 ,'PLANT CAPACITY (MMSCFD)',Q
490 PRINT USING 1620 ,'PLANT CAPACITY (MOLS/HR)',Q(7)
500 PRINT USING 1620 ,'PLANT CAPACITY (LB/MIN)',Q(8)
510 PRINT USING 1620 ,'GAS MOL WEIGHT',M2
520 PRINT USING 1610 ,'RATIO OF SPECIFIC HEATS,K3,',K3
530 PRINT USING 1610 ,'COMPRESSION EXPONENT,S,',S
540 PRINT USING 1640 ,'SPEED, R4, (RPM)',R4
550 PRINT USING 1620 ,'ROTOR DIAMETER (INCHES)',D
560 PRINT USING 1630 ,'SUCTION PRESS, P1, (PSIA)',P1
570 PRINT USING 1630 ,'DISCHARGE PR, P, (PSIA),',P
580 PRINT USING 1640 ,'SUCTION TEMP, T1, (°R),',T1
590 PRINT USING 1610 ,'COMPR Z-SUCTION, Z1,',Z1
600 PRINT USING 1610 ,'COMPR Z-DISCHARGE, Z2,',Z2
610 PRINT
620 PRINT
630 PRINT
640 PRINT USING 1600 ,'CALCULATED RESULTS'
650 PRINT
660 K8=K8+19
670 R=P/P1
680 H(1)=(R↑S-1)*772.5*(Z1+Z2)*T1/(M2*S)
```

Table 11.30 *(continued)*

```
690 S2=INT(H(1)/(15000-1500*(M2)↑0.35))+1
700 H(2)=H(1)/S2
710 V(1)=10.73*T1*Z1/(M2*P1)
720 Q(1)=1.83*Q*M2*V(1)
730 W(1)=Q*M2/32.8
740 Q(2)=V(1)*W(1)
750 N(3)=R4*Q(2)↑0.5/H(2)↑0.75
760 D(3)=D*H(2)↑0.25/(12*Q(2)↑0.5)
770 X=D(3)*N(3)
780 IF X>160 THEN 810
790 IF X<140 THEN 810
800 GO TO 910
810 PRINT 'SPECIFIC SPEED NOT COMPATIBLE. RUN STOPPED'
820 PRINT USING 1610 ,'CAPACITY (CFS)',Q(2)
830 PRINT USING 1630 ,'TOTAL ADIABATIC HEAD (FT-LB/LB)',H(1)
840 PRINT USING 1640 ,'NUMBER OF STAGES',S2
850 PRINT USING 1630 ,'ADIABATIC HEAD/STG (FT-LB/LB)',H(2)
860 PRINT USING 1620 ,'SPECIFIC SPEED', N(3)
870 PRINT USING 1610 ,'SPECIFIC DIAMETER', D(3)
880 PRINT USING 1620 ,'X=(NS*DS)',X
890 K8=K8+7
900 GO TO 1680
910 N(5)=N(3)
920 GO SUB 950
930 E(1)=E(5)
940 GO TO 1210
950 IF N(5)<90 THEN 990
960 IF N(5)>130 THEN 990
970 E(5)=82
980 RETURN
990 IF N(5)<75 THEN 1030
1000 IF N(5)>150 THEN 1030
1010 E(5)=80
1020 RETURN
1030 IF N(5)<60 THEN 1070
1040 IF N(5)>185 THEN 1070
1050 E(5)=75
1060 RETURN
1070 IF N(5)<40 THEN 1110
1080 IF N(5)>210 THEN 1110
1090 E(5)=70
1100 RETURN
1110 IF N(5)<30 THEN 1150
1120 IF N(5)>250 THEN 1150
1130 E(5)=60
1140 RETURN
1150 IF N(5)<25 THEN 1190
1160 IF N(5)>270 THEN 1190
1170 E(5)=50
1180 RETURN
1190 PRINT 'DYNAMIC EFFICIENCY < 50 PERCENT. RUN STOPPED.'
```

(Table 11.30 continued on p. 168)

Table 11.30 (*continued*)

```
1200 GO TO 820
1210 E(3)=E(1)*(0.95*(S2-1)+1)/S2
1220 T2=T1*(R↑(100*S/E(1)))
1230 T4=T2-460
1240 L(2)=W(1)*H(1)/550
1250 L(3)=100*L(2)/E(1)
1260 L(1)=L(3)+(L(3)↑0.4)
1270 Q(3)=Q(1)/R4
1280 K(1)=223.*((K3*T1/M2)↑0.5)
1290 K(2)=223.*((K3*T2/M2)↑0.5)
1300 D(5)=(Q(1)*144*4/(60*0.1*K(1)*3.1416))↑0.5
1310 U(4)=R4*D/229
1320 K(4)=U(4)/K(2)
1330 Q(3)=Q(1)/R4
1340 S3=H(2)*32.2/U(4)↑2
1350 PRINT USING 1640 ,'TOTAL ADIABATIC HEAD (FT-LB/LB)',H(1)
1360 PRINT USING 1640 ,'NUMBER OF STAGES',S2
1370 PRINT USING 1640 ,'ADIABATIC HEAD/STG (FT-LB/LB)',H(2)
1380 PRINT USING 1630 ,'ADIABATIC H P',L(2)
1390 PRINT USING 1630 ,'GAS H P',L(3)
1400 PRINT USING 1630 ,'BHP',L(1)
1410 PRINT USING 1620 ,'SPECIFIC DIA,DS,',D(3)
1420 PRINT USING 1620 ,'SPEC SPEED, NS,',N(3)
1430 PRINT USING 1610 ,'PRESSURE COEFF. S3',S3
1440 PRINT USING 1620 ,'X (=NS*DS)', X
1450 PRINT USING 1610 ,'FIRST STAGE DYN EFF (PERCENT)',E(1)
1460 PRINT USING 1610 ,'UNIT DYNAMIC EFF (PERCENT)',E(3)
1470 PRINT USING 1630 ,'INLET CAPACITY,(ACFM)',Q(1)
1480 PRINT USING 1620 ,'DIAMETER SUCT NOZZLE (INCHES)',D(5)
1490 PRINT USING 1610 ,'CAPACITY INDEX,Q/N, (ACFM/RPM),',Q(3)
1500 PRINT USING 1620 ,'DISCHARGE TEMP, T4,(°F)', T4
1510 PRINT USING 1620 ,'SUCTION SONIC VEL (FPS)',K(1)
1520 PRINT USING 1620 ,'DISCHARGE SONIC VEL (FPS)',K(2)
1530 PRINT USING 1620 ,'ROTOR TIP SPEED (FPS)',U(4)
1540 PRINT USING 1610 ,'MACH NUMBER OF TIP SPEED', K(4)
1550 PRINT
1560 PRINT
1570 PRINT
1580 PRINT'              --END OF CALCULATIONS--'
1590 K8=K8+24
1600:    ##########################    #####
1610:    ##############################          ###.###
1620:    ##############################         ####.##
1630:    ##############################        #####.#
1640:    ##############################       ######.
1650 GO TO 1680
1660 PRINT 'INPUT DATA FOR Q IS MISSING. PLEASE CHECK. RUN STOPPED.'
1670 GO TO 1720
1680 K7=66-K8+6
1690 FOR I=1 TO K7
1700 PRINT
1710 NEXT I
1720 END
```

Table 11.30 (*concluded*)

```
GIVEN DATA                              TEST

PLANT CAPACITY (MMSCFD)                      36.90
PLANT CAPACITY (MOLS/HR)                   4054.94
PLANT CAPACITY (LB/MIN)                    1350.54
GAS MOL WEIGHT                               20.00
RATIO OF SPECIFIC HEATS,K3,                   1.283
COMPRESSION EXPONENT,S,                       0.221
SPEED, R4, (RPM)                          18000.
ROTOR DIAMETER (INCHES)                       9.50
SUCTION PRESS, P1, (PSIA)                   200.0
DISCHARGE PR, P, (PSIA),                    900.0
SUCTION TEMP, T1, (°R),                     520.
COMPR Z-SUCTION, Z1,                          1.000
COMPR Z-DISCHARGE, Z2,                        0.940


CALCULATED RESULTS

TOTAL ADIABATIC HEAD (FT-LB/LB)           69498.
NUMBER OF STAGES                              7.
ADIABATIC HEAD/STG (FT-LB/LB)              9928.
ADIABATIC H P                              2843.1
GAS H P                                    3467.2
BHP                                        3493.3
SPECIFIC DIA,DS,                              1.41
SPEC SPEED, NS,                             101.39
PRESSURE COEFF. S3                            0.573
X (=NS*DS)                                  143.01
FIRST STAGE DYN EFF (PERCENT)                82.000
UNIT DYNAMIC EFF (PERCENT)                   78.486
INLET CAPACITY,(ACFM)                      1883.9
DIAMETER SUCT NOZZLE (INCHES)                 6.69
CAPACITY INDEX,Q/N, (ACFM/RPM),               0.105
DISCHARGE TEMP, T4,(°F)                     319.32
SUCTION SONIC VEL (FPS)                    1287.97
DISCHARGE SONIC VEL (FPS)                  1576.74
ROTOR TIP SPEED (FPS)                       746.72
MACH NUMBER OF TIP SPEED                      0.474


        --END OF CALCULATIONS--
```

Table 11.31
Gas Compression
Glossary of Symbols Used in Computer Programs

SYMBOL	DESCRIPTION	UNITS	PROGRAM
A(J,1)	Piston/Valve—Area Ratio	decimal	All Piston
A(J,2)	Adjusted P/V—Area Ratio	decimal	RKBCS
A(J,3)	Net Eye Area	square inches	NSG,NSQ
A(J,4)	Area (Hub)	square inches	NSG,NSQ
A(J,5)	Area (Eye + Hub)	square inches	NSG,NSQ
B(J,1)	Bore of Cylinder	inches	All Piston
B(J,2)	Suction Valve Loss	decimal	All Piston
B(J,3)	Discharge Valve Loss	decimal	All Piston
B(J,(3 + N1))	Discharge Valve Loss	decimal	RKCPC
B(J,4)	Intrinsic Loss Factor	decimal	All Piston
B(J,(8 + N1))	Intrinsic Loss Factor	decimal	RKCPC
B(J,5)	Temporary Bore (Parallel Cylinders)	inches	RKBCS
B(J,6)	Final Adjusted Cylinder Bore	inches	RKBCS
B(J,7)	Temporary Bore (Two Parallel)	inches	RKBCS
B(J,8)	Adjusted Suction Valve Loss	decimal	RKBCS
B(J,9)	Adjusted Discharge Valve Loss	decimal	RKBCS
B(J,10)	Adjusted Intrinsic Factor	decimal	RKBCS
B(J,11)	Temporary Bore (Three Parallel)	inches	RKBCS
B(J,12)	Finalized Bore	inches	RKBCS
C(J,1)	Cylinder Clearance	decimal	PBS,RKBCS,RKCPC
C(J,2)	Adjusted Cylinder Clearance	decimal	RKBCS
C(J,4)	Meridional Velocity	fps	NSG, NSQ
C(J,6)	Cylinder Clearance	decimal	CPC, RKCPC
C(J,7)	Required Pressure Coefficient	decimal	All Centrifugal
C2	Critical Pseudo Pressure	psia	RKBCS, RKCPC
D	Impeller Diameter	inches	All Centrifugal
D1, D(1)	Adjusted Impeller Diameter	inches	CKNP
D2	Critical Pseudo Temperature	°R	RKBCS, RKCPC
D8	Hub Diameter	inches	NSQ,NSG
D(3)	Specific Diameter		All Centrifugal
D(4)	Adjusted Specific Diameter		CKNP
D(5)	Diameter Suction Nozzle	inch	All Centrifugal
D(J,1)	Critical Pressure Array	psia	All Piston
D(J,2)	Critical Temperature Array	°R	RKBCS,RKCPC
D(J,3)	Specific Diameter		NSG,NSQ
D(J,4)	Impeller Diameter	inches	NSG,NSQ
D(J,7)	Eye Diameter	inches	NSG,NSQ
E(1)	Dynamic Efficiency	percent	CKNP,CCA
E(2)	Adjusted Dynamic Efficiency	percent	CKNP
E(3)	Composite Multistage Efficiency	percent	CKNP,CCA
E(4)	Adjusted Composite Efficiency	percent	CKNP
E(5)	Temporary Efficiency	percent	CKNP,CCA
E(6)	Adjusted First Stage Efficiency	percent	CKNP
E(J,1)	Compression Efficiency	decimal	All Piston
E(J,N1)	Array of Compression Efficiency	decimal	RKCPC,P6S
E(J,1)	First Stage Efficiency	percent	NSG,NSQ
E(J,2)	Adjusted Compression Efficiency	percent	All Centrifugal
E(J,3)	Composite Multistage Efficiency	percent	NSG,NSQ
E(J,4)	Adjusted Multistage Composite Efficiency	percent	NSG,NSQ
E(J,5)	Sub-Routine First Stage Efficiency	percent	NSG,NSQ
E(J,6)	Adjusted First Stage Efficiency	percent	NSG,NSQ
F(J,1)	Volumetric Efficiency	decimal	All Piston

Table 11.31 (*continued*)

F(J,N1)	Volumetric Efficiency	decimal	RKCPC,P6S
F(J,2),(J,3)	Temporary Displacement 2 or 3 Par Cylinder	cfm/unit	RKBCS
F(J,7)	Adjusted Volumetric	decimal	RKBCS
F(J,8)	Adiabatic Horsepower	hp	UOHP
F(J,9)	Gas Horsepower	hp	UOHP
G(J,1)	Gas Number in Reference Data File	digit	All RK
H2	Mol Specific Heat Compression Gas	Btu/mol	All RK
H(J,1)	Adiabatic Head	feet	All Units
H(J,2)	Mol Specific Heat, Each Gas	Btu/mol	RKZ
H(J,2)	Adiabatic Head/Stage	ft lb/lb	MUCS
H(J,3)	Adiabatic Head	ft lb/lb	PBS
H(J,(3 + N1))	Adiabatic Head Array	ft lb/lb	RKCPS,P6S
H(J,(8 + N1))	General Form 0.8 to 1.2 Design Point	digit	RKCPC,P6S
I,J,K	Counters	digit	General
I5	Indicator	digit	P6S,RKBCS
I(J)	Counter	digit	RKBCS
J(J)	Counter	digit	RKBCS
K1	Number of Gases	digit	All RK
K2	Gas Counter		All RK
K3	Ratio of Specific Heats		All
K4	Stage Counter		All RK
K5	Mechanical Efficiency	decimal	All
K5	Casing Counter	digit	MUCS
K6	Page Counter	digit	RKCPC
K6	Line Counter	digit	RKBCS
K7,K8	Page, Line Counter	digit	All
K9	$K9 = (1/K3)$ Input		P6S
K(1)	Suction Sonic Velocity	fps	CKNP,CCA
K(2)	Discharge Sonic Velocity	fps	CKNP,CCA
K(4)	Mach Number of Tip Speed	decimal	RKBCS
K(J)	Bore Change Counter		RKBCS
K$	Literal Identification		All Centrifugal
K(8,1)	Mach Suction Velocity	decimal	NSQ,NSG
K(J,2)	Mach Discharge Velocity	decimal	NSQ,NSG
K(J,3)	Mach Tip Speed	decimal	NSQ,NSG
K(J,4)	Adjusted Impeller Tip Speed	fps	NSQ
K(8,4)	Suction Mach Velocity	decimal	NSQ
L1	Total Station Bhp	Bhp	All Piston
L2	Adjusted Total Station Bhp	Bhp	RKBCS
L3	Adiabatic Horsepower	hp	CKNP
L4	Gas Horsepower	hp	CKNP
L(1)	Brake Horsepower	Bhp	CKNP,CCA
L(2)	Adiabatic Horsepower	hp	CKNP,CCA
L(3)	Gas Horsepower	hp	CKNP,CCA
L(J,1)	Tension Rod Thrust	pounds	All Piston
L(J,2)	Calculated Bhp/Cyl	Bhp	All Piston
L(J,2)	Adiabatic Horsepower	hp	NSQ
L(8,2)	Adiabatic Horsepower	hp	NSG
L(J,3)	Calculated Bhp/unit	Bhp	RKBCS
L(J,3)	Calculated Bhp	Bhp	
L(J,4)	Adiabatic hp/Stage	hp	CPC,PBS,RKBCS
L(J,5)	Adjusted Bhp/Stage	hp	RKBCS

(*Table 11.31 continued on p. 172*)

Table 11.31 (*continued*)

L(J,6)	Adjusted Rod Thrust	M-lb	RKBCS
L(J,7)	Gas Horsepower	hp	NSG
L(J,8)	Unit Stress-Piston Rod	psi	All Piston
L(J,N1)	Tension Rod Thrust	M-lb	RKCPC,P6S
L(7,N1)	Total Station Array	Bhp	RKCPC,P6S
L(7,1)	Total Bhp—All Stages	Bhp	UOHP
L(J,(5 + N1))	Calculated Bhp/Cyl Array	Bhp	RKCPC,P6S
L(J,(10 + N1))	Calculated Bhp/Stage Array	Bhp	RKCPC
L(J,(15 + N1))	Unit Stress-Piston Rod	psi	RKCPC,P6S
M2	Mol Weight of Gas Handled	digit	All
M(J,1/10)	Compressibility Z Calculation (Array)	digit	All RK
M(J,11)	Mol Fraction	decimal	All RK
N1	Counter-H(*J*,*N*1) Series	digit	P6S,RKCPC
N(3)	Specific Speed	digit	CKNP,CCA
N(4)	Adjusted Specific Speed	digit	CKNP
N(5)	Temporary Specific Speed	digit	CKNP
N(J,1/4)	Compressibility Array		All RK
N(J,3)	Specific Speed		NSQ,NSG
N(J,5)	Sub-routine Specific Speed		NSQ,NSG
N(J,6)	Adjusted Specific Speed		NSQ,NSG
O$(J)	Literal Name Variable		All RK
O(I,J)	Gas Data Array		All RK
O(G(J,1),2)	Gas Mol Weight Variable		All RK
O(G(J,1),3)	Gas Critical Temperature Variable	°R	All RK
O(G(J,1),4)	Gas Critical Pressure Variable	psia	All RK
O(G(J,1),5)	Gas Mol Specific Heat Variable	digit	All RK
P	Final System Pressure	psia	All
P1	Ambient Pressure	psia	P6S
P1	Suction Pressure	psia	All
P2	Discharge Pressure	psia	All
P(J,1)	Suction Pressure	psia	All Piston
P(J,2)	Discharge Pressure	psia	All Piston
P(J,4)	Discharge Pressure (+2,000)	psia	All RK
P(J,(5 + N1))	Suction Pressure Array	psia	RKCPC,P6S
Q	Plant Capacity	MMscfd	All
Q(1)	Actual Capacity	acfm	CKNP,CCA
Q(2)	Actual Capacity	cfs	CKNP,CCA
Q(3)	Capacity Index	cf/rev	CKNP,CCA
Q(4)	Plant Capacity	MMscfd	CKNP,CCA
Q(5)	Plant Capacity	mols/hr	CKNP,CCA
Q(6)	Plant Capacity	lb/min	CKNP,CCA
Q(7)	Plant Capacity~*Q*5	mols/hr	CKNP,CCA
Q(8)	Plant Capacity~*Q*6	lb/min	CKNP,CCA
Q(J,1)	Cylinder Displacement/Unit	cfm/cyl	All Piston
Q(J,1)	Actual Capacity	acfm	NSQ
Q(6,1)	Actual Capacity	acfm	NSG
Q(J,2)	Total Capacity/Unit	acfm	All Piston
Q(J,2)	Capacity	cfs	NSQ
Q(6,2)	Capacity	cfs	NSG
Q(J,3)	Each Cylinder Capacity	acfm	RKBCS,CPC
Q(J,4)	Adjusted Capacity Each Cylinder	acfm	RKBCS
Q(J,6)	Capacity Stage	MMscfd	PBS,UOHP

Table 11.31 (*continued*)

Q(J,7)	Flow Index, Q/N	cfm/rpm	NSQ,NSG
Q(J,8)	Capacity	MMscfd	NSQ
Q(6,8)	Plant Capacity	MMscfd	NSG
Q(6,9)	Plant Capacity	mols/hr	NSG
Q(6,10)	Plant Capacity	lb/min	NSG
Q(J,11)	Plant Capacity	mols/hr	NSQ
Q(6,11)	Plant Capacity	mols/hr	NSG
Q(J,12)	Plant Capacity	lb/min	NSQ
Q(6,12)	Plant Capacity	lb/min	NSG
Q(J,14)	Choke Capacity	cfs	NSQ,NSG
Q(J,15)	Surge Capacity	cfs	NSQ,NSG
R	Ratio of Compression	digits	All
R1	Compression Ratio/Stage, Casing #1	digits	MUCS
R2	Compression Ratio/Stage, Casing #2	digits	MUCS
R3	Compression Ratio/Stage, Casing #3	digits	MUCS
R4	Revolution/Minute	rpm	All
R5	Rod Diameter	inches	All Piston
R6	Piston Stroke	inches	All Piston
R(6)	Adjusted Speed	rpm	NSQ
R(J,1)	Ratio of Compression	digits	All Piston
R(1,1)	Applied Compression Ratio/System	digits	MUCS
R(2,1)	Applied Compression Ratio/System	digits	MUCS, after CSG #1
R(3,1)	Applied Compression Ratio/System	digits	MUCS, after CSG #2
R(J,(1 + N1))	Compression Ratio Array	digits	RKCPC
R(J,(5 + N1))	Compression Ratio Array	digits	P6S
R(J,4)	Calculated Speed	rpm	NSG
R(J,5)	Rod Diameter	inches	P6S
R(J,6)	Adjusted Impeller Speed	rpm	NSG
S	Adiabatic Exponent	decimal	All
S2	Number of Stages	digit	All
S3	Pressure Coefficient	decimal	All Centrifugal
S4	Gas Head Slip	decimal	All Centrifugal
S(J,K)	Number of Stages and Stage/Casing	J = 1—3, K = 1—6	MUCS
S(J,5)	Ratio Impeller/Eye Diameter	digits	NSG,NSQ
T1	Initial Suction Temperature	°F	All RK
T1	Initial Suction Temperature	°R	All
T2	Discharge Temperature	°R	All
T3	Suction Temperature, Casing #2	°R	MUCS
T4	Discharge Temperature	°F	CKNP,CCA
T5	Suction Temperature, Casing #3	°R	MUCS
T(1,1)	Initial Suction Temperature	°R	CPC
T(J,1)	Suction Temperature/Stage	°R	All Piston & RK
T(J,2)	Discharge Temperature	°R	All RK
T(J,3)	Suction Temperature/Stage	°F	All RK
T(J,4)	Discharge Temperature/Stage	°F	All Piston
T(J,6)	Discharge Temperature	°F	RKBCS,RKZ
T(J,6)	Discharge Temperature	°R	NSQ,NSG
T(J,8)	Discharge Temperature	°F	NSQ,NSG
T(J,(8 + N1))	Discharge Temperature	°F	P6S,RKCPC

(*Table 11.31 continued on p. 174*)

Table 11.31 (*concluded*)

U	Piston Speed and Rotor Tip Speed	fps	All
U1	Bhp/Unit	Bhp	All Piston
U3	Number of Units/Station	digit	All Piston
U4	Number of Cylinders/Unit	digit	All Piston
U5	Rated hp/Cylinder	hp	All Piston
U6	Trial Number of Units	digit	RKBCS
U7	Adjusted Trial Number of Units	digit	RKBCS
U8	Adjusted Trial Number of Units	digit	RKBCS
U(4)	Rotor Tip Speed	fps	All Centrifugal
U(J,1)	Cylinder in Parallel	digit	RKCPC
U(J,1)	Piston Speed	fps	P6S
U(J,2)	Alternate Cylinder Displacement	cfm	RKBCS
U(J,3)	Impeller Tip Speed	fps	NSG
U(1,3)	Impeller Tip Speed	fps	NSQ
U(2,3)	Adjusted Impeller Tip Speed	fps	NSQ
U(J,4)	Adjusted Impeller Tip Speed	fps	NSG
V(1)	Suction Specific Volume	cf/lb	All Centrifugal
V(J,1)	Suction Specific Volume	cf/lb	All Piston
V(J,N1)	Suction Specific Volume	cf/lb	RKCPC,P6S
W	Weight Flow/Stage	lb/min	RKBCS
W(1)	Weight Flow	lb/sec	CKNP,CCA,NSG
W(J,1)	Weight Flow/Stage	lb/min	CPC,UOHP
W(J,1)	Weight Flow/Stage	lb/sec	NSQ
W(J,3)	Weight Flow/Cylinder	lb/min	CPC
W(J,N1)	Weight Flow/Stage	lb/min	RKCPC,P6S
W(J,6)	Weight Flow/Stage	lb/min	PBS
X	Product *Ns*Ds*	digits	All Centrifugal
X1	Compression Ratio, Casing #1	digits	MUCS
X2	Compression Ratio, Casing #2	digits	MUCS
X3	Compression Ratio, Casing #3	digits	MUCS
X(J,1)	Compression Ratio/Casing	digits	MUCS
X(J,1)	Product *NS*DS*	digits	NSG,NSQ
X(J,1)	Discriminate in *Z*1	digits	All RK
X(J,2)	Discriminate in *Z*2	digit	All RK
Y1	Cumulative Gas Composition	percent	All RK
Y(J,1)	Mol Fraction Array	decimal	All RK
Y(J,2)	Mol Fraction Array	percent	All RK
Z1	System Initial Compressibility	decimal	All
Z2	System Final Compressibility	decimal	All
Z(J,1)	Suction Compressibility/Stage	decimal	All Piston
Z(J,2)	Discharge Compressibility/Stage	decimal	All Piston

Computer Program Index

Computer Program Index continued on p. 176

Computer Program Index (concluded)

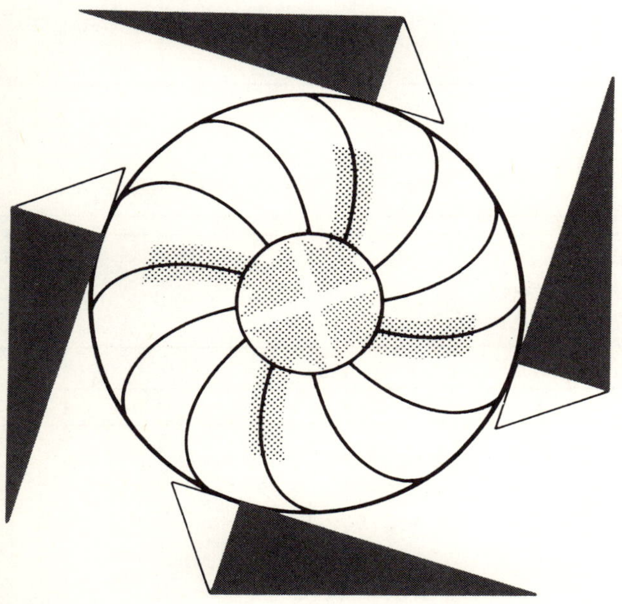

12 Application Calculation Sheets

The following seven charts, Tables 12.1 through 12.7, outline the sequence of calculations to size or determine the capabilities of piston, centrifugal and rotary compressors. This chapter also includes a method for sizing a *radial inflow* (RIF) turbo-expander and a method of sizing nominal pulse dampers.

The technique for these procedures are generally remote and obscure. Making such applications occurs so seldom that the technician may become inept. These procedural outlines offer a method for resolving these problems.

Equations used in more than one chart are given in the Calculation Glossary, Table 12.8. The identifying symbols and equation styles are in the same glossary. They closely follow the pattern used in the computer glossary.. The equations are given in a direct comprehensive form. The derivation or development for most of this material is given in earlier chapters.

Blank form calc sheets are included for those who may want to use them for application evaluations. Detailed calculations are shown adjacent to the example case in column 1 on the second calc sheet.

Calc Sheet No. 12.1

Table 12.1 develops the capacity of a 30-inch cylinder having a 20-inch stroke as 9.51 MMscfd when running 277 rpm. The valve losses bring the ideal adiabatic horsepower from 575 to 733 bhp for the load on the driver. The analysis continues by developing the piston rod tension thrust load and the rod stress. The analysis is completed with the discharge temperature.

Calc Sheet No. 12.2

Table 12.2 develops the number of parallel and size cylinders required for a given assignment. The example shows the method of reducing large cylinder requirement to a commercial size. The compromise selection is an array of 12 parallel 27.5-inch cylinders. While a 30-inch bore is presumed to be the largest diameter cylinder capable of 100 psig operation, a 27.5-inch bore was selected to avoid a non-extendable installation. The remaining details are determined from the glossary equations and in descriptive notes following the calc sheet.

Table 12.1a
Piston Compressor Capabilities

Item	Calculations	Case 1	2	3	4
01	*R* (*P*/*P*1)				
02	*F*1 (*Vol. Eff.*)				
03	*Q*1 (*Displ.*), cfm				
04	*Q*3 (*Q*1 x *F*1), acfm				
05	*W*3 (*Q*3/*V*1), ppm				
06	*Q* (0.547 x *W*3/*M*2)				
07	*B*2 (*Suct. Valve Loss*)				
08	*B*3 (*Disch. Valve Loss*)				
09	*B*/*E*3 (*Intrinsic Factor* / *Compr. Eff.*)				
10	*H* (*Head*), ft				
11	*L*2/bhp				
12	*L*1 (*Rod Thrust*), lb				
13	*L*6 (*Rod Stress*), psi				
14	*T*2 (*Disch. Temp.*), °F				
15	Remarks				

Note:
Gas: ____ M2, ____ K3, ____ S, ____ Z1, ____ Z, ____ T1, ____ R5, ____ R6
Terminal: ____ P1, ____ P, ____ V1, ____ A1, ____ U, ____ R4, ____ B1, ____ C1,

Table 12.1b
Piston Compressor Capabilities
(85 psig Wet Gas Booster Power Analysis)

Item	Calculations	Case 1	2	3	4
01	*R* (*P*/*P*1)	3.33	*100/30 (Expanded calcs)*		
02	*F*1 (*Vol. Eff.*)	72.0	*115-15 (1.1) 2.62*		
03	*Q*1 (*Displ.*), cfm	4,520	*0.327 (900-4.5) 15.4*		
04	*Q*3 (*Q*1 x *F*1), acfm	3,260	*4,500 (0.72)*		
05	*W*3 (*Q*3/*V*1), ppm	350	*3,260/9.33*		
06	*Q* (0.547 x *W*3/*M*2)	9.56	*0.547 (350)/20*		
07	*B*2 (*Suct. Valve Loss*)	0.131	*144 (237) 20/520 (10)*^{4}		
08	*B*3 (*Disch. Valve Loss*)	0.103	*0.131/1.272*		
09	*B*/*E*3 (*Intrinsic Factor* / *Compr. Eff.*)	1.27/81.5	*1.103/0.869; .272/.324*		
10	*H* (*Head*), ft	54,600	*0.272 (1545) 520/20* 0.2*		
11	*L*2/bhp	579/737	*350 (54,600)/33,000*		
12	*L*1 (*Rod Thrust*), lb	48,900	*707 (100-30) —7.1 (100-15)*		
13	*L*6 (*Rod Stress*), psi	6,900	*48,900/7.1*		
14	*T*2 (*Disch. Temp.*), °F	201	*1.272 (520)—460*		
15	Remarks		Load and stress are nominal		

Note:
Gas: 20 M2, 1.250 K3, 0.200 S, 1.00 Z1, 1.00 Z, 520 T1, 3 R5, 20 R6,
Terminal: 30 P1, 100 P, 9.33 V1, 12 A1, 15.4 U, 277 R4, 30 B1, 15 C1,

Table 12.2a
Piston Compressor Sizing

Item	Calculation	Case 1	2	3	4
01	R, ($P/P1$)				
02	$F1$, (*Vol. Eff.*)				
03	$W3$, (1.83 x Q x $M2$), ppm				
04	$Q2$, ($W3$ x $V1$), acfm				
05	$Q1$, ($QZ/F1$), cfm				
06	$B1$, (*Cyl. Bore*), in.				
07	$B6/U4$, (*Adj. Cyl. Bore, In.* / *No Parallel Cyl.*)				
08	H, (*Head*), ft				
09	$B2/B3$, (*Suct. Valve Loss* / *Disch. Valve Loss*)				
10	$B/E3$, (*Intrinsic Factor* / *Compr. Eff.*)				
11	$L1$, (*Rod Thrust*), lb				
12	$L2/L5$, (*Ad. hp.* / *Bhp*)				
13	$L6$, (*Rod Stress*), psi				
14	$T2$, (*Disch. Temp.*), °F				
15	Remarks				

Note:

Gas: ____M2, ____K3, ____S, ____Z, ____Z, ____T1,
____P1, ____P, ____Q, ____A1, ____U, ____R4,
____U1, ____R5, ____C1, ____V1, ____R6

Table 12.2b
Piston Compressor Sizing
(Selection of Cylinders for Gas Booster)

Item	Calculation	Case 1	Case 2	Case 3	Case 4
01	*R*, (*P*/*P*1)	3.33	*100/30 (expanded calcs)*		
02	*F*1, (*Vol. Eff.*)	75.5	*113–13 (1.1) 2.62*		
03	*W*3, (1.83 x *Q* x *M*2), ppm	3,660	*1.83 (100) 20*		
04	*Q*2 (*W*3 x *V*1), acfm	34,200	*3,660 (9.33)*		
05	*Q*1 (*Q*2/*F*1), cfm	45,250	*34,200/0.755*		
06	*B*1 (*Cyl. Bore*), in.	95	$((45{,}250 * 3.06/15.4) + 4.5)^{0.5}$		
07	*B*6, (*Req* 10 *Max WP B*1 30 / *U*4 12 *Adj Q*2 3770)	27.5	$((3{,}770/5.04) + 4.5)^{0.5}$		
08	*H*, (*Head*), ft	54,500	*0.272 (1,545) 520/0.2 (20)*		
09	*B*2/*B*3, (*Suct. Valve Loss* / *Disch. Valve Loss*)	091/.072	*100 (237) 20/520* $(10)^4$		
10	*B*/*E*3, (*Intrinsic Factor* / *Compr. Eff.*)	1.18/86.5	*1.072/0.909; .272/.315*		
11	*L*1, (*Rod Thrust*), lb	40,900	*593 (100 − 30) − 7.1 (100 − 15)*		
12	*L*2/*L*5, (*Ad. Hp.* / *bhp*)	6050/7000	*54,500 (3660)/33,000*		
13	*L*6, (*Rod Stress*), psi	5,770	*40,900/7.1*		
14	*T*2, (*Disch. Temp.*), °F	201	*1.272 (520) − 460*		
15	Remarks				

Note:

Gas:	20 M2,	1.250 K3,	0.200 S,	1.00 Z,	1.00 Z,	520 T1,
Terminal:	30 P1,	100 P,	100 Q,	10 A1,	15.4 U,	277 R4,
		5000 U1,	3.0 R5,	13 C1,	9.33 V1,	20 R6,

Item Description: Table 12.2

06 A 44-inch compressor cylinder is as large as practical to manufacturer. These large cylinders are usually restricted to vacuum service where the pressure differential does not exceed 25 psia. The cylinder working pressures are usually established as a result of hydrostatic testing. There is no design criteria that governs the maximum working pressure. The cylinder bore for 100-psig service is usually limited to about 30 inches and smaller. Whereas this example calls for a 95-inch diameter cylinder, it is an unrealistic requirement.

07 A 30-inch cylinder has a displacement of 4,490 cfm at the specified conditions. It requires 10 cylinders to satisfy the system requirement. It is good engineering practice to provide a margin for expansion of services. In lieu of providing an exact minimum cylinder count, it is better practice to provide 12 cylinders, which may be distributed among two, three, four or six units. A 5 percent clearance bottle attached to the head end of the compressor cylinder provides 5 percent unloading. The estimated power requirement for the 8.33 MMscfd per cylinder is evaluated from the approximate equation. Estimated bhp is $22\ R\ Q = 22\ (3.33)\ 8.33 = 610$ bhp.

08 The equations for all of the necessary design criteria are included in the calc sheet glossary, Table 12.28.

Calc Sheet No. 12.3

The capability of the example six-stage, 24-inch diameter centrifugal compressor in Table 12.3 is 236 MMscfd at 8,250 rpm and for the given terminal conditions. The ratio of the impeller diameter to the eye diameter is 1.86. This ratio is within good design limits. When this ratio is less than 1.3, vane lengths are short and permit excessive slippage. Conversely, when the diameter ratio is greater than 2.3, the vanes are too long and involve excessive friction. Calculation Sheet No. 12.4 illustrates this point.

Table 12.3a
Centrifugal Compressor Capabilities

Item	Calculation	Case			
		1	2	3	4
01	*R*, (*P*/*P*1)				
02	*H*, (*Ad. Head*), ft				
03	*H*3, (*Where S*2 =), ft/stg				
04	*S*4, Adj (*Where S*3)				
05	*U*, (*SQR*(32.2 x *H*3/*S*4)), fps				
06	*N*, (*U* x 229/*D*), rpm				
07	*Q*3/*W*, (*Comp. Inlet, cfs* / *Flow, pps*)				
08	*Q*/*E*1, (*Compr. Cap., MM/D* / *Dyn. Eff.*)				
09	*A*3, (144 x *Q*3/*C*4 x 0.85), sq. in.				
10	*A*5, (*A*3 + *A*4), sq. in.				
11	*D*4/*F*, (*Eye diam., in.* / *Ratio D*/*D*4)				
12	*U*/*C*4				
13	*E*2/*L*3, (*Compr. Eff.* / *Ad. hp.*)				
14	*L*5, (*bhp*)				
15	*T*2, (*Disch. Temp.*), °F				

Note:

Gas: ___M2, ___K3, ___S, ___Z1, ___Z, ___T1,

Terminal: ___P1, ___P, ___D, ___C5, ___V1, ___N3

Table 12.3b
Centrifugal Compressor Capabilities
(Wet Natural Gas Transmission, Power and Hardware)

Item	Calculation	Case 1	Case 2	Case 3	Case 4
01	R, ($P/P1$)	4.00	*1,000/250 (Expanded calcs)*		
02	H, (*Ad. Head*), ft	62,500	*0.32(1,545)520/0.2(20)*		
03	$H3$ (*Where* $S2 = 6$), ft/stg	10,400	*62,500/6; 15,000—1,500*$(20)^{0.35}$		
04	$S4$, Adj (*Where* $S3 = 0.53$)	0.45	*((0.53(.83).988)5 + 0.53)/6*		
05	U, (*SQR* (32.2 x $H3/S4$)), fps	863	$(32.2(10{,}400)/.45)^{0.5}$		
06	N, (U x 229/D), rpm	8,250	*863(229)/24*		
07	$Q3/W$, (*Comp. Inlet, cfs* / *Flow, pps*)	155/144	$(102{,}500/8{,}250)^2$ *; *155/1.08*		
08	$Q/E1$, (*Compr. Cap.* MM/D / *Dyn. Eff./Baljé*)	236/83	*144(32.8)/20*		
09	$A3$, (144 x $Q3/C4$ x 0.85), sq. in.	101	*155(144)/0.3(.863)0.85*		
10	$A5$, ($A3 + A4$), sq. in.	130	*101 + 29*		
11	$D4/F$, (*Eye diam., in* / *Ratio D/D4*)	12.9/1.86	$(1.273(/30))^{0.5}$; *24/12.9*		
12	$E2$, (*Adj. Dyn. Eff.*)	76.0	*83((.90(6—1) + 1)/6)*		
13	$L3$, (*Ad. hp.*)	16,350	*144(62,500)/550*		
14	$L5$, (*bhp*)	21,610	*(16,350/.76) + 60*		
15	$T2$, (*Disch. Temp.*), °F	289	$(520(4)^{0.263})-460$		

*$Q3$ is derived from $N3^* H^{0.75}/N = Q3^{0.5}$

Note:

Gas:	20 M2,	1.250 K3,	0.200 S,	0.95 Z1,	0.99 Z,	520 T1,
Terminal:	250 P1,	1000 P,	24.0 D,	0.30 C5,	1.08 V1,	100 N3

Calc Sheet No. 12.4

The analysis in Table 12.4 shows a wide choice of three five-stage centrifugal compressors. One machine has a low specific speed of 70 and a potential dynamic efficiency of 73 percent. A second machine has an optimum specific speed of 100 and a potential dynamic efficiency of 83 percent. A third machine has a high specific speed of 150 and 80 percent efficiency. Diameters range from 31 to 21.8 and 14.5 inches, after correcting the pressure coefficients. The diameter ratios indicate that the first and last case have poor geometry. The resultant low efficiencies mean 13 percent more power is required to operate the first case than is required for the second case. The performance data developed in the calc sheet refers to the optimum design conditions. The dynamic efficiency is at a peak for the design flow rate. The minimal or *surge* flow is generally realized at two-thirds of the design rate for machines having a tip speed of approximately 900 fps and a vane backward lay of 40°, which is the most popular design. The pressure coefficient (*S*3 or *qad*) is expected to rise about 6 percent and the efficiency drop off about 4 percent at this minimal flow. The maximum choke or Stonewall flow is expected at 1.35 to 1.45 times the design flow rate. The pressure coefficient is off about 40 percent and the dynamic efficiency is off about 25 percent from the design values.

The slope of these performance curves is steepened by a greater degree of the vane back-lay. The gas mol weight and the impeller tip speed also tend to steepen the slope of the basic characteristic curves. A flattening of these curves is the result of a lesser back-lay vane angle, a lighter gas and/or a lower tip speed. The advantage of the machine having the optimized geometry of N3 = 100 is quite obvious.

Table 12.4a
Centrifugal Compressor Sizing

Item	Calculation	Case 1	Case 2	Case 3	Case 4
01	*R*, (*P*/*P*1)				
02	*H*, (*Ad. head*), ft				
03	*H*3, (*where S*2 =), ft/stg				
04	*W*, (0.0305 x *Q* x *M*2), pps				
05	*Q*3,(*V*1 x *W*), cfs				
06	*N*3/*E*1 (*Specific Speed* / *Dyn. Eff.*)				
07	*N*, (*N*3 x *H*3↑0.75/*Q*3↑0.5), rpm				
08	*D*, (1,800 x *H*3↑0.5/*N*), in.				
09	*U*, (*Tentative*), fps				
10	*S*4, Adj. (*where S*3 =)				
11	*S*5, (*g* x *H*3/*U*↑2)				
12	*D*1, (*S*5/*S*4)↑0.5)**D*, in.				
13	*E*2, (*Corrected Eff.*)				
14	*U*, (*Corrected*), fps				
15	*A*3, (144 x *Q*3/*C*4 x 0.85), sq. in.				
16	*A*5, (*A*3 + *A*4), sq. in.				
17	*D*4/*F*, (*Eye Diam.* / *Ratio D/D*4)				
18	*L*3, (*Ad. hp.*)				
19	*L*5, (*bhp*)				
20	*T*2, (*Discharge Temp.*) °F				
21	Remarks				

Note:
Gas: ____M2, ____K3, ____S, ____Z1, ____Z, ____T1,
Terminal: ____P1, ____P, ____Q, ____V1,

Table 12.4b
Centrifugal Compressor Sizing
Sulphur Plant Furnace Air

Item	Calculation	Case 1	Case 2	Case 3	Case 4
01	*R*, (*P*/*P*1)	4.0	*56/14 (Expanded calcs)*		
02	*H*, (*Ad. Head*), ft	47,000	*0.486 (1,545) 520/29 (.286)*		
03	*H*3, (*where S*2 = 5), ft/stg	9,400	*47,000/5; 996/98/9.9*		
04	*W*, (0.0305 x *Q* x *M*2), pps	9.0	*(.0305 (10.15) 29)*		
05	*Q*3, (*V*1 x *W*), cfs	123.7	*9 (13.75); Q3↑0.5 = 11.1*		
06	*N*3/*E*1 (*Specific Speed* / *Dyn. Eff. Baljé*)	70/73	100/83	150/80	
07	*N*, (*N*3 x *H*3↑0.75/*Q*3↑0.5), rpm	6,280	8,950	13,450	
08	*D*, (1,800 x *H*3↑0.5/*N*), in.	28.1	19.8	13.2	
09	*U*, (*Tentative*), fps	770	773	777	
10	*S*4, *Adj.* (*Where S*3 = .53)	0.46	*(0.53 (.83) 4 + 0.53)/5*		
11	*S*5, (*g* x *H*3/*U*↑2)	0.506	*32.2 (9,400)/773(773)*		
12	*D*1, (*S*5/*S*4)↑0.5)**D*, in.	29.5	20.7	13.8	
13	*E*2, (*Corrected Eff.*)	67.0	76.4	73.6	
14	*U*, (*Corrected*), fps	810	*20.7 (8,950)/229*		
15	*A*3, (144 x *Q*3/*C*4 x 0.85), sq. in.	86	*144(123)/0.3(810)0.85*		
16	*A*5, (*A*3 + *A*4), sq. in.	112	*(86 + 26)*		
17	*D*4/*F*, (*Eye Diam.* / *Ratio D/D4*)	12/2.4	12/1.7	12/1.15	
18	*L*3, (*Ad. hp.*)	770	*9(47,000)/550*		
19	*L*5, (*bhp*)	1170	1025	1065	
20	*T*2, (*Disch. Temp.*) °F	480	415	430	
21	Remarks	Low Eff. Long Vanes	Best	Short Vanes	

Note: **H↑0.75, .5 & .25*

Gas:	29 M2,	1.40 K3,	0.286 S,	1.00 Z1,	1.00 Z,	520 T1
	14.0 P1,	56 P,	10.15 Q,	13.75 V1		

Calc Sheet No. 12.5

Table 12.5 undertakes a new procedure for sizing a rotary compressor. The methods given in this text are used to evaluate the adiabatic head and the capacity requirements. The charging loss *B*2 requires two velads of static energy to match the smaller rotor tip speed. The equation for the charging loss is

*B*2 = 2.5* *U*↑2* *M*2 / *T*1* 10↑5.

The exhaust loss is *B*2/*R*↑*S*. The intrinsic effect that these two losses have on the basic *R*, ratio of compression, is

B = (1 + *B*3) / (1 − *B*2).

The dynamic efficiency is

*E*3 = (*R*↑*S* − 1) / (*B** *R*↑*S* − 1).

The rotary compressor experiences a good deal of by-pass leakage unless the rotors are flooded with a sealant oil. The slippage efficiency for a dry rotary unit is *E*5 = (1. − *Ws*/*Q*1). Means for determining *Ws* is given in Table 12.5.

The slippage gas warms the incoming gas, thereby adding an extra thermal burden to the power of compression. It is evaluated by

*E*7 = 1.0 − (1. − *E*5) (0.15*R*)

The volumetric efficiency for the rotary compressor is *F*3 = *E*5**F*2. The latter is determined by

*F*2 = 1.00 − (1. − *C*1 (*R* ↑ (1/*K*3)))

The displacement of rotary compression is

*Q*1 = *D***L***U***X* = *D*8**L*8**U***X* cfm

The diameter, *D*8, and rotor length, *L*8, are both

used in inches. $D8$ is usually equal to $L8$ in high pressure casings. U is the tip speed, fps, and the displacement factor is X. The diameter for the square ($D8 = L8$) model is determined from

$$D8 = (Q1^* / U^*X)\uparrow 0.5 = L8,$$

where $L9 = 1.5\ D9$, $D9 = ((Q1^* / U^* X) / 1.5) \uparrow 0.5$ and $L9 = 1.5\ D9$. The brake horsepower is a product of

$$\text{Bhp} = H^* W3 / 33{,}000^* E3^* E5^* E7^* 0.90.$$

This sizing program was designed primarily for the helical-screw type compressor. The analysis and the equations should have equal application value for the other four types of rotary units given in Chapter 5. Because of the discretionary limitations of tip speeds to 0.12 Mach and 3.5 R for the spiral-axial machines and 0.05 Mach and 1.7 for the straight-lobe machines, the following two equations better approximate the overall efficiency, including the mechanical losses:

$E8$ (SPIRAL-AXIAL) = 1.00 − 0.15 R
$E9$ (STRAIGHT-LOBE) = 1.00 − 0.2 R
$E0$ (LIQUID-LINER) = CE (0.80 − 0.5 UO) / $SG\uparrow 2$

where UO is the tip speed, fps, and SG is the specific gravity.

The dominate friction in a liquid-liner type gas compressor is the thrashing of the sprocket rotor in the liquid. As the viscosity is increased, the CE viscosity correction is applied. The CE values given in Table 5.7 were derived from the Hydraulic Institute as indices of viscous drag for pump impellers. The frictional drag for sliding-vane compressors does not follow a consistent pattern. It is best to assume that it has a flat efficiency of 70 percent.

The discharge gas temperature from a liquid-liner compressor should not exceed the temperature of the entering coolant by more than 50° F. The discharge temperature from an oil-quenched compressor is usually limited to a 90° F rise. The discharge temperature from dry rotary compressors should reflect all of the inefficiencies except the mechanical losses.

Table 12.5a
Rotary Compressor, Sonic Slip Rating

Object: *D, L, N, HP* & *T2*

Item	Calculations:	Case: Type 1	2	3	4
01	$R = P/P1$, ($R \uparrow S$ = ______) PE				
02	$W3 = (1.83^*Q^*M2)$ ppm				
03	$Q2$, $W3^*V1$ acfm				
04	$H1$, $(PE - 1)1545^*T1/M2^*S$				
	$B2$, $2.5(M2)\ U^2/T(10)^5$				
05	$B3$, $B2/PE$				
	B, Intrinsic Factor				
06	$E3$, $(PE - 1)/(BR \uparrow S - 1)$				
	$Ws = RZ^*36^*A(K3^*T2/M2)^{0.5}$				
07	$E5$, *SLIP EFF*, $(1 - Ws/Q1)$				
08	$E7$, *Thermal Effect*				
09	$F2$, $(1 - C(R \uparrow 1/K3)$				
10	$F3$, *Overall Vol Eff* $(E5^*F2)$				
11	$Q1$, $(Q2/F3)$ *Displ.* cfm				
12	*D, Rotor L = D*				
13	*D, Rotor L* = 1.5*D*				
14	*N cor, with STD D* x *L*				
15	$L3$, $H1^*W3/33{,}000$ hp				
16	$L4$, *Gas* hp. $L3/E3^*E5^*E7$				
17	$L5$, *Brake* hp				
18	$T2$, $(T1\ (R \uparrow S - 1)/E3^*E7) + t1$				

Note:

Gas ___Q ___M2 ___K3 ___S ___V1
___P1 ___P ___T1 ___U ___A ___G
___X ___N ___D ___L ___C

$Rz = R((K3 + 1)/2) \uparrow (1/S)$
$A = 2\ G^*L + 15\ G^*D/nc$, where nc = number of rotor cells,
nc = 4 (SRM), 12 (R − V), 6 to 24 (S − V)

Table 12.5b
Rotary Compressor, Sonic-Slip Rating
Object: *D, L, N, HP* & *T2*

Item	Calculations	Case: Type: Helical Screw 1	2	3	4
01	*R* = *P/P*1, (*R* ↑ *S* = 1.272) *PE*	3.33	*(50/15) Example*		
02	*W*3, (1.83* *Q***M*2), ppm	350	*1.83 (9.57) 20*		
03	*Q*2, *W*3**V*1, acfm	6500	*1.86 (350)*		
04	*H*1, (*PE* - 1) 1545* *T*1/*M*2**S*	54,600	*.272 (1545) 520/20 (.20)*		
05	*B*2, 2.5 (*M*2)U^2/*T*1$(10)^5$	0.091	*2.5 (20)* 307^2*/520*$(10)^5$		
	*B*3, *B*2/*PE*	0.071	*0.091/1.272*		
06	*B, Intrinsic Factor*	1.18	*1.071/0.909*		
	*E*3, (*PE* - 1)/(*BR* ↑ *S* - 1)	86.3	*(1.272 - 1)/(1.314 - 1)*		
07	*Ws, Rz**36**A* (*K*3**T*2/*M*2)$^{0.5}$	1420	*(3.33/1.80) 36(3.18) (1.25*700/20)*$^{0.5}$		
	*E*5, *SLIP EFF*, (1 - *Ws/Q*1)	85.0	*1. - (1420/9500)*		
08	*E*7, *Thermal Effect*	92.5	*1. - (1. - E5) (0.15R)*		
09	*F*2, (1 - *C* (*R* ↑ 1/*K*3)	80.0	*1. - (1. - C(R ↑ (1/K3)))*		
10	*F*3, *Overall Vol. Eff,* (*E*5**F*2)	68.0	*0.85 * 0.80*		
11	*Q*1, *Q*2/*F*3, *Displ, cfm*	9500	*6500/.68*		
12	*D, Rotor, L* = *D*	22.5	*(9500/307* .0612)*$^{0.5}$		
13	*D, Rotor, L* = 1.5*D*	18.5*27.5	*(505/1.5)*$^{0.5}$		
14	*Ncor, with STD D* x *L*	2950	*505/Std(20*30)(3510)*		
15	*L*3, *H*1* *W*3/33,000 hp	580	*350*54,600/33,000*		
16	*L*4, *Gas hp, L*3/*E*3**E*5**E*7	855	*580/.863*.85*.925*		
17	*L*5, *Brake hp*	884	*L4 + SQR*L4*		
18	*T*2, (*T*1(*R* ↑ *S* - 1)/*E*3**E*7) +*t*1	237	*(520*(1.272 - 1))/.863*925+60*		

Note:

Gas	Nat Gas	Q	9.57	M2	20	K3	1.25	S	0.20	V1	18.6
P1	15.0	P	50.0	T1	520	U	307	A	3.18	G	0.030
X	0.0612	N	3510	D	—	L	—	C	7.5		

↓ Rz = R((K3 + 1)/2) ↑ (1/S)
A = 2G*L + 1SG*D/nc, where nc = number of rotor cells,
nc = 4(SRM), 12(R — V), 6 to 24 (S — V)

Table 12.6a
Radial Inflow Turboexpanders

Item	Calculation	Case 1	2	3	4
01	*R*, (*P*/*P*1)				
02	*H*, (778*Δ*H*) ft				
03	*T*6, (*Mollier*), °F				
04	*T*4, (*Polytropic*), °F				
05	*Q*1, (*V*1**W*), cfs				
06	*N*3/*D*3				
07	*N*, (*N*3, *etc.*), rpm				
08	*D*, (*D*3, *etc.*), in.				
09	*E*1/(*UX*/*S*6)				
10	*UX* (*Tip Speed*), fps				
11	*No. of Noz* (*A* = ____)				
12	*B*7 (*Noz bore*), in.				
13	*A*5/*D*4, in.				
14	*E*6/*F*				
15	*Bhp* (*Adhp* ____)				

Note:
Gas: ____M2, ____K3, ____S, ___Z1, ____Z2, ____T1,
Terminal: ____P1, ____P, ____Q, ___W, ____V1, ____ΔH

Table 12.6b
Radial Inflow Turboexpanders
(An Array of Cryogenic Expander Performance)

Item	Calculation	1	2	3	4
01	R, (*P/P1*)	5.0			
02	H, (778*Δ*H*), ft	40,450			
03	T6, (*Mollier*), °F	−197	−200	−192	−182
04	T4, (*Polytropic*), °F	−190	−192	−183	−168
05	Q1, (*V1*W*), cfs	105	105	110	115
06	N3/D3	80/1.3	60/1.6	40/2.2	30/2.8
07	N, (*N3, etc.*), rpm	22,200	16,700	10,600	7,770
08	D, (*D3, etc.*), in.	11.25	14.0	20.0	25.75
09	E1/(*UX/S6*)	80/.81	81/.76	73/.68	65/.63
10	UX, (*Tip Speed*), fps	1,090	1,020	925	850
11	No. Noz. (*A* = *2.00*)	9	12	16	18
12	B7, (*Noz. Bore*), in.	0.531	0.460	0.400	0.375
13	A5/D4, in.	67/9.2	70/9.5	77/10	80/10
14	E6/F	94/1.2	95/1.5	96/2.0	96/2.6
15	Bhp, (*Adhp 2,860*)	2,135	2,175	1,975	1,765

Note:

Gas:	16 M2,	1.355 K3,	0.262 S,	1.00 Z1,	1.00 Z2,	365 T1,
Methane:	60 P1,	300 P,	80 Q,	39 W,	2.7 V1,	52 ΔH

Item Description: Table 12.6

02 This example concerns the performance of a methane cryogenic expander. The gas is supplied to the turbine at 300 psig and minus 95° F, where the enthalpy is 162 Btu/lb and specific volume is 0.72 cf/lb. The terminal pressure is 60 psia where the isentropic temperature is minus 222° F and the enthlapy is 110 Btu/lb. The expansion head (*H*) released by this function is 778 (162 − 110) = 40,450 ft-lb/lb. The 0.75 and the 0.25 *head roots* produce values of 2,850 and 14.2. These values are used in Items 07 and 08.

03 The exhaust temperature (*T*6) from the turbine which presumably has a dynamic efficiency of 81 percent is read from Mollier Chart 30 in the appendix. The dynamic and exhaust losses are 25 percent and represent 13 Btu/lb, which is added to the exhaust enthlapy of 110, making the 60 psia temperature minus 203° F.

04 If the gas had been a heterogeneous mixture rather than a 90 percent or greater purity of a single gas, the approximate exhaust temperature is determined from the following equation:

*T*4 = *T*1 (1 − 1/*R*↑*S*) *E*1.

Applying the same physical properties, *T*4 = 365 (1 − 1/1.525) = (365 − 126) = 239 Δt or 460 − 239 = minus 221° F. This adiabatic solution is a very close check on the isentropic exhaust temperature of minus 222° F. Applying the same 81 percent dynamic efficiency to the 126° F temperature drop makes the realistic adiabatic exhaust temperature of minus 197°F, i.e., −95° F − (126* 0.81). Designers as a whole agree that the temperatures taken from a Mollier Chart are more reliable than those derived from the adiabatic equation.

05 *Q*1 = 24.4* 2.7 = 66 cfs at the realistic exhaust conditions of 60 psia and minus 203° F. The *SQR* of *Q*1 is 8.12.

06 An array of four sets of specific speeds and specific diameters are exhibited to show the effect on the turbine efficiency and hardware dimensions.

07 The head roots from Item 02 and the *SQR Q*1 of 8.12, Item 05, are applied to the specific speed equation to obtain the optimum turbine speed for each case.

N = *N*3* *H*↑0.75/*Q*1↑0.5.

08 The same root data is applied to the specific diameter equation, which is

$$D = 12^* D3^* Q1\uparrow 0.5/H\uparrow 0.25$$

for *inches in diameter*.

09 The dynamic efficiencies are apparent for each point on the Baljé chart, Figure 6.8, for RIF turbines.

10 The impeller tip speed *UX* is found by the equation, *D** *N*/229. Much of the impeller geometry is related to the tip speed. The meridional velocity flowing through the peripheral passage is about 25 percent of *UX*. The meridional velocity leaving the exhaust eye or annulus is about 30 percent of *UX*.

11 The spouting velocity Mach number is the key item to check in analyzing a turboexpander. There is no problem of developing a damaging wave front when the spouting velocity is less than 0.93 Mach. This number depends upon the flow friction through the blading or the vane drag in a RIF turbine. Supersonic flow requires that the gas contain a persistent accelerating force. Any supersonic flow of equal or a decelerating intensity provokes an unstable performance. The method of circumventing this trouble is to provide a reaction balance of 50 to 70 percent of the enthalpy released through the primary nozzles. The balance of the energy is released in passing through the rotor blades or impeller vanes. In this instance the sonic velocity is *SQR* (1.355* 22.8)* 224 = 1,250 fps. The full enthalpy drop can cause a potential spouting velocity of *SQR* (52)* 224 = 1,615 fps or 1.3 Mach. The corrective measure for this case is to dilute the spouting velocity with a 69 percent reaction. The corrective spouting velocity is 1,344 fps. This velocity is 1.075 Mach, but the reaction feature acts as a cushion to absorb any deceleration in flow. The impeller volute should hold at or about 100 psia and minus 170° F for normal design conditions. The enthalpy is 162 − (52*0.93) = 126 Btu/lb. The *UX*/*S*6 or *U*/*Co* range from 0.71 to 0.76, which is consistent with the Baljé chart, Figure 6.8. The total nozzle area is *A* = *V*2* *W** 144/0.94* *S*6 = 0.72* 24.4* 144/0.94* 1344 = 2.0 sq in. The number of nozzles (*n*) can be approximated from the factor, 0.85* *D*. This provides the necessary arc for a nozzle set at about 25° from a peripheral tangent.

12 The bore of the primary nozzles (*B*7) is determined from *SQR* (1.275* *A*/n). The breadth of vanes at the periphery is about 0.075 larger than the nozzle bore.

13 The eye or annulus area (*A*5) is *V*1* *W** 144/0.85* *C*4 + 12.5 sq in when the shaft extends through the eye. This allowance is for a 4-inch diameter hub. The 0.85 factor allows for vane, etc., obstructions. The eye diameter *D*4 is *SQR* (1.273* *A*5).

14 The exhaust efficiency (*E*6) is readily evaluated by the equation,

$$1 - ((C4/S6)\uparrow 2).$$

Where *C*4 ≅ 0.3* *UX*, the exhaust loss is approximately 5 percent of the available energy. The ratio (*F*) is the quotient of *D*/*D*4.

15 The adiabatic hp is *H** *W*/550 = 40,450* 39/550 = 2,860. The gas hp = 2,860* 0.75 = 2,160. The bhp = 2,160 − (2,860 ↑ 0.4) = 2,135 bhp.

Table 12.7a
Pulse Damper Evaluation and Sizing

Item	Calculation	Stage 1	Stage 2
01	$R/Adhp.$ *Index* $(P/P1/(R\uparrow S-1)(h1)$		
02	$Q4$, Piston Sweep, cu in.		
03	AF, $(1/(1+AQ))$ Where $AQ = 11.0$, Figure 3.2		
04	$a/U, \left(\frac{\text{Piston/Valve Area Ratio}}{\text{Avg. Piston Speed, fps.}}\right)$		
05	$B2/B3, \left(\frac{\text{Suction Valve Loss}}{\text{Disch. Valve Loss,}}\right)$ dec		
06	B/Power Index, $\left(\frac{\text{Intrinsic Fact.}}{B\text{x}R\uparrow S-1}\right)$ $(h2)$		
07	Θ_{sp}/Θ_{ap} *(Peak Pulses)*, dec		
08	$B_p/(B_p \text{ x } R\uparrow S-1)$, $(h3)$		
09	Θ_{sn}/Θ_{an} *(Nominal Piping)*, dec		
10	$B_n/(B_n \text{ x } R\uparrow S-1)$, $(h4)$		
11	Θ_{sa}/Θ_{da}, $(AQ = 11.0)$*(Attenuated)*		
12	$B_a/(B_a \text{ x } R\uparrow S-1)$, $(h5)$		
13	$Q5/Q6$, Suction/Disch. Vessel, cu in.		
14	Suction Vessel, ID x L, tan. x tan.		
15	Discharge Vessel, ID x L, tan. x tan.		
	Remarks:		

Note:

Gas: ___M2, ___K3, ___S, ___T1, ___P1, ___P,
Hdw: ___B1, ___R4, ___R5, ___R6, ___U, ___V1,

Table 12.7b
Pulse Damper Evaluation and Sizing
(Propane–Propylene Refrigeration)

Item	Calculation	Stage 1	Stage 2
01	*R*/Ad Power, (*P*/*P*1/(*R*↑*S*–1)(*h*1)	3.83/0.167	928 hp*
02	*Q*4, Piston Sweep cu in.	1,330	
03	*AF*, (1/(1 + *AQ*)) Where *AQ* = 11.0 Figure 3.2	0.0833	
04	*a*/*U*, (*Piston/Valve Area Ratio* / *Avg. Piston Speed, fps.*)	10.6/11.0	
05	*B*2/*B*3, (*Suction Valve Loss* / *Disch. Valve Loss*)	0.0955/0.082	
06	*B*/Power Index, (*Intrinsic Fact.* / *B* x *R*↑*S*–1) (*h*2)	1.20/0.192	1068 hp†
07	Θ*sp*/Θ*dp* (*Peak Pulses*)	0.191/0.086	
08	*Bp*/(*Bp* x *R*↑*S*–1) (*h*3)	1.38/0.211	1172 hp**
09	Θ*sn*/Θ*dn* (*Nominal Piping*)	0.048/0.041	
10	*Bn*/(*Bn* x *R*↑*S*–1) (*h*4)	1.09/0.179	1000 hp††
11	Θ*sa*/Θ*da*, (*AQ* = 11.0) (*Attenuated*)	0.008/0.007	
12	*Ba*/(*Ba* x *R*↑*S*–1) (*h*5)	1.015/0.1695	945 hp***
13	*Q*5/*Q*6, Suction/Disch. Vessel, cu in.	14,630/4,450	
14	Suction Vessel, ID x L, tan. x tan.	19.25 x 50.5	
15	Discharge Vessel, ID x L, tan. x tan.	12.0 x 39.5	

Note:

Gas:	41 M2,	1.13 K3,	0.115 S,	590 T1,	15.7 P1,	60 P
Hdw:	12.5 B1,	360 R4,	2 R5,	11.0 R6,	11.0 U,	9.83 V1

*Intrinsic Load, No Losses
†Intrinsic and Valve Loss Load
**Instantaneous Peak Pulse Power
††Basic Min. Load With Nominal Piping, *AQ* = 1
***Intrinsic and Valve Load Idealized, *PD* and *AQ* = 11

Θ*dp* = 0.086 is an arbitrary value and is applied to all power evaluations. Calculated value of 0.325 is not compatible for heavy gases.

Item Description: Table 12.7

01 The expressions (*R*↑*S* – 1) and values tagged (*h*1), (*h*2), etc., are power indices for each of the five cases analyzed. The above factor represents the minimal power index which is experienced with adiabatic compression without any valve and pipeline surge losses.

02 The piston sweep is the volume displaced in each action stroke. A double acting cylinder has two sweeps per revolution. The sweep volume is

0.875* (*B*↑2 – 0.5* *R*5↑2) *R*6 = *Q*4.

03 The design objective is presumed to limit the pulse intensity to 1.5 percent (or less) of the respective pressure system. Figure 3.2 shows that an *attenuation quotient* value of 11 is required, and the *attenuation factor* is 0.0833.

04 The valve area is 3.85 square inches. There are three valves per corner. The piston area is 122.5. The piston/valve area is 122.5/3*3.85 = 10.6. The piston speed *U* is *R*4* *R*6/360 = 11.0 fps.

05 $B2$, the suction valve loss, is developed from the equation:

$(a^*U)\uparrow 2^* M2 / T1^* 10\uparrow 4.$

The value is a decimal percentage of the suction pressure. The discharge valve loss, $B3$, is $B2 / R\uparrow S$.

06 B is the intrinsic factor which is applied to R as a correction for the friction of the suction and discharge valves. The power index ($h2$) is 15 percent greater than the basic adiabatic power. This loss persists and can never be avoided. It cannot be altered by size of the attenuation vessel, the form of the choke tube or such sonic filters. This dynamic loss can be altered by changing the compressor speed or the valve area. The valve losses are reduced by higher gas inlet temperature and from a lower gravity gas.

07 The peak pulse experienced in the suction stroke is about twice the magnitude of the average $B2$ value. The discharge peak pulse is about $4^* B2 / (R \uparrow 1/K3)$.

08 If the watt meter was to register the peak power requirements, the intrinsic power index would be 4 percent greater than the average value ($h2$); $(h3 / h1) - (h2 / h1) = 4$ percent.

09 The typical piping connections provide an attenuation quotient of one and a factor of 0.5. The valve losses for this condition are equal to half of the average intrinsic valve loss given as Item 05.

10 The power index ($h4$) for this condition is 7.2 percent greater than the $h2$ value.

11 The effect of the prescribed pulse vessels reduces the average intrinsic loss values by the amount of the attenuation factor, which is 0.0833 for the example shown in the text. This factor is applied to Item 05.

12 The power index, $h5$, for the attenuated case is 1.3 percent greater than case $h2$ or $h3$. The power economy that might be credited to the pulse vessels is $(h2)^* (h4) - (h2)^* (h5) = 7$ percent.

13 The volume of the suction pulse vessel is $AQ^*Q4 = Q5$. For the example given, 11^* 1,330 = 14,630 cubic inches. The discharge vessel ($Q6$) is sized by dividing the volume of the suction vessel by $(R\uparrow 1/K3)$.

14 The length of the pulse vessels are usually proportional to the size of the unit. A small unit such as concerned in this example would have a pulse vessel 3 to 5 feet long. The necessary diameter for the suction vessel is 19.25 ID and 50.5 inches tangent to tangent in length.

15 The discharge bottle is fabricated from 12-inch pipe 39.5 inches, tangent to tangent in length. Figure 3.4 gives the Shell Oil Company's acceptable pulse intensities for the full array of operating pressures. Figure 3.2 provides an equivalent power evaluation and the prescription for a more effective damper.

Notes on Pulse Dampers

The numbers in Column 2 represent the relative power load to render a common service. The minimum intrinsic adiabatic power requirement is 928 hp. This case presumes no valve or attenuation losses.

Adding just the valve losses, the power is enhanced 15 percent to 1,068 hp. Adding pulse bottles that have an $AQ = 11$ and the same valve losses, the idealized power is reduced to 942 hp.

The effect of the nominal piping sizes produces an $AQ = 1$, and the reference load is 1,000 hp. While the 942 hp appears unrealistic and optimistic, the margin between the 942 and the 1,000 hp load offers a hopeful evaluation.

The 1,172 hp load represents the instantaneous power demand to accommodate the peak pulse. The rest of the power valves represent the average recordable load.

Table 12.8
Glossary for Calculation Sheets

Symbol	Description	Units
A	Total nozzle or valve area per quadrant	sq in.
$A1$	Piston/valve area ratio, also identified by "a"	
$A3$	Net suction (ROF) or exhaust (RIF) eye area, $Q3*144*229/0.3*D1*N*0.85 = Q3*144/C4*0.85$	sq in.
$A4$	Hub area, usually about 5.75 in. or 26 sq in.	sq in.
$A5$	Gross, eye plus hub, $A3 + A4$, area	sq in.
Ad	Adiabatic head, hp or efficiency, see specific item	
B	Intrinsic power factor, to include valve losses, $(1 + B3) / (1 - B2)$	
Bhp	Brake horsepower requirement of driver	hp
$B1$	Bore of cylinders, $SQR\ [3.06*Q1/U + (0.5*R5\uparrow 2)]$	in.
$B2$	Suction valve loss, $(A1*U)\uparrow 2*M2/T1*10\uparrow 4$ Rotary, $2.5*M2*U\uparrow 2/T1*10\uparrow 5$	dec.
$B3$	Discharge valve loss, $B2/R\uparrow S$	dec.
$B5$	Breadth of vane at impeller tip	
$B6$	Adjusted cylinder bore	in.
$B7$	Bore of expander nozzles, $SQR\ (1.275\ A/n)$	in.
$C1$	Clearance, decimal percent of cylinder volume	dec.
CX	Adjusted effective clearance	dec.
$C2$	Meridional velocity in (RIF) or out of impeller periphery	fps
$C3$	Eventual $C2$ check velocity; $V2*W*144/D*B5*3.14$	fps
$C4$	Meridional velocity in (ROF) or out (RIF) of impeller eye	fps
$C5$	Eye flow coefficient, $C4/U$, $C5$ usually is +/— 0.3	
D	Diameter of impeller $D = 1{,}800*H\uparrow 0.5/N$, for back-lay vane, $S3 = 0.53$ and $D3*N3 = 150$	in.

	D = 1,980*N↑0.5/N, for radial vanes, S3 = 0.63 and D3*N3 = 165	
D1	Corrected impeller diameter, SQR (S5/S4)D	in.
D3	Specific diameter, 12*D*H↑0.25/Q1↑0.5	
D4	Eye diameter, SQR (1.273*A5)	in.
E	Overall efficiency of a machine, (E1 or E2 or E3) E4	dec.
E1	Dynamic compression efficiency, from Baljé Charts, etc.	dec.
E2	Multistage efficiency, E1*([0.90(S2—1)+1]/S2)	dec.
E3	Compression efficiency, (R↑S-1)/(B*R↑S-1)	dec.
E4	Mechanical efficiency, 1 — (L4↑0.5)/L4 for piston machine; use exponent 0.4 for turbomachines	dec.
E5	Slip efficiency, QS/Q1	dec.
E6	Exhaust efficiency, 1 — (C4/S6)↑2	dec.
E7	Thermal efficiency, QS/Q1 (0.15)R	dec.
F	Ratio of impeller diameters, D/D4; 3.0>F>1.3	
F1	Volumetric efficiency, 1.0 + C1 — 1.1*C1*R(1/K3), for piston machines	dec.
F2	Volumetric efficiency for rotary machines, 1.0 — QS/Q1	
h	Power indices for evaluating pulse dampers, etc.	
H	Adiabatic head, (R↑S—1)*772.5*(Z1+Z)*T1/(M2*S) or 778*(H2-H1) at constant entropy for an expander, (1-1/R↑S)*772.5*(Z1+Z)*T1/(M2*S)	ft.
H1	Enthalpy at the lower pressure (P1) condition	Btu/lb
H2	Enthalpy at the higher pressure (P) condition and same entropy as H1	Btu/lb
H3	Adjusted head per stage, H/S2	
H5	Empirical head capability per stage, 15,000 — 1500*(M2↑0.35) S2 = H/H5 to nearest practical integer	ft

(Table 12.8 continued on p. 194)

Table 12.8 (continued)

hp	Horsepower, 550 ft-lb/sec	hp
$K3$	Ratio of specific heats, CU&CU	
L	Length, usually as a multiple of D	in.
$L1$	Piston rod thrust load = $B1\uparrow 2*0.785*(P—P1)$ $-RS\uparrow 2*0.785*(P-15)$	lb
$L2$	Adiabatic hp/cylinder, $W3*H/33{,}000$	hp
$L3$	Adiabatic hp, $W*H/550$	hp
$L4$	Gas horsepower, $L3/E1$ or $E2$ or $E3$	hp
$L5$	Brake horsepower, $L4/E4$	bhp
$L6$	Rod stress, $L1/(R5\uparrow 2*0.785)$	psi
M	Thousand of units	1,000
MM	Million of units	1,000,000
$M1$	Mols per hour, $110*Q$	
$M2$	Molecular weight of gas	
n	Number of parallel expander nozzles	
N	Revolutions per minute, $N3*H\uparrow 0.75/Q\uparrow 0.5$	rpm
$N3$	Specific speed, $N*Q3\uparrow 0.5/H\uparrow 0.75$	
P	Higher system pressure, compressor or expander	psia
$P1$	Lower system pressure	psia
$P2$	Discharge of compressor	
Q	System capacity, MMscfd, millions of standard cubic feet, $32.8*W/M2$	
$Q1$	Cylinder displacement, cfm/cyl Piston, $0.327*(B1\uparrow 2-0.5*R5\uparrow 2)U$, Rotary, = $D*L*U*X$, cfm	cfm
$Q2$	Cylinder capacity, $Q1*F1$	acfm
$Q3$	Compressor inlet or expander outlet $(N3*H\uparrow 0.75/N)\uparrow 2$	cfs
$Q4$	Expander supply or compressor discharge volume	cfs

Symbol	Description	Units
R	Ratio of compression	
*R*4	Revolutions per minute	rpm
*R*5	Piston rod diameter	in.
*R*6	Piston stroke	in.
S	Adiabatic exponent, (*K*3—1) / *K*3	
*S*2	Number of stages, *H* / *H*5	ft
*S*3	Nominal pressure coefficient, 0.53 for back-lay vanes and *D*3**N*3 = 150, for radial vanes, *S*3 = 0.63 and *D*3*N*3 = 165.	
*S*4	Multistage correction of nominal coefficient [*S*3*0.83([(*M*2+51)/80]↑0.1)*(*S*2—1)+*S*3]/*S*2	dec.
*S*5	Dynamic pressure coefficient required, (*g***H*/*S*2)/(*N***D*/229)↑2 or (*g***H*3/*U*↑2)	dec.
*S*6	Spouting velocity, *SQR* (*H*)224 or *SQR* (*H*/778)224 where *H* is the differential enthalpy expended in the nozzle. *H* = (*H*2—*H*1)	fps
T	Discharge or higher pressure system temperature, calculation in Rankine, other data shown Adiabatic rise; (*R*↑*S*—1)*T*1, Polytropic rise; *T*1(*R*↑*S* — 1)/*E*1, or *E*2, or *E*3	°F °F °F
*T*1	Supply temperature for compressor, expander	°R
*T*2	Discharge temperature for compressor	°R
*T*4	Estimated discharge temperature by polytropic	°F
*T*6	By Mollier Chart (see Notes on Table 12.6)	°R
U	Impeller tip or piston speed, *D***N*/229; *SQR*[*g***H*/*S*2*(*S*4 or *S*5)]; or *R*4**R*6/360	fps
UX	Expander tip speed, *D***U*/229	fps
*U*1	Horsepower per unit	hp
*U*4	Minimum number of parallel cylinders per system	
*V*1	Lower pressure specific volume, 10.73**T*1**A*1/*M*2**P*1	cf/lb
*V*2	Higher pressure specific volume	cf/lb

(Table 12.8 continued on p. 196)

Table 12.8 (concluded)

W	Weight flow, 0.0305*Q*$M2$	lb/sec or pps
$W1$	Weight flow	lb/min or ppm
$W3$	Weight flow per cylinder, 1.83*Q*$M2$	ppm
W_s	By-pass slippage for rotary unit	dec.
Z	Compressibility of gas in higher pressure system	dec.
$Z1$	Suction or lower pressure system compressibility factor	dec.
$Z2$	Discharge compressibility	dec.
θ	All theta values are equivalent to B2 and B3 values in Sheet No. 7; other special designations on this sheet are described therein	dec.
*	Instruction to multiply	
↑	Indicated an exponential function	

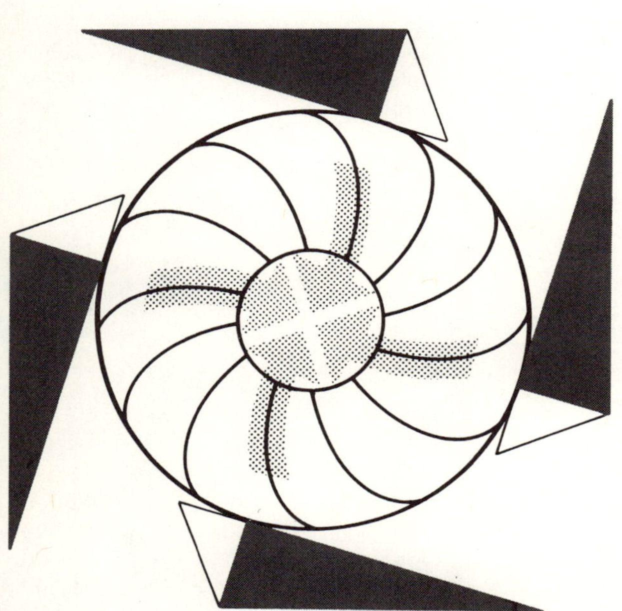

13 Machinery and Economics

The most popular type of piston gas compressor that is used in the refinery and petro/chem industry is the *balanced-opposed* frame, illustrated in Figure 1.13. A list of these frame sizes for each of the principal manufacturers is shown in Table 13.1. The frames carry from two to 10 cylinders. The true balanced-opposed effect is obtained by having the crank throws 180° apart for each pair of identical cylinders. The crankshafts of some designs are sufficiently rigid to avoid the need for intermediate bearing. This further reduces the unbalanced force couples.

The next most popular compressor is the *integral* type gas engine direct driven gas compressor. It is used primarily in the operation of natural gas gathering and transmission. The power cylinders are in the vertical position having a pair of "V" power cylinders or one vertical power cylinder per single "L" or per pair of balanced-opposed compressor cylinders. A list of these integral units are given in Table 13.2.

The *straight-line* compressors are perhaps the oldest form of gas machines. They are still indispensable for small quantity recycle and recovery processes. This model was the outgrowth of the old single cylinder steam engine. The piston rod was extended through the steam cylinder head into a horizontal cylinder supported by a distance piece attached to the steam cylinder. Similar single, tandem or triple extensions are frequently assembled on one piston rod and driven through "V" belts or coupled through a gear which is driven by a multiple cylinder gas engine, electric motor or steam turbine.

The *duplex* frame unit is another outgrowth of the two-stage steam engine. It was also a forerunner of the integral design in that it employed a complex crankshaft that served a pair of straight-line frames. A torque balancing flywheel was mounted on the main shaft between the two frames. As the steam power was replaced with electrical power, the flywheel was replaced with a synchronous motor. Still earlier duplex frames were belt driven to flat sheaves in lieu of flywheels. They in turn were replaced with the direct driven gas engine compressors by means of a connecting yoke and crosshead.

The articulated *cast en bloc* design is used primarily for air conditioning and refrigeration purposes up to 400 hp. Beyond this rating the canned motor driven centrifugal compressor has a distinct market advantage. The piston machines that use ammonia gas are driven with four-pole motors. The heavy fluorocarbon refrigerants use six and eight-pole motors to avoid excessive valve losses.

Table 13.1
Balanced-opposed Crankshaft-driven Compressors

Manufacturer and Model	Frame Stroke in.	Rated Speed, rpm	Number of Cylinders	Rod Dia.	Nominal Frame Hp/Cyl	Piston Rod Rating, bhp	Thrust Tension, lb
Cooper-Bessemer							
AMA	5	1,000	2 to 6		350	2,000	20,000
AMC	5	1,000	2 to 8		400	3,000	30,000
EM	9	514	2 to 4	2.25	200	1,000	20,000
EMA	9	514	2 to 6	2.25	200	1,000	25,000
FM	10 1/2	450	2 to 8	2.25	250	1,875	40,000
JM	14	327	2 to 10	3.00	500	4,000	75,000
KM	14	327	2 to 10	3.50	750	6,000	100,000
LM	20	257	2 to 10	4.00	1,250	10,000	150,000
LMA	20	257	2 to 10	4.00	2,000	16,700	150,000
Worthington							
BDC	9 1/2	514	2 to 4		250	600	22,000
BDC	12	400	2 to 6	2.50	300	1,250	30,000
BDC	14	327	2 to 6	3.00	500	1,750	45,000
BDC	16	300	2 to 10	3.25	1,250	7,000	100,000
BDC	18	277	2 to 10		2,000	16,000	15,000
Ingersoll-Rand							
HHE	10	600	1 to 7	2.00	250	1,625	25,000
HHE	11	450	2 to 6	2.50	425	2,500	40,000
HHE-S	15	327	2 to 10	3.25	400	4,000	72,000
HHE-H	15	327	2 to 10	3.50	800	8,000	90,000
HHE-SL	15 1/2	300	2 to 8	5.00	2,000	10,000	180,000
Clark							
CMA	8	600	2 to 6	1.50	100	600	18,000
CMB	8 7/8	600	2 to 6	2.00	200	1,250	22,000
CJA	11	450	2 to 6	2.75	200	1,750	42,000
CLRA	14	327	2 to 8	3.00	400	4,000	70,000
CLBA	17	300	4 to 10	4.00	750	8,000	90,000
CLBA	19	300	4 to 12	5.00	1,000	12,000	125,000
Chicago-Pneumatic							
FE	12	400	2 to 8	3.25	250	1,800	30,000
FE	14	327	2 to 10	4.00	1000	5,000	100,000
Joy							
WFB	7	900	2 to 4	1.75	325	1,300	20,000
White							
W6	6	900	2 to 8	2.0	250	2,000	35,000
HW6	6	900	2 to 6	3.0	750	4,500	60,000
HW8	8	720	2 to 6	3.0	750	4,500	60,000

Table 13.2
Integral Gas Engines for Compressors

Manufacturer and Model	Bore and Stroke, in.	Piston Rod Thrust Tension	Speed, rpm	Brake Mean Effective Pressure	Rated bhp Min.	Rated bhp Max.	Comp. Nominal Bhp/Cyl.	Number Comp. cyl.
Cooper-Bessemer								
GMXH	9 3/4 x 10 1/2	40,000	450	93	500	1,000	250	3 to 6
GMVA	14 x 14	75,000	300	82	775	1,550	350	3 to 6
GMVC	14 x 14	75,000	300	102	2,000	2,000	400	6
GMVH	14 x 14	75,000	330	111	1,200	2,400	500	3 to 6
V-250	18 x 20	115,000	250	105	2,060	5,500	700	3 to 8
W-330	18 x 20	115,000	330	107	3,625	7,250	900	4 to 8
Z-330	20 x 20	180,000	330	120	5,000	12,500	1,250	4 to 10
Worthington								
SLHC	11 1/2 x 12	30,000	450	135	475	1,000	250	3 to 10
SLHCA	12 x 12	30,000	450	143	440	1,100	300	3 to 10
MLI	16 x 16	75,000	320	108	1,250	2,800	600	3 to 10
MLV	16 x 18	150,000	310	104	3,400	5,500	800	3 to 9
MLV	17 x 19	175,000	330	111	4,000	8,000	1,200	3 to 10
Ingersoll-Rand								
TVR	11 1/2 x 13	35,000	475	124	400	1,200	100	4 to 12
KVS	15 1/4 x 18	60,000	330	122	500	2,000	170	3 to 12
KVR	17 x 22	150,000	350	153	1,400	5,500	700	3 to 12
Clark								
TVM	8 5/8 x 9	32,000	600	104	825	1,000	100	10 to 12
HRA-T	14 x 14	50,000	330	101	1,100	1,440	200	6 to 8
TLA	17 x 19	90,000	300	103	1,700	3,400	350	5 to 10
TCY	17 x 19	115,000	300	103	3,400	5,500	350	10 to 16
TCVD	17 3/4 x 19	115,000	330	108	5,100	7,250	450	12 to 16
TCVE	22 1/2 x 22		330	116		12,500	800	16

Most large European designed compressors carry vertical cylinders. They are usually synchronous motor driven. They feature intricate staging combinations on a single piston rod. Several such combinations are assembled on a common crankshaft. About 12 years ago, the largest compressor on either continent was 6,000 hp per frame and 1,200 hp per piston rod. The capabilities of these units have been more than doubled in many categories.

The parity energy cost for natural gas is $1 per Mscf, *electricity* is one cent per KWH. This is realized with a fuel rate of 7,400 net Btu per brake horsepower and a thermal efficiency of 31 percent. Natural gas at the source is worth about 15¢/Mscf at the Texas well head. A recent FPC hearing favors a price hike to 25¢/Mscf. For this reason the Big Inch pipeliners can afford to consume one-half more fuel for a gas turbine, because of its lower initial and mechanical maintenance cost.

The gas industry has long sought a dual-fuel, diesel or gas engine. There are stationary dual-fuel engines available for interruptible gas fuel services. This engine can be operated as a full diesel. It can also be operated as a gas-diesel engine on proportions as high as 90 percent gas fuel without electric ignition. The engine can easily be converted to an Otto cycle gas engine in a matter of 10 man-hours time. The fuel injection nozzles are replaced with an electrical ignition system. The diesel engine compression is reduced for Otto cycle operation by means of a dual camshaft.

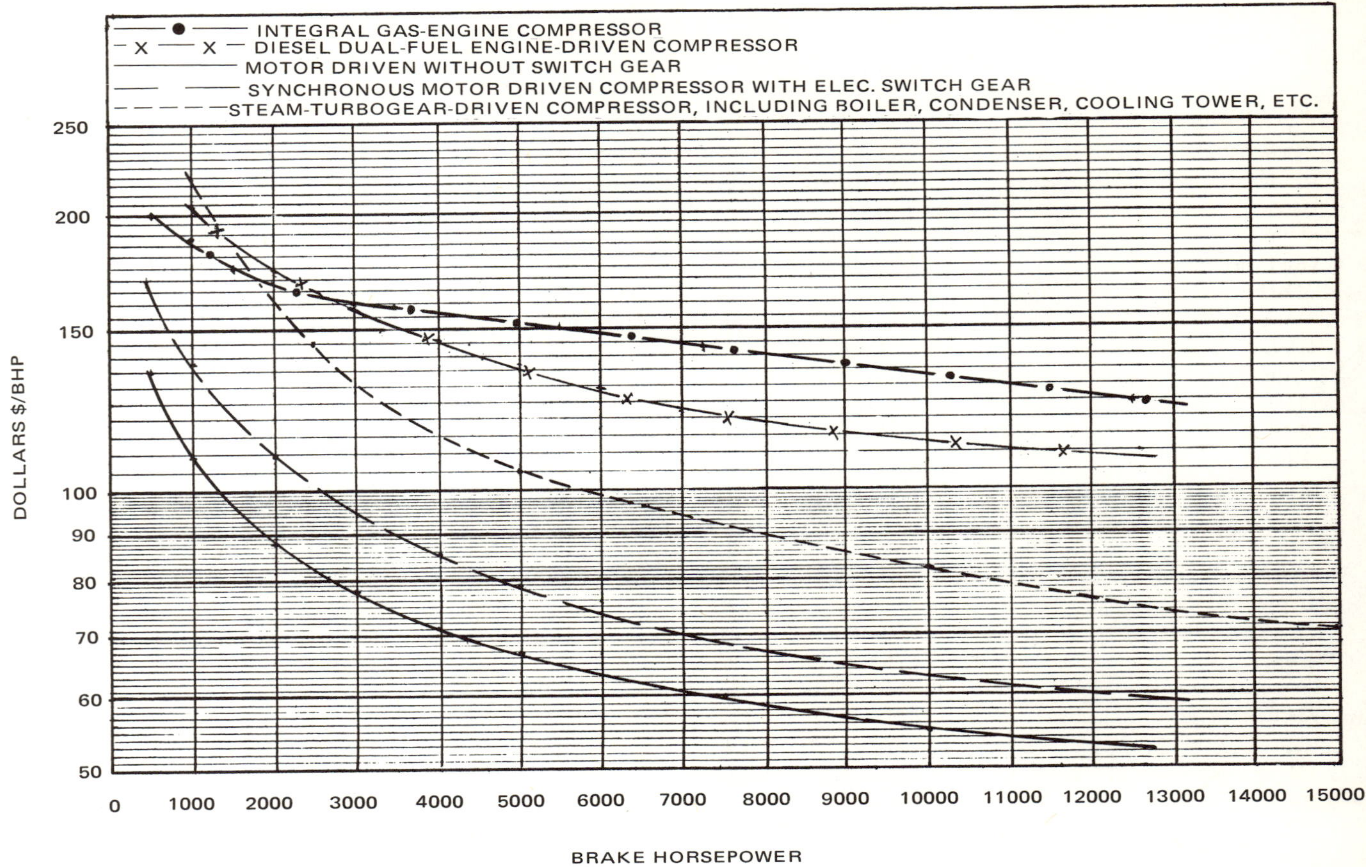

Figure 13.1 (above). Unit cost per brake horsepower of reciprocating compressors with various types of prime mover.

Cost of Reciprocating Compressor Stations

Compressor facilities are appraised in terms of installed cost per brake horsepower. The cost of a piston compressor and five of the most popular types of drivers are given on Figure 13.1. The Nelson Construction Indices show an average annual price increase of gas compressors is 4.3 percent (27). The labor cost has increased 7.4 percent over the same 13 years. The cost of the incidental material components has increased 4 percent. The combined labor, equipment and miscellaneous construction costs have increased 4.7 percent. Technological advances in manufacturing, construction metnods and engineering designs contribute to the low combined plant unit cost. This is the resultant of greater power capabilities per unit, less cylinders to pipe, etc. Table 13.3 gives the break-down cost of constructing three stations that contained 3, 8 and 19 units, each 1,500 hp in size. Table 13.4 gives the cost of installing 1,500-hp units using competing motive power. The prime movers are synchronous motor, steam turbogear and a dual-fuel engine-gear drive. Figure 13.2 shows the influence that the number of units installed at one time has on the unit cost.

Table 13.5 gives the approximate cost of the various incidental components used in a compressor station. All stations require the first eight items; the next five items are required in the engine driven stations. In the instance of multistage plants, an extra $4 is added for each additional stage over the first stage which is priced at $5 per hp. The cost of a fully automated plant is between $20 and $25 per hp. Acoustical treatment adds about $6 per hp. This provides a suspended ceiling with asbestos-composition siding or brick-wall construction.

Table 13.3
Unit Cost of Integral Reciprocating Compressor Stations

Prime Units in Station	3		8		19	
Total Horsepower of Station	3,960		10,560		25,550	
Item of Cost	Labor	Mat'l	Labor	Mat'l	Labor	Mat'l
Grading, Excavating, etc.	$ 15.65	$ 6.85	$9.20	$ 6.07	$ 7.65	$ 2.95
Concrete and Foundations	18.60	11.10	16.25	10.40	14.90	9.18
Equipment, Less Prime Units	8.65	52.90	9.60	51.30	7.05	49.80
Piping	30.40	58.90	29.40	50.00	23.50	44.40
Building and Painting	11.75	10.90	11.75	9.22	10.20	5.98
Electrical Work	7.65	10.00	7.65	7.00	4.31	5.25
Steel Work	2.95	6.50	2.35	4.81	1.57	3.00
Instruments and Controls	2.35	6.47	1.96	4.32	1.37	3.23
Subtotal, $/bhp	$ 98.00	$ 163.62	$ 88.16	$ 143.12	$ 70.55	$ 123.79
Labor (carried forward from subtotal)	$ 98.00		$ 88.16		$ 70.55	
Material (carried forward from subtotal)	163.62		143.12		123.79	
Equipment Rental	10.40		9.75		7.15	
Prime Unit	175.00		175.00		175.00	
Prime Unit Freight	5.00		5.00		5.00	
Field Supervision, etc.	16.25		15.65		10.60	
General Overhead, Engineering, and Fee	16.25		13.75		10.00	
Total Unit Cost, $/bhp	$484.52		$450.43		$402.09	

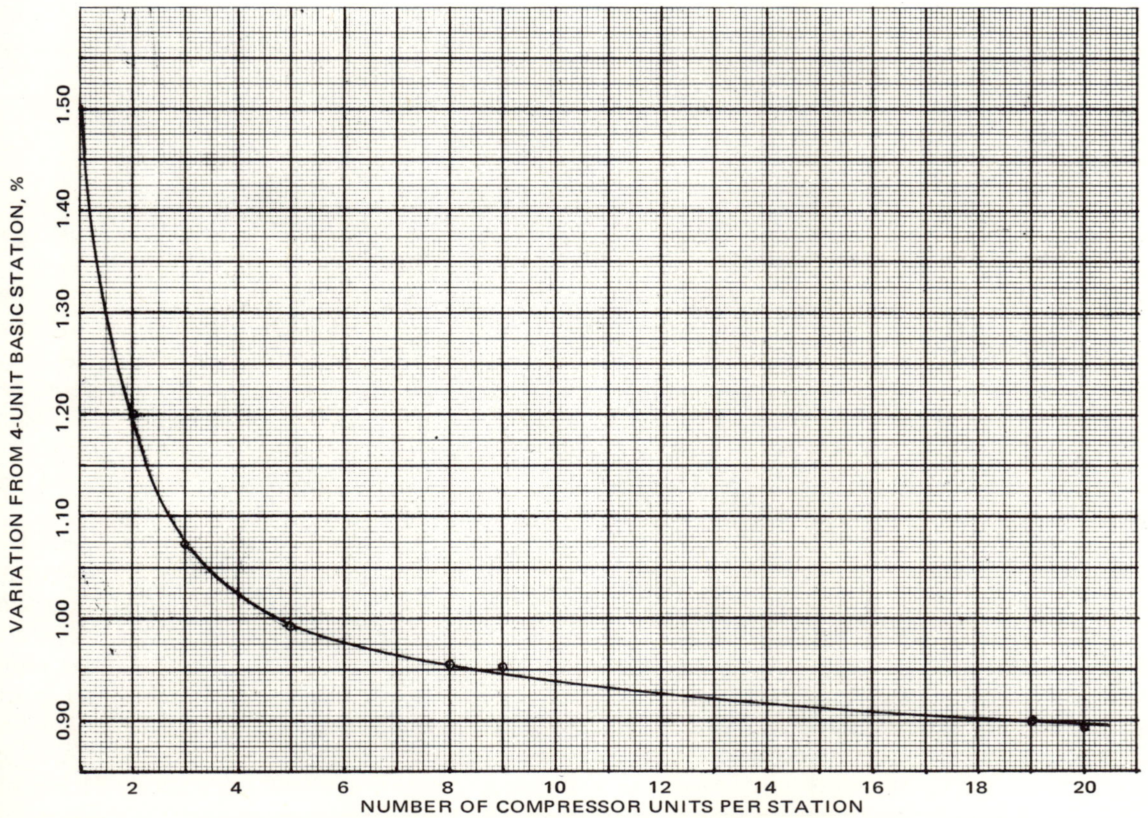

Figure 13.2. Influence of numbers of units installed in one station on unit cost

Table 13.4
Unit Cost of Reciprocating Compressor Stations with Various 1,500-bhp Drivers, Four Units per Station

Driver Connection	Electric Synchronous Shaft Motor		Steam-Turbine Gear Increaser		Dual-Fuel Engine Gear and Pneumatic Coupling	
Item of Cost	Labor	Mat'l	Labor	Mat'l	Labor	Mat'l
Grading, Excavation, etc.	$11.75	$ 6.22	$11.75	$ 6.22	$11.75	$ 6.22
Concrete and Foundations	5.88	4.17	7.85	5.55	11.75	6.95
Equipment Less Prime Units	5.88	35.70	5.88	35.70	11.75	54.50
Piping	27.50	41.70	27.50	41.70	29.40	44.50
Building and Painting	11.75	9.10	15.70	11.70	11.75	9.10
Electrical Work	23.50	25.60	7.85	10.50	7.85	10.50
Steel Work	3.90	6.50	3.90	7.80	3.90	7.80
Instrument and Controls	1.96	6.15	1.96	6.15	1.96	6.15
Subtotal, $/bhp	$92.13	$135.14	$82.39	$125.32	$90.11	$145.72
Labor (carried forward from subtotal)	$ 92.13		$ 82.39		$ 90.11	
Material (carried forward from subtotal)	135.14		125.32		145.72	
Equipment Rental	11.60		12.00		12.00	
Prime Unit	123.00		188.00		180.00	
Prime Unit Freight	5.35		8.00		6.65	
Field Supervision, etc.	21.00		21.00		22.50	
General Overhead, Engineering, and Fee	19.50		19.50		18.00	
Total Unit Cost, $/bhp	$407.72		$456.21		$474.98	

Table 13.5
Compressor Station Accessories

Items	$/hp
1. Gas separators	6.70
2. Boiler, fuel tank and space heaters	3.40
3. Emergency engine driven generators	7.80
4. Oil-cooler system, forced air	5.50
5. Lube oil storage and supply system	1.30
6. Pulse dampers	1.40
7. Utility and fire water system and tanks	1.90
8. Instrument-air compressors and receivers	1.50
9. Starting-air compressors and receivers	2.40
10. Jacket-water cooler, forced-air units	8.00
11. Jacket-water circulation system	2.50
12. Engine exhaust muffler	2.00
13. Engine air cleaners	0.60
14. Gas-cooler, forced-air units, $5 + $4 (>1 Stage)	5.00
Total, items 1 to 8 for all plants	$28.00
Total, gas engine plant, no gas coolers	$50.00

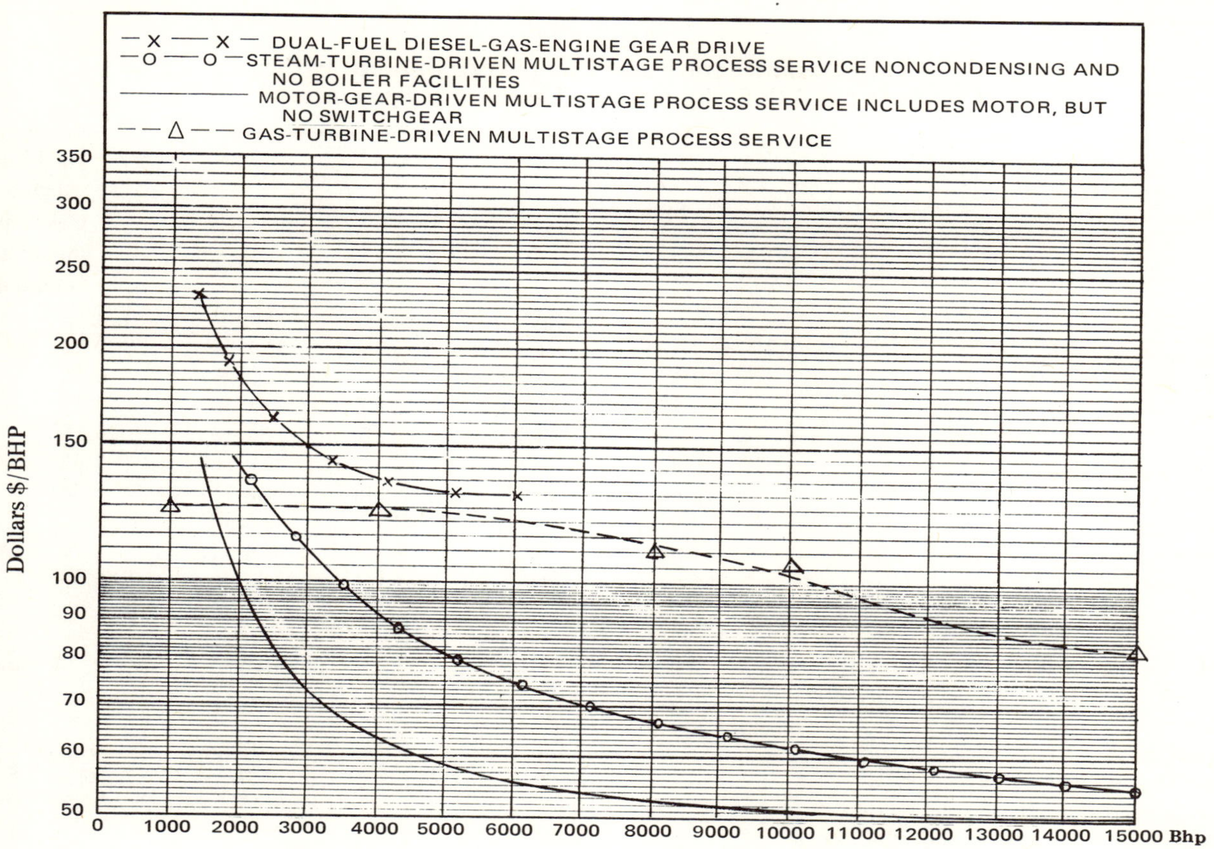

Figure 13.3. Unit cost per brake horsepower of centrifugal compressors with various types of prime movers.

Cost of Centrifugal Compressor Stations

The cost of centrifugal compressors driven by four of the most popular prime movers is given in Figure 13.3. The cost of installing four 4,000-hp units is shown in Table 13.6. The steam turbine is the most popular driver, especially where steam is available and where the load conditions are subject to change. It usually is the least costly prime mover when the boilers and condensing facilities are not required. Each individual case must be evaluated on the respective merits.

The cost of the high-speed, 8,000 to 15,000 rpm steam turbines is approximately

Steam Turbine Cost = \$50,000 + \$25/hp. (13-1)

The cost of condenser and water cooling system is

Condenser System Cost = \$35,000 + \$4/hp.(13.2)

The cost of 1,200 rpm synchronous motor is

Electric Motor Cost = \$5,000 + \$15/hp. (13.3)

Add \$20,000 for 300 rpm synchronous brushless motors. Add 15 percent for switch gear.

The above described tubular condenser system is capable of 27 inches of mercury vacuum. The system includes a vacuum and water circulating pump, a 20°F differential induced draft cooling tower having a 12°F approach to wet bulb and the associated piping. The power range for this class of equipment is 5,000 to 20,000 hp.

The cost of steam generation facilities is estimated from the following equation for a popular sized boiler.

For 900 psi, 750°F boilers,

Boiler Plant Cost = \$20,000 + \$1.50 per lb/hr.(13.4)

An estimated cost of speed increasers and reducers is

Gear, Base and Couplings Cost = \$15,000 + \$5/hp. (13.5)

The boiler plant cost includes the feed-water treating facilities, pumps, forced draft blowers, oil burners and the usual boiler accessories. The equipment

Table 13.6
Unit Cost of Centrifugal Compressor Stations with Various 4,000-bhp Drivers, Three Units Per Station

Driver Connection	Electric Synchronous Motor, Gear		Steam Turbine Gear		Dual-Fuel Engine Gear and Pneumatic Coupling	
Item of Cost	Labor	Mat'l	Labor	Mat'l	Labor	Mat'l
Grading, Excavating, etc.	$ 2.35	$ 1.55	$ 2.35	$ 1.55	$ 2.35	$ 1.55
Concrete and Foundations	3.53	2.09	3.53	2.09	6.46	3.48
Equipment Less Prime Unit	5.87	31.00	5.87	31.00	13.75	49.00
Piping	19.60	34.70	19.60	34.70	23.50	39.60
Building and Painting	8.82	6.50	11.75	9.10	8.82	6.50
Electrical Work	18.60	20.40	5.88	8.15	5.88	8.15
Steel Work	3.92	5.20	3.92	5.85	3.92	5.85
Instruments and Controls	1.96	4.62	1.96	4.62	1.96	4.62
Subtotal $/bhp	$64.65	$106.06	$54.86	$97.06	$66.64	$118.75
Labor (carried forward from subtotal)	$ 64.65		$ 54.86		$ 66.64	
Material (carried forward from subtotal)	106.06		97.06		118.75	
Equipment Rental	11.70		7.15		9.10	
Prime Unit	63.00		91.00		135.00	
Prime Unit Freight	3.75		4.38		5.00	
Field Supervision, etc.	10.00		8.75		10.00	
General Overhead, Engineering, and Fee	10.00		8.75		11.25	
Total Unit Cost $/bhp	$269.16		$271.95		$355.74	

is packaged and requires a minimum installation labor. The plant size limitation is about 250,000 lb/hr. The turbine water rates used in these tables are based on a 70 percent Rankine efficiency. The water rate is 11 lb/hp-hr for 10 psig exhaust and 8 lb/hp-hr for 27 inches of mercury vacuum exhaust. The many small 25 to 200-hp turbines used in refineries have an average Rankine efficiency of about 40 percent and a steam rate of 40 lb/hp-hr. The minimal Rankine efficiency is about 25 percent. The economy of the steam turbine is relatively poor for these small turbines.

Power Costs

Table 13.7 shows that the gas engine is the most economical prime mover to operate. It is in some disfavor on account of the amount of maintenance and the frequency of shut-downs necessary for repair. The next closest competitor is the gas turbine. Industrial applications have been restricted by the 50 percent greater fuel requirement and by reason of the limited number of power denominations. There are several hundred gas turbines, 1,000 hp in size, that have been rendering excellent service for the past five years. There are several new models in the 3,000-hp range which have excellent possibilities. There are several other large industrial gas turbines that range up to 15,000 hp. The parity price of gas turbine fuel is 17.3¢/Mscf to balance the $8.03 per hp-yr (supplies and repair) for the gas engine in comparison with the $1.81 per hp-yr for the gas turbine Ref. Table 13.7. The parity equation is

$$\$8.03 + 72\,n\ (\text{¢ per Mscf}) = \$1.81 + 108n\ (\text{Turbine}) \tag{13.6}$$

Then 36n = $6.22, the parity price is 17.3¢/Mscf.

The electric motor is the most popular prime motor because of its dependability and low installed cost. The parity price for gas engine fuel to match the 0.87¢ KWH motor power is $1.53/Mscf, $0.88 + (8750* 1.34* $.0087/0.87) = $8.03 + (8750* 8.22) n (¢/Mscf). Then, 72n = $110.27; n = 153¢/Mscf. (13.7)

Table 13.7
Unit of Operating Compressor Prime Movers

Prime Mover	Gas Engine	Diesel Engine	Gas Turbine	Steam Turbine	Electric Motor
Item of Cost	Dollars per Horsepower-year				
Operating Labor	$ 5.32	$ 5.32	$ 3.55	$ 7.10	$ 1.78
Operating Energy or Fuel	a) 36.00*	68.00†	b) 54.00*	c) 76.20*	65.00**
Operating Lubricant	1.56	4.55	.13	.13	.07
Operating water	.39	.39	.03	5.75	.03
Other Operating Supplies	.25	.44	.16	.28	.16
Repair Labor	3.37	4.88	.71	1.33	.36
Repair Material	2.46	3.23	.78	2.07	.26
Total Cost	$49.35	$86.81	$59.36	$92.86	$67.66
Less Operating Labor	44.03	81.49	55.81	85.76	65.88
Less Energy or Fuel	$ 8.03	$13.49	$ 1.81	$9.56	$ 0.88

*Natural Gas at 50 cents per MSCF, $\eta_a = 34\%$, $\eta_b = 23\%$, $\eta_c = 13\%$
†Diesel Oil at 15 cents per gal, $\eta = 36\%$
**Electricity at 0.87 cent per Kwhr, $\eta = 87\%$

The steam turbine is the most reliable prime mover. It is used extensively in refinery and petro/chem plants where variable quantities, compression heads and gas densities are encountered. The variable speed feature of the turbine provides a means of coping with this complication. The parity price of boiler fuel gas for the large steam turbine is 37¢/Mscf in contrast with 0.87¢/KWH motor power. The price of boiler fuel gas must be 10¢/Mscf or less to justify small turbines having a 40 percent Rankine efficiency in contrast with electric motor operation. The cost of the steam turbine power can be sharply reduced to about $50 per hp-yr by using Bunker C fuel oil. The cost of Bunker C fuel oil is about half the cost of No. 5 fuel oil. There is a substantial increase in maintenance associated with burning the heavy Bunker C fuel oil.

The diesel engine is used primarily for emergency standby services. It has a high fuel cost and a high maintenance cost. It is subject to a less favorable continuity record than the gas engine.

Operating Supplies

The principal consumable supply required by the diesel and gas engines is the lubrication oil. Table 13.8 gives the quantity of lube oil required by reciprocating gas compressors. This data was compiled from the record of 10 plants over a period of 20 years. This experience indicated that the crankcase system could serve 20,000 hp-hr per gallon of lubricant. The gas engine cylinders experience normal service of 4,000 to 8,000 hp-hr per gallon.

The compressor cylinders and most power cylinders are lubricated by means of mechanical forced-feed lubricators. Each point is served with an individual minute plunger pump that is cam-activated through gearing from the engine or by means of a separate fractional horsepower motor driven lubricator. The gearing provides 2 to 20 drops per minute per point. There are two types of lubricators. The vacuum type has two plungers that operate in series. The first metering plunger feeds into a visible cup, the second plunger charges these droplets into the tubing that serves the application point. The feed rate for the vacuum type lubricator is 12,500 drops per pint.

The pressure-type lubricator has a single plunger that feeds directly into the tubing. The discharge line contains a pressurized sight fixture where the feed rate may be observed. This type of lubricator passes about 4,000 drops per pint. An oil consumption of 6,000 hp-hr per gallon is used for the gas engine in Table 13.7. The diesel engine oil consumption of 2,000 hp-hr per gallon is used. The cost of oil used in Table 13.7 is $1 per gallon.

Water is an essential item to consider in evaluating the cost of prime movers. The water requirements in Table 13.9 were observed from several plant records.

Table 13.8. Quantity of Lube Oil Required for Compressor Cylinder Lubrication

Range of cylinder Diameters, in.	Nominal Discharge Pressure, psig	Film Thickness, microinches	bhp-hr per gal (thousands)	Compressor Lube Oil, pints/(cyl)(100 hp load)	
				Pints/day	Drops/min*
24-36	80	.03	13	1.5	13
15-23	200	.03	19	1.0	9
10-14	800	.04	24	0.9	8
7-9	2,000	.06	24	0.8	7
4-7	8,000	.09	24	0.8	7
3-5	20,000	.11	27	0.7	6

*Based on 12,500 drops per pint of 5,000 SSU SAE 40 lube oil at 75°F with vacuum-type lubricator. Feed includes rod packers, e.g., 300-hp, 16-in. cylinder requires 3 pints/day and 27 drops/min. Reduce drop count to one-third of above for pressure-type lubricator.

Table 13.9
Water Requirements

Service	bbl/hp yr.
Open natural draft cooling tower	500
Induced-draft	
20°ΔF cooling tower	400
10°ΔF cooling tower	200
Jacket-water closed system makeup	20
Turbine leakage and boiler blowdown	20

Table 13.10
Cost of Operating Compressor Units

Type of Unit	Reciprocating	Centrifugal
Item of Cost	Dollars per Horsepower-year	
Operating Lubricant	$ 0.40	$ 0.07
Other Operating Supplies	0.13	0.07
Repair Labor	0.80	0.18
Repair Materials	0.47	0.28
Total Cost	$ 1.80	$ 0.60

The value of raw water may vary from 0.1¢ to 5¢ per barrel depending upon the locale. A reasonable average price is 0.5¢/bbl, which is equivalent to 12¢/ M-gal, 1.44¢/ M-lb or 90¢/Mcf. The cost of steam is usually given in terms of cents per 1,000 pounds. The most frequently quoted value is 50¢/M-lb. This price includes fuel gas at 27¢/Mscf. A better evaluation is 16¢ + 1.6 (¢/Mscf). The price of $92.86 per horsepower year quoted in Table 13.7 is predicated on a steam rate of 11 lb/hp-hr. This rate is equivalent to 96,250 pounds per year, which is approximately $1.00 per thousand pounds of steam.

Table 13.10 gives an expense breakdown for maintenance of the piston and centrifugal compressor proper. Both expenses are typical for 1,000 to 10,000-hp sizes with a normal amount of repair. The maintenance expense for small 100-hp reciprocating machines that operate at excessive piston speed can exceed the normal quoted figure of $1.80 by as much as tenfold. This abnormal expense can mean a valve replacement every other week, which frequently does occur.

Glossary

a	Piston/valve area ratio
acfs	Actual cubic feet per second
acfm	Actual cubic feet per minute
A	Area of flow passage, usually square inch
Ad	Adiabatic function
B	Intrinsic R_c Factor
bhp	Brake horsepower
cfm	Cubic feet per minute
C	Clearance volume retained in cylinder, percentage
C_p	Specific heat at constant pressure
C_v	Specific heat at constant volume
C_{pm}	Molal specific heat of gas
d	Diameter of cylinder bore, impeller, etc., inches
D	Diameter of cylinder bore, impeller, etc., feet
D_s	Specific diamter, feet
E_{dy}	Dynamic efficiency
E_m	Mechanical efficiency
E_v	Volumetric efficiency
f	Feet in compund abbreviates fps.
f	Frictional factor, $f\ (L/D) = K$
fps	Feet per second, velocity
ft	Foot, feet
F	Flow coefficient
F	Temperature ambient, usually degrees Fahrenheit
F	Force of thrust, pounds
g	Acceleration due to gravity, 32.2 ft/sec/sec
G	Gap, clearances
H	Enthalpy, Btu/lb
Hg	Mercury, usually inches of vacuum
icfm	Inlet cubic feet per minute
in.	Inch, inches
J	Mechanical heat equivalent, 778 ft-lb/Btu
k	Ratio of specific heats, C_p/C_v
K	Velocity heads of resistance, ft/velad and other assigned constants
L	Head of gas, ft-lb/lb
m	m = m
m	Molecular weight of gas
M	Mach number
M	Thousand
MM	Million
MMscfd	Million standard cubic feet per day
Mcfm	Thousand cubic feet per minute, inlet conditions
N	Speed of rotation, rpm
N_s	Specific speed
θ	Valve loss, decimal fraction of system pressure
psi	Pounds per square inch, Pressure
psia	Pounds per square inch absolute
psig	Pounds per square inch gauge
ppm	Pounds per minute, flow
pps	Pounds per second, flow
P	Pressure in system, usually discharge, psia
P_d	Suffix d, 2 or any even number, denotes discharge condition; P, V, T or Z
P_s	Suffix s, 1 or any odd number denotes suction condition; P, V, T or Z
Q	Volume, cfm
R	Rankine absolute temperature, 460 + °F
R	Universal gas constant, 1,545/m
R_c	Ratio of compression, P_d/P_s
σ	$(k-1)\ /\ k$
ρ	Density, lb/cu ft
scf	Standard cubic feet
scfm	Standard cubic feet per minute
SP	Static pressures, usually inches of water
t	Thickness, usually inches or decimal thereof
T	Temperature, usually absolute degrees Rankine (R)
U	Average piston or impeller tip speed, fps
V	Velocity, fps
velad	Velocity head $V^2/2g$
V_s	Specific volume, 10.73 ZT/m P_s, cf/lb
W	Flow rate, lb/min
w	Flow rate, lb/sec
Σ	Brake horsepower per MMscfd

Appendix

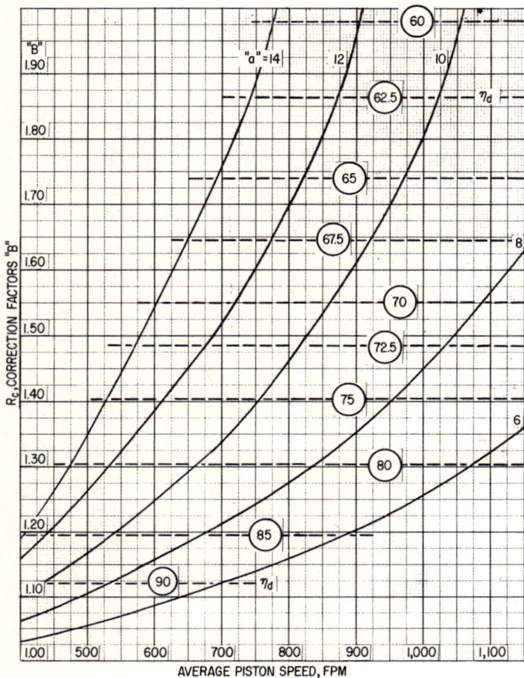

Chart 1. Correction factor B to convert the visual Rc to the intrinsic Rc for butane and other gases having a sonic velocity of τ = 700 fps.

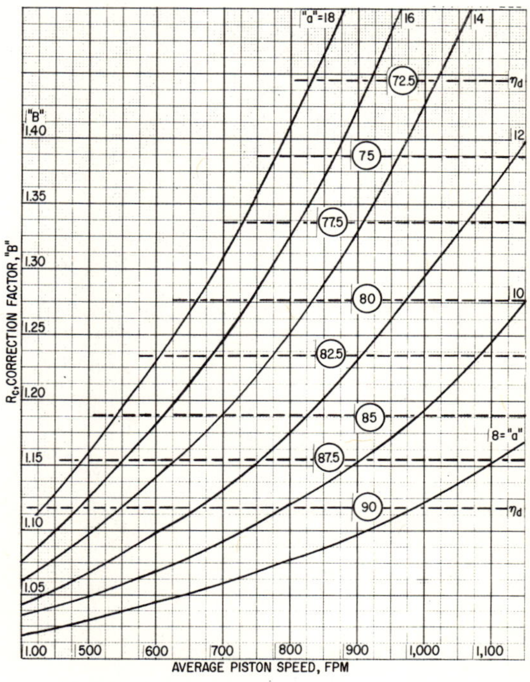

Chart 3. Correction factor B to convert the visual Rc to the intrinsic Rc for natural gas and other gases having a sonic velocity of 1,320 fps.

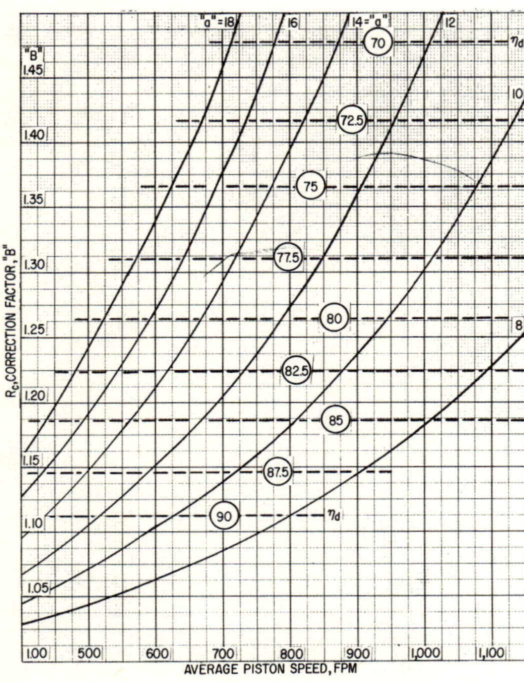

Chart 2. Correction factor B to convert the visual Rc to the intrinsic Rc for air and other gases having a sonic velocity of 1,140 fps.

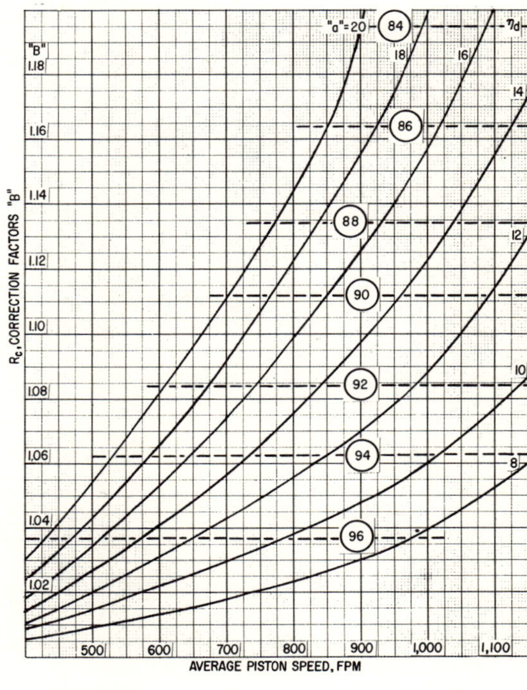

Chart 4. Correction factor B to convert the visual Rc to the intrinsic Rc for a 6.5 m hydrogen mixture with τ = 2,400 fps.

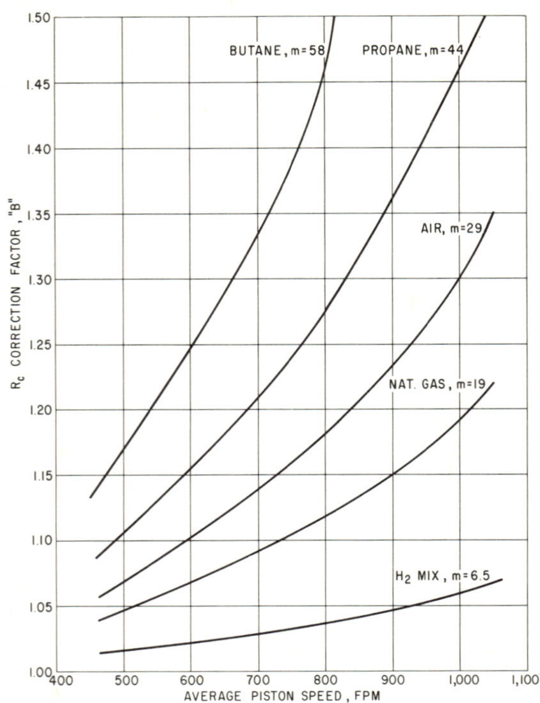

Chart 5. Correction factor B to convert the visual Rc to the intrinsic Rc for and array of gases, m = 6.5 to 58 and the piston/valve area = 10.

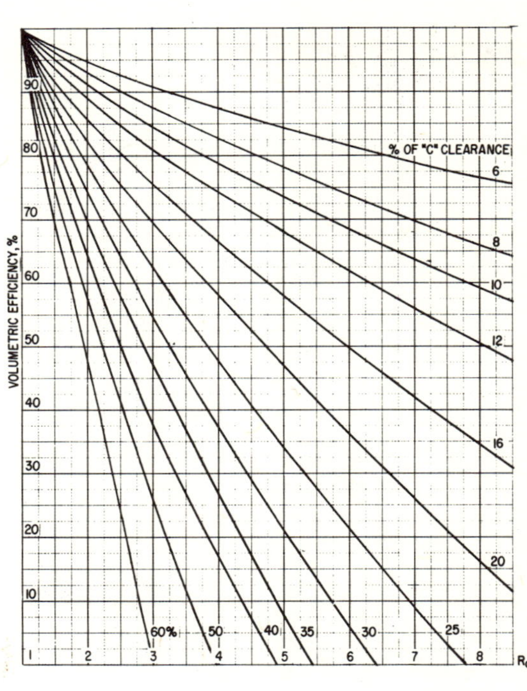

Chart 7. Volumetric efficiency versus Rc for an array of normal clearances for air and diatomic gases. The clearance leakage λ = 1.10, k = 1.40, n = 1.35

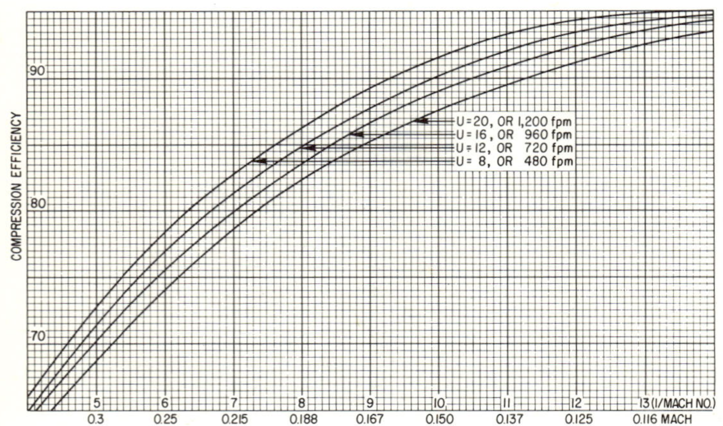

Chart 6. The compression efficiency as a function of the average piston speed and valve velocity Mach Number. Note: The Mach Number carries a 1.5 acceleration factor for the central piston speed over the average speed.

Chart 8 (right). Volumetric efficiency versus Rc for dry natural gas. λ = 1.10 k = 1.28, n = 1.25.

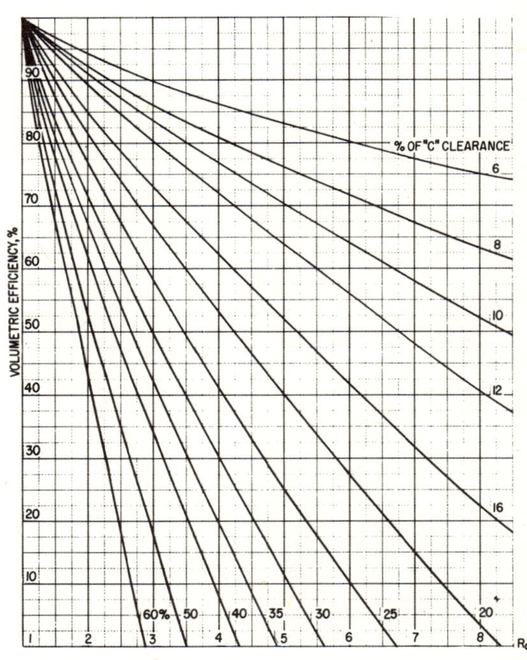

VOLUMETRIC EFFICIENCY, %
% OF "C" CLEARANCE
6
8
10
12
16
20
25
30
35
40
50
60%
Rc
1
2
3
4
5
6
7
8
10
20
30
40
50
60
70
80
90

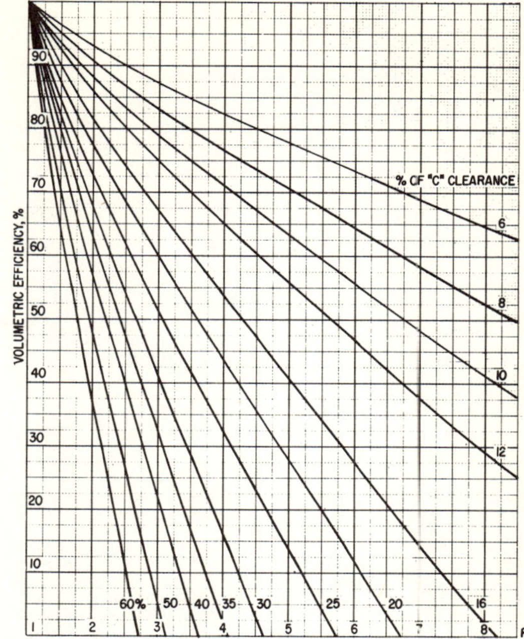

Chart 9. Volumetric efficiency versus Rc for LPG *gases:* $\lambda = 1.1$, $k = 1.13$, $n = 1.10$.

Chart 10. Molal heat capacity of gases. (Compiled by W.C. Edminster, Applications of Thermodynamics to Hydrocarbon Processing.*)*

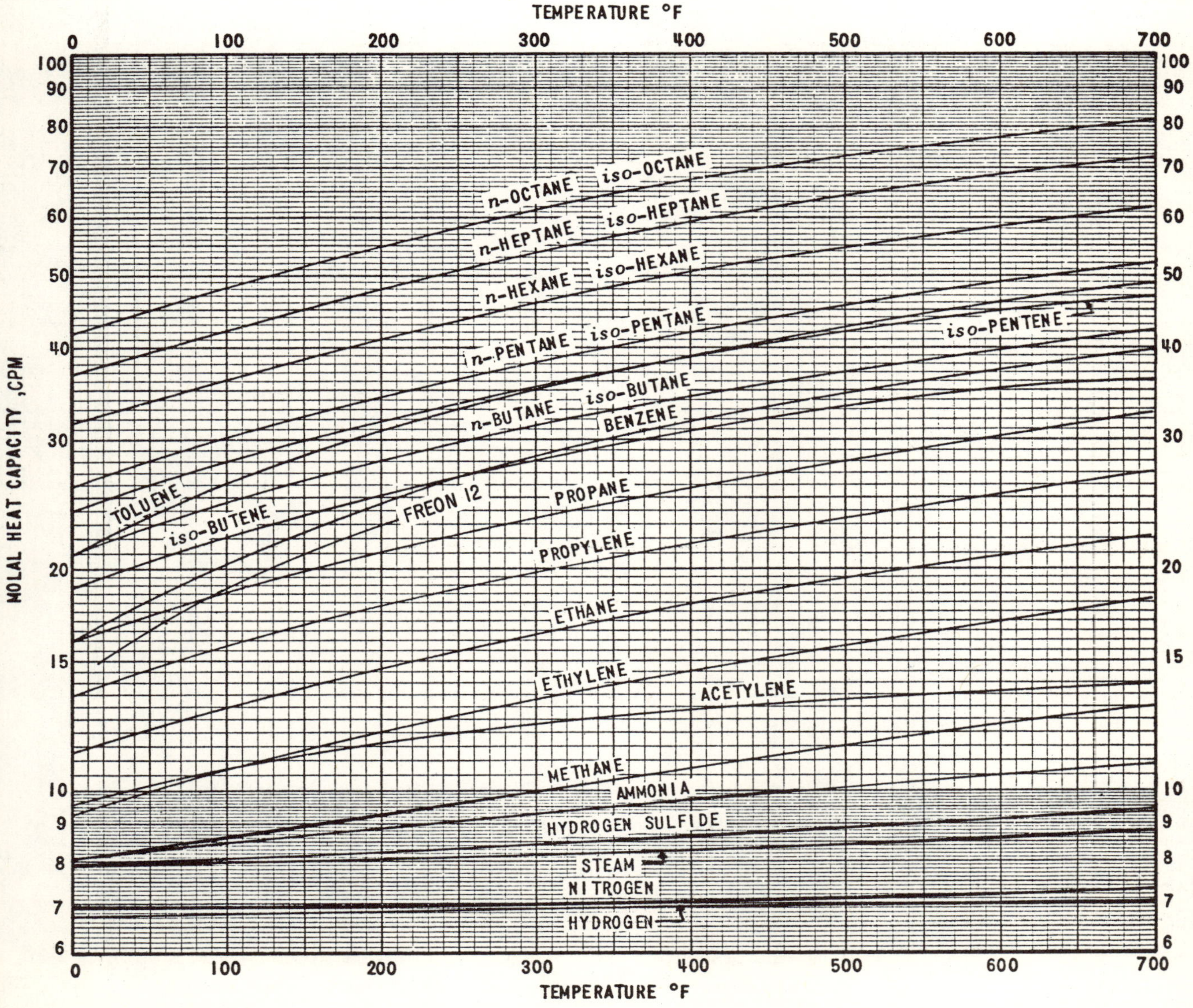

TEMPERATURE °F
0
100
200
300
400
500
600
700
MOLAL HEAT CAPACITY ,CPM
6
7
8
9
10
15
20
30
40
50
60
70
80
90
100
n-OCTANE
iso-OCTANE
n-HEPTANE
iso-HEPTANE
n-HEXANE
iso-HEXANE
n-PENTANE
iso-PENTANE
iso-PENTENE
n-BUTANE
iso-BUTANE
BENZENE
TOLUENE
iso-BUTENE
FREON 12
PROPANE
PROPYLENE
ETHANE
ETHYLENE
ACETYLENE
METHANE
AMMONIA
HYDROGEN SULFIDE
STEAM
NITROGEN
HYDROGEN
TEMPERATURE °F

PHYSICAL CONSTAN

No.	Compound	Formula	Molecular Weight	Boiling Point °F., 14.696 psi., abs.	Vapor Pressure 100°F., psi., abs.	Freezing Point, °F., 14.696 psi., abs.	Critical Constants: Pressure, psi., abs.	Critical Constants: Temperature, °F.	Critical Constants: Volume, cu. ft. per lb.	Density of Liquid; 60°F., 14.696 psi., abs.: Specific Gravity 60°F./60°F. a,b	lb. per gal.* a (Wt. in Vacuum)	lb. per gal.a,c * (Wt. in Air)	Gal/Lb. Mol*	Temperature Coefficient
1	Methane	CH_4	16.042	−258.68	—	−296.46[d]	673.1	−115.78	0.0991	0.3[i]	2.5[i]	2.5[i]	6.4[i]	—
2	Ethane	C_2H_6	30.068	−127.53	—	−297.89[d]	709.8	+90.32	0.0788	0.3771[h]	3.144[h]	3.144[h]	9.56[h]	—
3	Propane	C_3H_8	44.094	−43.73	190	−305.84[d]	617.4	206.26	0.0728	0.5077[h]	4.233[h]	4.220[h]	10.42[h]	0.001
4	n-Butane	C_4H_{10}	58.120	+31.10	51.6	−217.03	550.7	305.62	0.0702	0.5844[h]	4.872[h]	4.865[h]	11.93[h]	0.001
5	2-Methylpropane (isobutane)	C_4H_{10}	58.120	+10.89	72.2	−255.28	529.1	274.96	0.0724	0.5631[h]	4.695[h]	4.686[h]	12.38[h]	0.001
6	n-Pentane	C_5H_{12}	72.146	96.93	15.570	−201.50	489.5	385.5	0.0690	0.6312	5.262	5.253	13.71	0.000
7	2-Methylbutane (isopentane)	C_5H_{12}	72.146	82.13	20.44	−255.82	483.	369.0	0.0685	0.6248	5.209	5.199	13.85	0.000
8	2,2-Dimethylpropane (neopentane)	C_5H_{12}	72.146	49.10	35.9	+2.21	464.0	321.08	0.0674	0.5967[h]	4.975[h]	4.965[h]	14.50[h]	0.001
9	n-Hexane	C_6H_{14}	86.172	155.73	4.956	−139.63	440.0	454.1	0.0685	0.6640	5.536	5.526	15.57	0.000
10	2-Methylpentane	C_6H_{14}	86.172	140.49	6.767	−244.61	440.1	435.7	0.0681	0.6579	5.485	5.476	15.71	0.000
11	3-Methylpentane	C_6H_{14}	86.172	145.91	6.098	—	453.1	448.2	0.0681	0.6690	5.578	5.568	15.45	0.000
12	2,2-Dimethylbutane (neohexane)	C_6H_{14}	86.172	121.53	9.856	−147.77	450.7	420.1	0.0667	0.6540	5.453	5.443	15.80	0.000
13	2,3-Dimethylbutane	C_6H_{14}	86.172	136.38	7.404	−199.37	455.4	440.2	0.0665	0.6664	5.556	5.546	15.51	0.000
14	n-Heptane	C_7H_{16}	100.198	209.17	1.620	−131.10	396.8	512.62	0.0682	0.6882	5.738	5.728	17.46	0.000
15	2-Methylhexane	C_7H_{16}	100.198	194.09	2.271	−180.90	400.	495.	0.0685	0.6830	5.694	5.685	17.60	0.000
16	3-Methylhexane	C_7H_{16}	100.198	197.33	2.130	—	413.	504.4	0.0668	0.6915	5.765	5.758	17.38	0.000
17	3-Ethylpentane	C_7H_{16}	100.198	200.26	2.012	−181.49	420.	513.8	0.0665	0.7026	5.858	5.849	17.10	0.000
18	2,2-Dimethylpentane	C_7H_{16}	100.198	174.55	3.492	−190.86	417.	477.9	0.0646	0.6783	5.655	5.645	17.72	0.000
19	2,4-Dimethylpentane	C_7H_{16}	100.198	176.90	3.292	−182.64	403.	476.9	0.0671	0.6772	5.646	5.637	17.75	0.000
20	3,3-Dimethylpentane	C_7H_{16}	100.198	186.92	2.773	−210.03	440.	505.0	(0.067)	0.6977	5.817	5.807	17.23	0.000
21	2,2,3-Trimethylbutane (triptane)	C_7H_{16}	100.198	177.59	3.374	−12.84	437.2	497.0	0.0631	0.6945	5.790	5.782	17.31	0.000
22	n-Octane	C_8H_{18}	114.224	258.20	0.537	−70.23	362.1	563.7	0.0682	0.7068	5.893	5.883	19.38	0.000
23	2,5-Dimethylhexane (diisobutyl)	C_8H_{18}	114.224	228.39	1.101	−132.16	362.	530.	0.0676	0.6980	5.819	5.810	19.63	0.000
24	2.2.4-Trimethylpentane ("isooctane")	C_8H_{18}	114.224	210.63	1.708	−161.28	374.7	520.1	0.0676	0.6963	5.805	5.795	19.68	0.000
25	n-Nonane	C_9H_{20}	128.250	303.44	0.179	−64.33	332.	610.5	0.0679	0.7217	6.017	6.008	21.31	0.000
26	n-Decane	$C_{10}H_{22}$	142.276	345.42	0.073	−21.39	304.	651.9	0.0679	0.7341	6.120	6.112	23.25	0.000
27	Cyclopentane	C_5H_{10}	70.130	120.67	9.914	−136.96	654.7	461.48	0.0593	0.7505	6.257	6.247	11.21	0.000
28	Methylcyclopentane	C_6H_{12}	84.156	161.26	4.503	−224.42	549.0	499.30	0.0607	0.7535	6.282	6.274	13.40	0.000
29	Cyclohexane	C_6H_{12}	84.156	177.33	3.264	+43.80	591.5(19)	535.6(19)	0.0586(17)	0.7834	6.531	6.522	12.89	0.000
30	Methylcyclohexane	C_7H_{14}	98.182	213.68	1.609	−195.87	504.4	570.2	0.0562	0.7740	6.453	6.443	15.21	0.000
31	Ethene (ethylene)	C_2H_4	28.052	−154.68	—	−272.47[d]	742.1	49.82	0.0706	0.35(16)	—	—	—	—
32	Propene	C_3H_6	42.078	−53.86	226.4	−301.45[d]	667.	197.4	0.0689	0.5220[h]	4.352[h]	4.343[h]	9.67[h]	0.001
33	1-Butene	C_4H_8	56.104	+20.73	63.05	−301.63[d]	583.	295.6	0.0689	0.6013[h]	5.013[h]	5.004[h]	11.19[h]	0.001
34	cis-2-Butene	C_4H_8	56.104	38.70	45.54	−218.04	600.	324.3	0.0503	0.6271[h]	5.228[h]	5.219[h]	10.73[h]	0.000
35	trans-2-Butene	C_4H_8	56.104	33.58	49.80	−157.99	600.	311.9	0.0503	0.6100[h]	5.086[h]	5.076[h]	11.03[h]	0.001
36	2-Methylpropene (isobutene)	C_4H_8	56.104	19.58	63.40	−220.63	579.8	292.5	0.0513	0.6004[h]	5.006[h]	4.996[h]	11.21[h]	0.001
37	1-Pentene	C_5H_{10}	70.130	85.94	19.12	−265.40	586.	376.9	(0.0672)	0.6457	5.383	5.374	13.03	0.000
38	1,2-Butadiene	C_4H_6	54.088	51.53	(20)	−213.14	(653)	(339)	(0.0649)	0.658[h]	5.486[h]	5.47[h]	9.86[h]	0.000
39	1,3-Butadiene	C_4H_6	54.088	24.06	(60)	−164.05	628.	306.	(0.0654)	0.6861	5.229[h]	5.220[h]	10.34[h]	0.001
40	2-Methyl-1,3-butadiene (isoprene)	C_5H_8	68.114	93.32	16.672	−230.71	(558)	(412)	(0.0650)	.6861	5.720	5.711	11.91	0.000
41	Ethyne (acetylene)	C_2H_2	26.036	−119[e]	—	−114[d]	905.	97.4	0.0695	0.6150[k]	—	—	—	—
42	Benzene	C_6H_6	78.108	176.18	3.224	+41.96	714.	552.0	0.0535	0.8845	7.374	7.365	10.59	0.000
43	Toluene	C_7H_8	92.134	231.12	1.032	−138.98	590.	605.5	0.0564	0.8719	7.269	7.260	12.67	0.000
44	Ethylbenzene	C_8H_{10}	106.160	277.13	0.371	−138.96	540.	651.2	0.0556	0.8717	7.268	7.259	14.61	0.000
45	1,2-Dimethylbenzene (o-xylene)	C_8H_{10}	106.160	291.94	0.264	−13.33	530.	674.8	0.0584	0.8848	7.377	7.367	14.39	0.000
46	1,3-Dimethylbenzene (m-xylene)	C_8H_{10}	106.160	282.39	0.326	−54.17	510.	649.9	0.0584	0.8687	7.243	7.234	14.66	0.000
47	1,4-Dimethylbenzene (p-xylene)	C_8H_{10}	106.160	281.03	0.324	+55.87	500.	649.4	0.0556	0.8657	7.218	7.209	14.71	0.000
48	Styrene (Phenyl Ethylene)	C_8H_8	104.144	293.25	(0.24)	−23.13	580.	706.0	0.0541	0.9111	7.596	7.586	13.71	0.000
49	Isopropylbenzene (cumene)	C_9H_{12}	120.186	306.31	0.188	−140.86	460.	724.5	0.0591	0.8663	7.223	7.213	16.64	0.000
50	Methyl Alcohol	CH_4O	32.042	148.1(2)	4.4(7)	−144.0(2)	1154.(17)	464.0(17)	0.0590(17)	0.796(3)	6.64	6.63	4.83	—
51	Ethyl Alcohol	C_2H_6O	46.069	173.3(2)	2.3(7)	−179.1(2)	926.(17)	469.(17)	0.0582(17)	0.794(3)	6.62	6.61	6.96	—
52	Carbon Monoxide	CO	28.010	−313.6(2)	—	−340.6(2)	507.(17)	−220.(17)	0.0532(17)	0.801[m](8)	—	—	—	—
53	Carbon Dioxide	CO_2	44.010	−109.3(2)	—	—	1071.(17)	87.8(17)	0.0343(17)	0.827[h](6)	6.90[h]	6.89[h]	6.38[h]	—
54	Hydrogen Sulfide	H_2S	34.076	−75.3(7)	394.0(6)	−117.2(7)	1306.(17)	212.7(17)	0.0461(17)	0.79[h](6)	6.59[h]	6.58[h]	5.17[h]	—
55	Sulfur Dioxide	SO_2	64.060	+14.0(7)	88.(7)	−103.9(7)	1143.(17)	315.5(17)	0.0304(17)	1.397[h](14)	11.65[h]	11.63[h]	5.50[h]	—
56	Ammonia	NH_3	17.032	−28.0(2)	212.(7)	−107.9(2)	1636.(17)	270.1(17)	0.0681(17)	0.6173(11)	5.15	5.14	3.31	—
57	Air	N_2O_2	28.966	−317.6(2)	—	—	547.(2)	−221.3(2)	0.0517(3)	0.856[m](8)	—	—	—	—
58	Hydrogen	H_2	2.016	−422.9(2)	—	−434.5(2)	188.1(17)	−399.8(17)	0.5159(17)	0.07[m](3)	—	—	—	—
59	Oxygen	O_2	32.000	−297.4(2)	—	−361.1(2)	736.(17)	−181.1(17)	0.0375(17)	1.14[m](3)	—	—	—	—
60	Nitrogen	N_2	28.016	−320.4(2)	—	−345.7(2)	492.(17)	−232.6(17)	0.0514(17)	0.808[m](3)	—	—	—	—
61	Chlorine	Cl_2	70.914	30.3(2)	158.(7)	−150.9(2)	1120.(17)	291.(17)	0.0281(17)	1.414(14)	11.79	11.78	6.02	—
62	Water	H_2O	18.016	+212.0	0.9492(12)	32.0	3208.(17)	705.6(17)	0.0500(17)	1.000	8.337	8.328	2.16	—
63	Hydrogen Chloride	HCl	36.465	−121(16)	925(7)	−173.6(16)	1198.(17)	124.5(17)	0.0208(17)	0.8558(14)	7.135	7.127	5.11	0.003

Chart 11. Physical constants of hydrocarbons. (Courtesy of the Natural Gas Processors Association.)

OF HYDROCARBONS[p] (1)

Pitzer Accentric Factor (18)	Compressibility Factor of real gas, Z 14,696 psia, 60° F	Gas Density, 60 F., 14.696 psi., abs. Ideal Gas*			Specific Heat; 60 F. 14.696 psi., abs.		Calorific Value; 60 F*				Heat of Vaporization, 14.696 psi., abs. at Boiling Point, Btu. per lb.	Refractive Index*, nD	Air Required for Combustion* Ideal Gas	Flammability Limits, Vol. % in Air Mixture		A.S.T.M. Octane Number		No.
					Cp Btu./lb/°F		Net	Gross										
		Specific Gravity Air = 1 *	cu. ft. Gas per lb. *	cu. ft. Gas per gal. Liquid *	Ideal Gas	Liquid	Btu. per cu. ft. Ideal Gas, 14,696 psi., abs. (20) *	Btu. per cu. ft. Ideal Gas, 14,696 psi., abs. (20) *	Btu. per lb. Liquid (Wt. in Vacuum) *	Btu. per gal. Liquid *		68° F.	cu. ft. per cu. ft.	Lower	Higher	Motor Method D-357	Research Method D-908	
013	0.9981	0.5538	23.66	59.[i]	0.5271		909	1010	—	—	219.2	—	9.55	5.0	15.0	—	—	1
105	0.9916	1.038	12.62	39.7[h]	0.4097	0.926[h]	1618	1769	—	—	210.4	—	16.71	2.9	13.0	+0.05[f]	+1.6[f,j]	2
152	0.9820	1.522	8.606	36.43[h]	0.3885	0.592[h]	2316	2517	21,498	—	183.1	—	23.87	2.1	9.5	97.1	+1.8[f,j]	3
201	0.9667	2.006	6.529	31.81[h]	0.3908	0.564[h]	3011	3262	21,136	102,980	165.7	1.3326[h]	31.03	1.8	8.4	89.6[j]	93.8[j]	4
192	0.9699	2.006	6.529	30.66[h]	0.3872	0.570[h]	3001	3253	21,086	98,990	157.5	—	31.03	1.8	8.4	97.6	+0.10[f,j]	5
252	0.9435	2.491	5.260	27.68	0.3883	0.542	3707	4009	20,926	110,120	153.6	1.35748	38.19	1.4	8.3	62.6[j]	61.7[j]	6
206	0.9482	2.491	5.260	27.40	0.3827	0.5353	3698	4000	20,887	108,800	147.1	1.35373	38.19	1.4	(8.3)	90.3	92.3	7
195	(0.95)	2.491	5.260	26.17[h]	0.3914	0.554	3685	3987	20,836	103,650	135.6	1.342[h]	38.19	1.4	(8.3)	80.2	85.5	8
290	—	2.975	4.404	24.38	0.3864	0.5333	4403	4756	20,784	115,060	144.0	1.37486	45.35	1.2	7.7	26.0	24.8	9
—	—	2.975	4.404	24.16	—	0.5264	4395	4747	20,756	113,850	138.7	1.37145	45.35	1.2	(7.7)	73.5	73.4	10
—	—	2.975	4.404	24.56	—	0.507	4398	4751	20,768	115,840	140.1	1.37652	45.35	(1.2)	(7.7)	74.3	74.5	11
—	—	2.975	4.404	24.01	—	0.5165	4382	4735	20,711	112,930	131.2	1.36876	45.35	(1.2)	(7.7)	93.4	91.8	12
—	—	2.975	4.404	24.47	—	0.5127	4391	4744	20,743	115,250	136.1	1.37495	45.35	(1.2)	(7.7)	94.3	+0.3[f]	13
352	—	3.459	3.787	21.73	0.3853	0.5276	5100	5502	20,681	118,660	136.0	1.38764	52.51	1.0	7.0	0.0	0.0	14
—	—	3.459	3.787	21.57	—	0.522	5092	5495	20,658	117,630	131.6	1.38485	52.51	(1.0)	(7.0)	46.4	42.4	15
—	—	3.459	3.787	21.83	—	0.511	5095	5497	20,668	119,160	132.1	1.38864	52.51	(1.0)	(7.0)	55.8	52.0	16
—	—	3.459	3.787	22.19	—	0.506	5098	5500	20,679	121,130	132.8	1.39339	52.51	(1.0)	(7.0)	69.3	65.0	17
—	—	3.459	3.787	21.42	—	0.517	5079	5482	20,620	116,610	125.1	1.38215	52.51	(1.0)	(7.0)	95.6	92.8	18
—	—	3.459	3.787	21.38	—	0.522	5084	5486	20,636	116,510	126.6	1.38145	52.51	(1.0)	(7.0)	83.8	83.1	19
—	—	3.459	3.787	22.03	—	.502	5084	5487	20,638	120,050	127.2	1.39092	52.51	(1.0)	(7.0)	86.6	80.8	20
—	—	3.459	3.787	21.93	—	.498	5081	5483	20,627	119,430	124.2	1.38944	52.51	(1.0)	(7.0)	+0.1[f]	+1.8[f]	21
—	—	3.943	3.322	19.58	0.3845	0.52.30	5796	6249	20,604	121,420	129.5	1.39743	59.67	0.96	—	—	—	22
—	—	3.943	3.322	19.33	—	0.5114	5780	6233	20,564	119,670	122.8	1.39246	59.67	(0.98)	—	55.7	55.2	23
—	—	3.943	3.322	19.29	—	0.4892	5778	6231	20,569	119,390	116.7	1.39145	59.67	1.0	—	100	100	24
—	—	4.428	2.959	17.80	—	0.5220	6493	6996	20,543	123,610	126.7	1.40542	66.83	0.87[s]	2.9	—	—	25
—	—	4.912	2.667	16.32	—	0.5207	7189	7743	20,495	125,440	118.7	1.41189	73.99	0.78[s]	2.6	—	—	26
193	—	2.421	5.411	33.86	—	0.4216	3512	3763	20,187	126,310	167.3	1.40645	35.80	(1.4)	—	84.9[j]	+0.1[f]	27
234	—	2.905	4.509	28.33	—	0.4407	4198	4500	20,129	126,450	147.8	1.40970	42.96	(1.2)	8.35	80.0	91.3	28
186	—	2.905	4.509	29.45	—	0.4332	4180	4482	20,038	130,880	153.7	1.42623	42.96	1.3	7.8	77.2	83.0	29
—	—	3.390	3.865	24.94	—	0.4397	4864	5216	20,003	129,080	138.9	1.42312	50.12	1.2	—	71.1	74.8	30
—	0.9940	0.9684	13.53	—	0.3622	—	1499	1599	—	—	207.6	—	14.32	2.7	34.0	75.6	+0.03[f]	31
143	0.9839	1.453	9.019	39.25[h]	0.3541	—	2183	2334	—	—	188.2	—	21.48	2.0	10.0	84.9	+0.2[f]	32
203	0.9694	1.937	6.764	33.91[h]	0.3548	—	2879	3081	20,679	103,670	167.9	—	28.64	1.6	9.3	80.8[j]	97.4	33
273	0.9653(15)	1.937	6.764	35.36[h]	0.3269	—	2872	3073	20,613	107,770	178.9	—	28.64	(1.6)	—	83.5	100.	34
234	0.9654(15)	1.937	6.764	34.40[h]	0.3654	—	2868	3068	20,584	104,680	174.4	—	28.64	(1.6)	—	—	—	35
201	0.9687(15)	1.937	6.764	33.86[h]	0.3701	—	2860	3061	20,548	102,860	169.5	—	28.64	(1.6)	—	—	—	36
—	(0.95)	2.421	5.411	29.13	—	0.5196	3576	3827	20,550	110,630	154.5	1.37148	35.80	(1.4)	8.7	77.1	90.9	37
—	(0.97)	1.867	7.016	38.49[h]	0.3458	—	2789	2940	—	—	(181)	—	26.25	(2.0)	(12)	—	—	38
—	0.975	1.867	7.016	36.69[h]	0.3412	—	2730	2881	20,039	104,790	(174)	—	26.25	2.0	11.5	—	—	39
—	(0.96)	2.352	5.571	31.87	—	0.5245	3411	3612	19,955	114,150	(153)	1.42194	33.41	(1.5)	—	81.0	99.1	40
186	0.9925	0.8988	14.58	—	0.3966	—	1422	1473	—	—	—	—	11.93	2.5	80	—	—	41
215	0.929(15)	2.697	4.859	35.83	—	0.4098	3591	3742	17,991	132,670	169.3	1.50112	35.80	1.3[g]	7.9[g]	+2.8[f]	—	42
252	0.903(21)	3.181	4.119	29.94	—	0.4017	4273	4475	18,251	132,670	156.2	1.49693	42.96	1.2[g]	7.1[g]	+0.3[f]	+5.8[f]	43
—	—	3.665	3.575	25.98	—	0.4114	4970	5222	18,494	134,110	145.7	1.49588	50.12	0.99[g]	6.7[g]	+97.9	+0.8[f]	44
—	—	3.665	3.575	26.37	—	0.4418	4958	5210	18,445	136,070	149.1	1.50545	50.12	1.1[g]	6.4[g]	100.	—	45
—	—	3.665	3.575	25.89	—	0.4045	4957	5208	18,441	133,560	147.4	1.49722	50.12	1.1[g]	6.4[g]	+2.8[f]	+4.0[f]	46
—	—	3.665	3.575	25.80	—	0.4083	4957	5209	18,445	133,130	146.1	1.49582	50.12	1.1[g]	6.6[g]	+1.2[f]	+3.4[f]	47
—	—	3.595	3.644	27.68	—	0.4122	4830	5031	18,150	137,870	(151)	1.54682	47.73	1.1	6.1	+0.2[f]	>+3[f]	48
—	—	4.149	3.158	22.81	—	(0.41)	5661	5963	18,653	134,710	134.3	1.49145	57.28	0.88[g]	6.5[g]	99.3	+2.1[f]	49
—	—	1.106	11.84	78.6	—	0.594(7)	—	—	9,760(16)	64,810	473(2)	1.3288(8)	7.15(7)	6.72(5)	36.50	—	—	50
—	—	1.590	8.237	54.5	—	0.562(7)	—	—	12,780(16)	84,600	367(2)	1.3614(8)	14.30(7)	3.28(5)	18.95	—	—	51
041	0.9995(15)	0.9670	13.55	—	0.2484(13)	—	—	321[r](13)	—	—	92.7(14)	—	2.38(7)	12.50(5)	74.20	—	—	52
225	0.9943(15)	1.519	8.623	59.4[h]	0.1991(13)	—	—	—	—	—	238.2[n](14)	—	—	—	—	—	—	53
100	0.9903(15)	1.176	11.14	73.4[h]	0.238(4)	—	588(16)	637(16)	—	—	235.6(7)	—	7.15(7)	4.30(5)	45.50	—	—	54
—	—	2.212	5.924	69.0[h]	0.145(7)	0.325[h](7)	—	—	—	—	166.7(14)	—	—	—	—	—	—	55
—	—	0.5880	22.28	114.7	0.5002(10)	1.114[h](7)	359(16)	434(16)	—	—	587.2(14)	—	3.57(7)	15.50(5)	27.00	—	—	56
—	0.9996(15)	1.000	13.10	—	0.2400(9)	—	—	—	—	—	92.(3)	—	—	—	—	—	—	57
—	1.0006	0.0696	188.2	—	3.408(13)	—	274[r](13)	324[r](13)	—	—	193.9(14)	—	2.38(7)	4.00(5)	74.20	—	—	58
0213	—	1.105	11.86	—	0.2188(13)	—	—	—	—	—	91.6(14)	—	—	—	—	—	—	59
040	0.9997(15)	0.9672	13.55	—	0.2482(13)	—	—	—	—	—	87.8(14)	—	—	—	—	—	—	60
—	—	2.448	5.351	63.1	0.119(7)	—	—	—	—	—	123.8(14)	—	—	—	—	—	—	61
—	—	0.6220	21.06	175.6	0.4446(13)	1.0009(7)	—	—	—	—	970.3(12)	1.3330(8)	—	—	—	—	—	62
—	—	1.259	10.41	74.3	0.190(7)	—	—	—	—	—	185.5(14)	—	—	—	—	—	—	63

Chart 11 (continued)

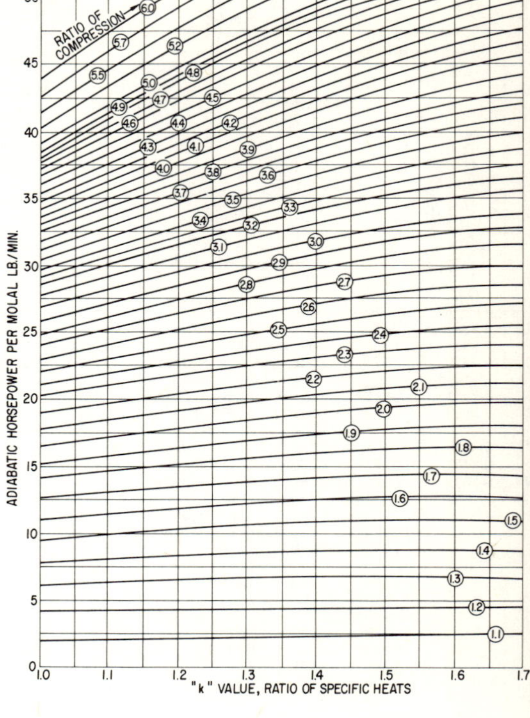

Chart 12. Adiabatic power per molal lb/min versus the ratio of specific heats and Rc.

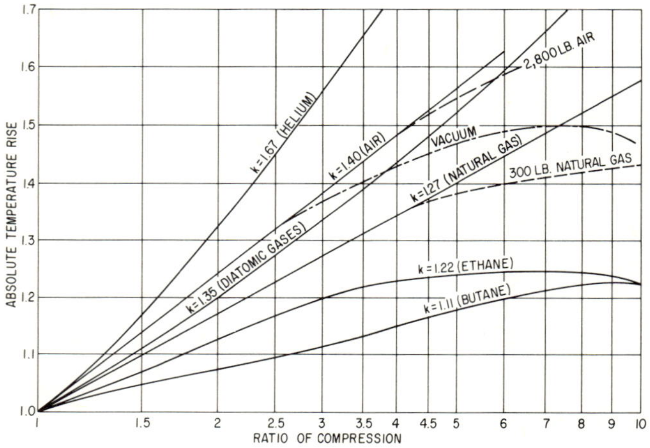

Chart 13. Absolute temperature rise versus Rc for estimating discharge temperature.

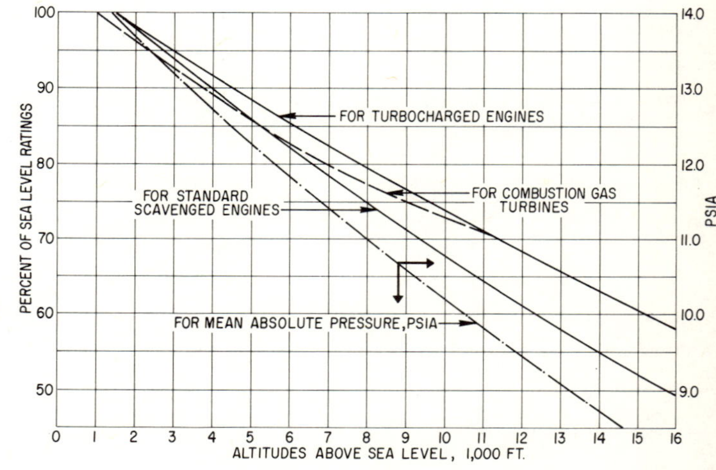

Chart 14. Elevation effect on psia ambient and engine derating.

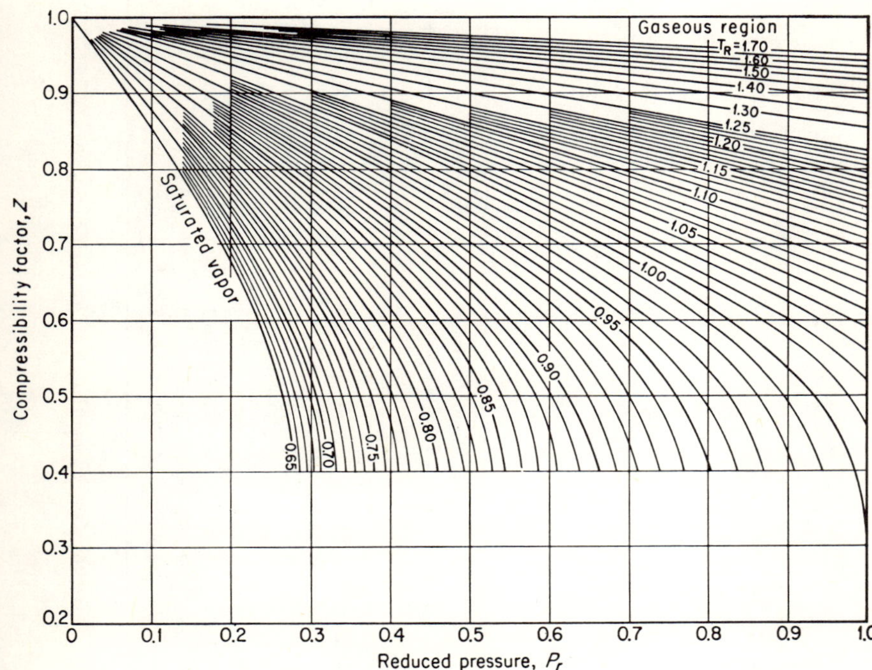

Chart 15a. Compressibility factors: low-pressure range (from Gas and Air Compression Machinery*).*

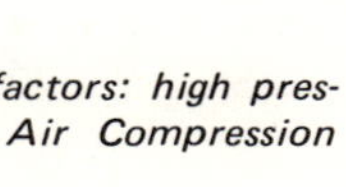

Chart 15b. Compressibility factors: high pressure range (from Gas and Air Compression Machinery).

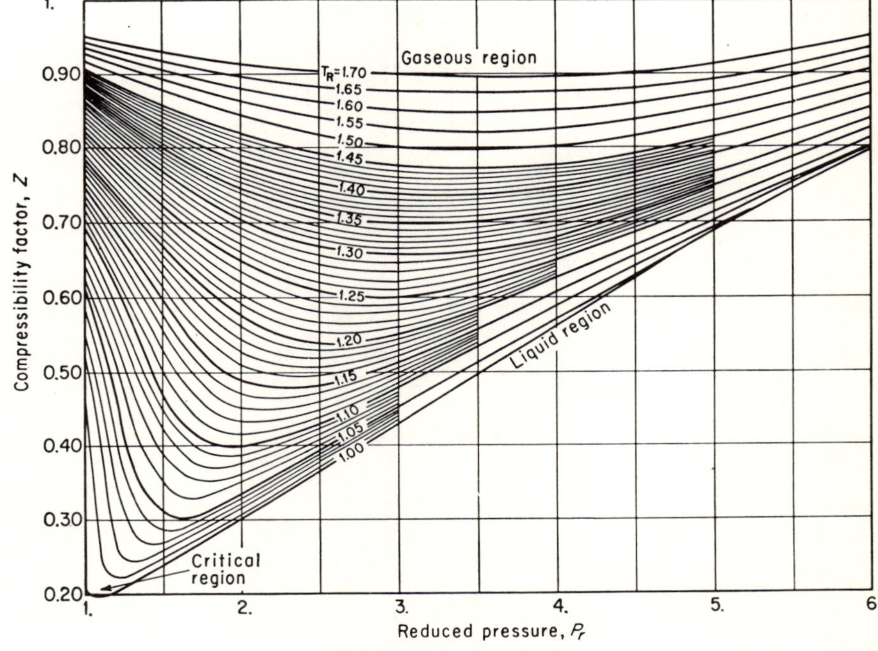

Diameter, in.	Piston area, sq in.	Displacement, cfm SA	Displacement, cfm DA	Clearance vol, %	Diameter, in.	Piston area, sq in.	Displacement, cfm SA	Displacement, cfm DA	Clearance vol, %
5	19.6	62.9	85.6	8.8	19	283	909	1,777	12.0
5½	23.8	76.2	111.9	7.4	19½	299	957	1,875	17.6
6	28.3	90.6	141.3	11.2	20	314	1,007	1,974	16.1
6½	33.2	106.4	172.3	10.3	20½	330	1,058	2,076	15.1
7	38.5	123.5	206.6	9.6	21	346	1,110	2,181	14.2
7¼	41.3	132.4	224.4	9.3	21½	363	1,165	2,288	13.4
7½	44.2	141.8	243.2	8.9	22	380	1,219	2,397	12.7
7¾	47.2	151.2	262.0	8.5	22½	397	1,275	2,510	12.0
8	50.3	161.2	282.0	8.2	23	415	1,332	2,624	20.3
8¼	53.5	171.2	302.5	14.7	23½	434	1,391	2,741	19.0
8½	56.8	181.7	323.5	13.4	24	452	1,450	2,861	17.8
8¾	60.0	192.8	345.1	12.3	24½	471	1,511	2,983	16.8
9	63.6	203.9	367.9	11.4	25	491	1,574	3,107	16.0
9¼	67.2	215.5	390.6	13.4	25½	511	1,638	3,234	15.4
9½	70.9	227.0	414.4	12.6	26	531	1,702	3,364	14.8
9¾	74.7	239.3	438.2	11.9	26½	552	1,768	3,496	13.2
10	78.5	252.1	463.1	11.3	27	573	1,836	3,631	12.3
10¼	82.5	264.8	488.6	18.9	27½	594	1,905	3,768	11.5
10½	86.6	277.6	514.7	17.6	28	616	1,974	3,908	10.7
10¾	90.8	290.9	541.8	16.4	28½	638	2,045	4,050	10.2
11	95.0	304.7	569.0	15.4	29	660	2,117	4,195	9.8
11¼	99.4	318.6	597.2	14.5	29½	684	2,192	4,342	9.5
11½	103.9	333.0	625.5	13.8	30	707	2,266	4,492	20.7
12	113.1	362.9	684.7	12.6	30½	731	2,342	4,645	19.7
12½	122.7	393.3	746.8	11.7	31	755	2,420	4,800	18.8
13	132.7	425.5	811.0	11.0	31½	779	2,499	4,957	17.9
13½	143.1	458.7	877.5	15.7	32	804	2,578	5,117	17.1
14	153.9	493.6	946.8	14.0	32½	830	2,659	5,279	16.3
14½	165.1	529.1	1,019	12.7	33	855	2,742	5,444	15.6
15	176.7	566.7	1,093	11.8	34	908	2,911	5,782	14.4
15½	188.7	605.0	1,170	11.0	35	962	3,085	6,128	19.8
16	201.1	644.9	1,249	10.4	36	1,018	3,264	6,486	18.1
16½	213.8	685.3	1,331	16.8	37	1,075	3,447	6,854	16.5
17	227.0	728.0	1,415	15.6	38	1,134	3,635	7,232	15.2
17½	240.5	771.2	1,502	14.5	39	1,195	3,830	7,620	13.6
18	254.5	816.0	1,591	13.5	40	1,257	4,029	8,016	12.4
18½	2,688	862.0	1,683	12.7	41	1,320	4,234	8,425	11.2

NOTES: Based on piston speed of 923 fpm and 3-in.-diameter piston rod. Compiled with permission and from data provided by Cooper-Bessemer Corporation.

The above clearance volume percentages represent typical variations in design. These percentages are applicable to only one manufacturer's line of heavy-duty cylinders. Clearances can be increased as desired. Substantial clearance percentage reduction generally involves a special cylinder design.

SA = single acting, DA = double acting.

Chart 16 (left). Data for 20-inch stroke x 277 rpm compressor cylinders

Stroke, in.	Multipliers for following speeds, rpm						
	396	432	500	585	715	875	1000
6	0.429	0.468	0.542	0.635	0.776	0.950	1.085
7	0.500	0.546	0.633	0.741	0.905	1.108	1.270
8	0.571	0.623	0.721	0.844	1.030	1.228	1.400
8⅞	0.634	0.693	0.803	0.940	1.150	1.365	1.560
9	0.643	0.702	0.812	0.950	1.160	1.420	1.620
10	0.715	0.780	0.903	1.055	1.290	1.580	1.800
10½	0.750	0.819	0.949	1.110	1.358	1.660	1.900
11	0.787	0.859	0.995	1.163	1.420	1.735	1.980
12	0.857	0.936	1.083	1.270	1.550	1.900	2.170

Chart 17. Capacity multipliers for short stroke cylinders

Stroke, in.	Multipliers for following speeds, rpm						
	240	257	277	300	327	360	396
12	0.520	0.557	0.600	0.650	0.707	0.780	0.857
14	0.606	0.650	0.700	0.758	0.825	0.910	1.000
15	0.650	0.697	0.750	0.812	0.885	0.975	1.070
16	0.693	0.743	0.800	0.866	0.943	1.040	1.142
17	0.736	0.790	0.850	0.920	1.002	1.105	1.213
18	0.780	0.836	0.900	0.975	1.150	1.170	1.286
19	0.823	0.882	0.950	1.029	1.212	1.235	1.357
20	0.866	0.929	1.000	1.083	1.180	1.300	1.430

Chart 18. Capacity multipliers for long stroke cylinders

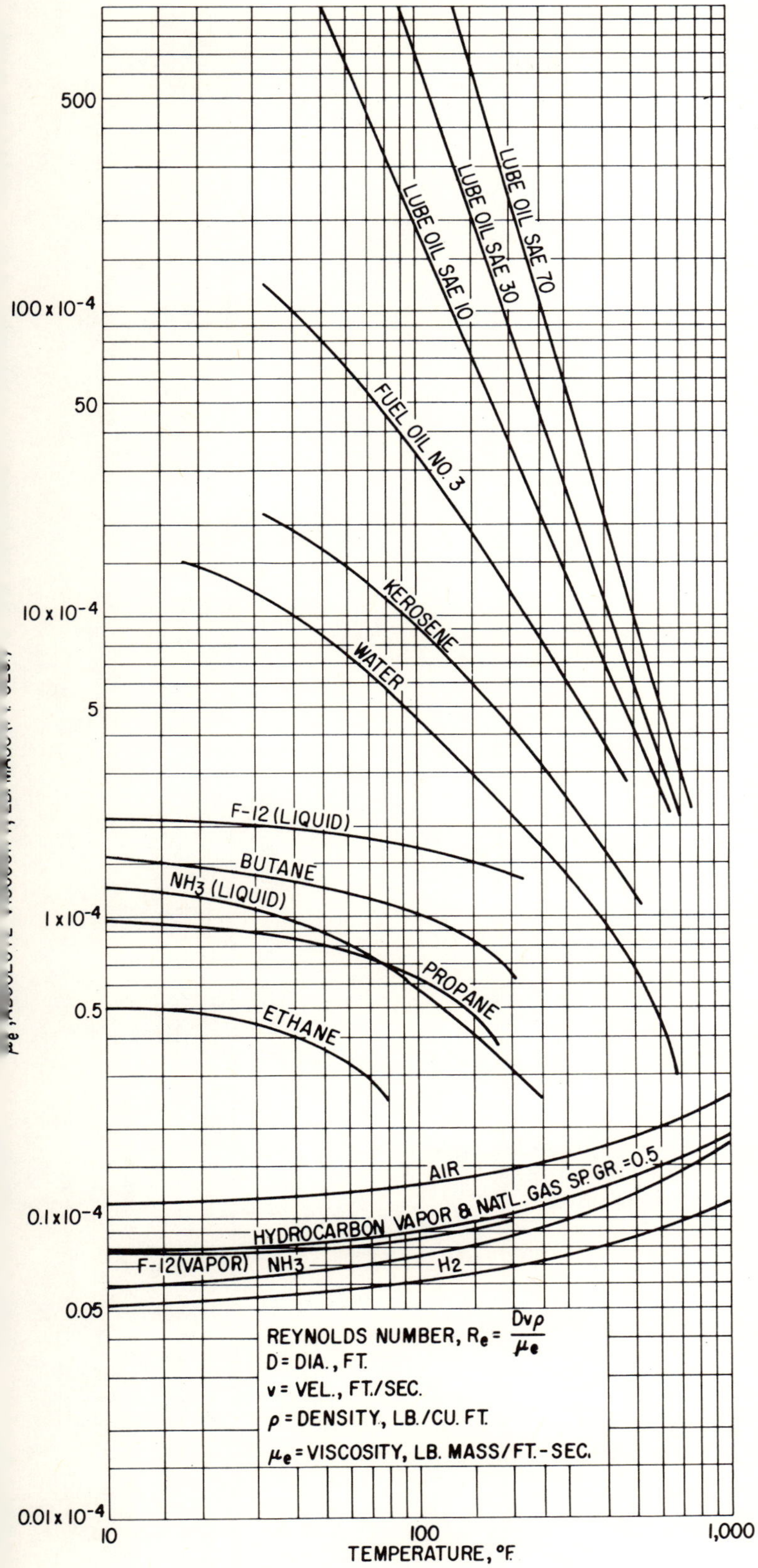

LUBE OIL SAE 70
LUBE OIL SAE 30
LUBE OIL SAE 10
FUEL OIL NO. 3
KEROSENE
WATER
F-12 (LIQUID)
BUTANE
NH_3 (LIQUID)
PROPANE
ETHANE
AIR
HYDROCARBON VAPOR & NATL. GAS SP. GR. = 0.5
F-12 (VAPOR)
NH_3
H_2

500, 100×10^{-4}, 50, 10×10^{-4}, 5, 1×10^{-4}, 0.5, 0.1×10^{-4}, 0.05, 0.01×10^{-4}

10, 100, 1,000

TEMPERATURE, °F

REYNOLDS NUMBER, $R_e = \frac{Dv\rho}{\mu_e}$

D = DIA., FT.

v = VEL., FT./SEC.

ρ = DENSITY, LB./CU. FT.

μ_e = VISCOSITY, LB. MASS/FT.-SEC.

Chart 19. Viscosity of fluids

Chart 20. Correction of polytropic efficiency to adiabatic efficiency.

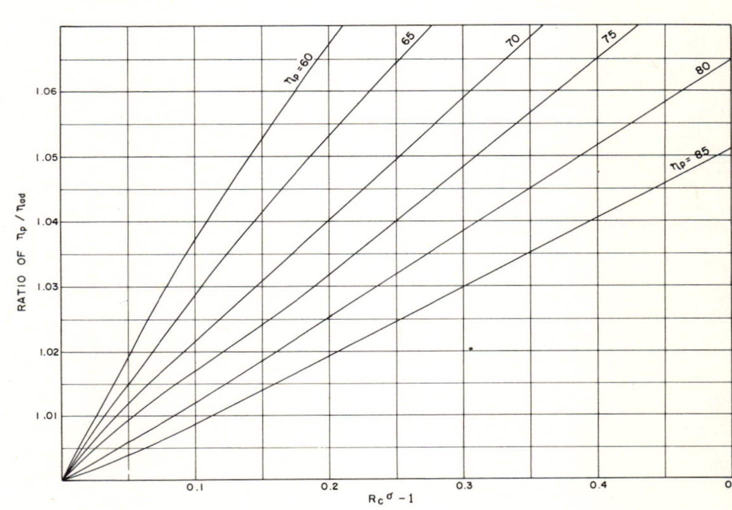

Chart 21. Compressibility nomograph for low-pressure hydrocarbons. (Courtesy of Hydrocarbon Processing.*)*

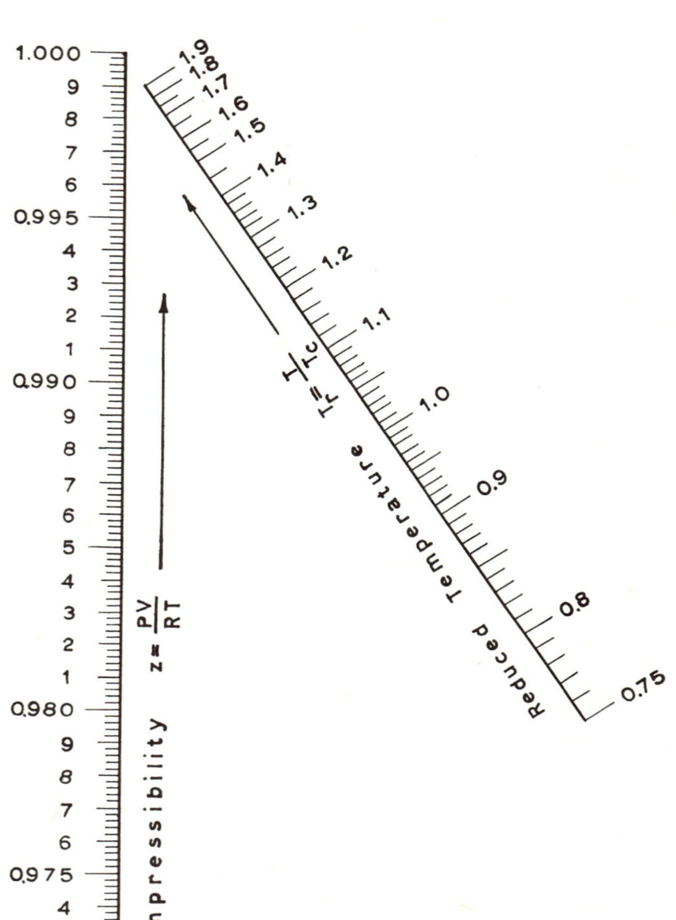

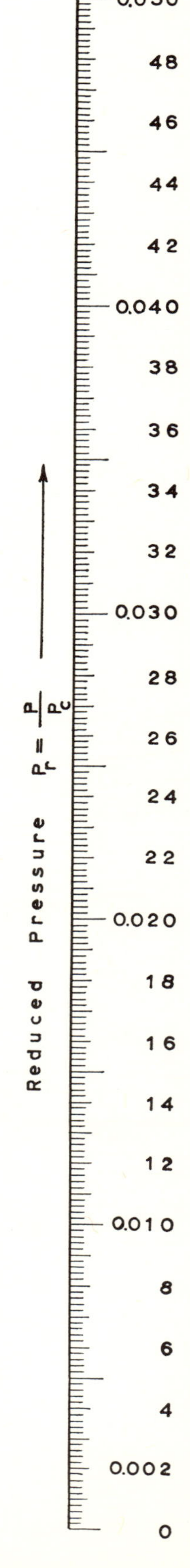

THE ACCOMPANYING LOW PRESSURE compressibility nomograph holds for the following hydrocarbons: methane, ethane, propane, *n*-butane, ethylene, propylene and *n*-butene. It has been constructed from the data of Pfennig and McKetta.[1]

In using the nomograph the following critical properties should be used.

	P_c atm	T_c°K
Methane	45.8	191.1
Ethane	48.2	305.5
Propane	42.0	370.0
n-Butane	37.5	425.2
Ethylene	50.0	282.4
Propylene	45.6	365
n-Butene	39.7	419.6

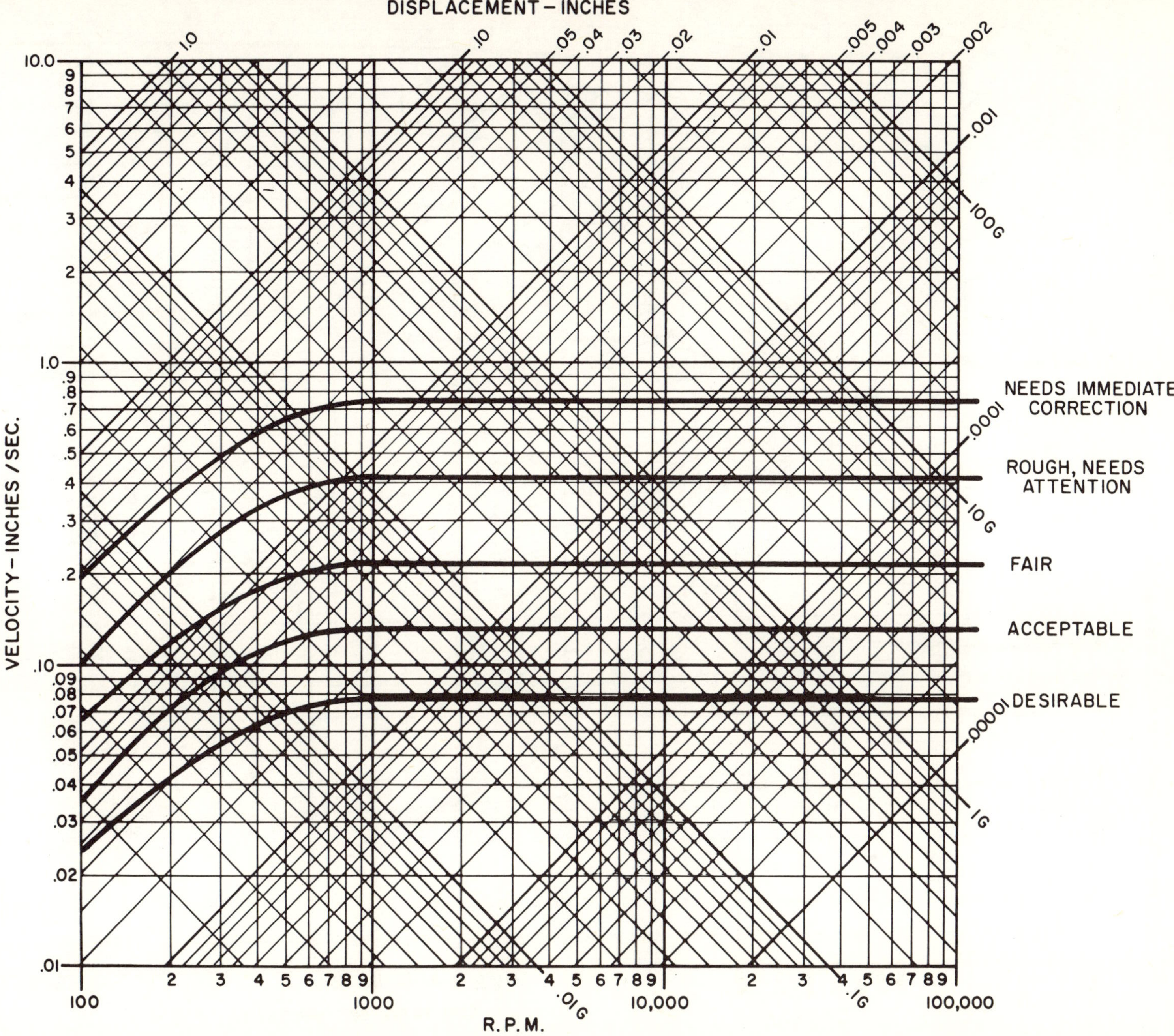

Chart 22. Vibration intensity evaluation.

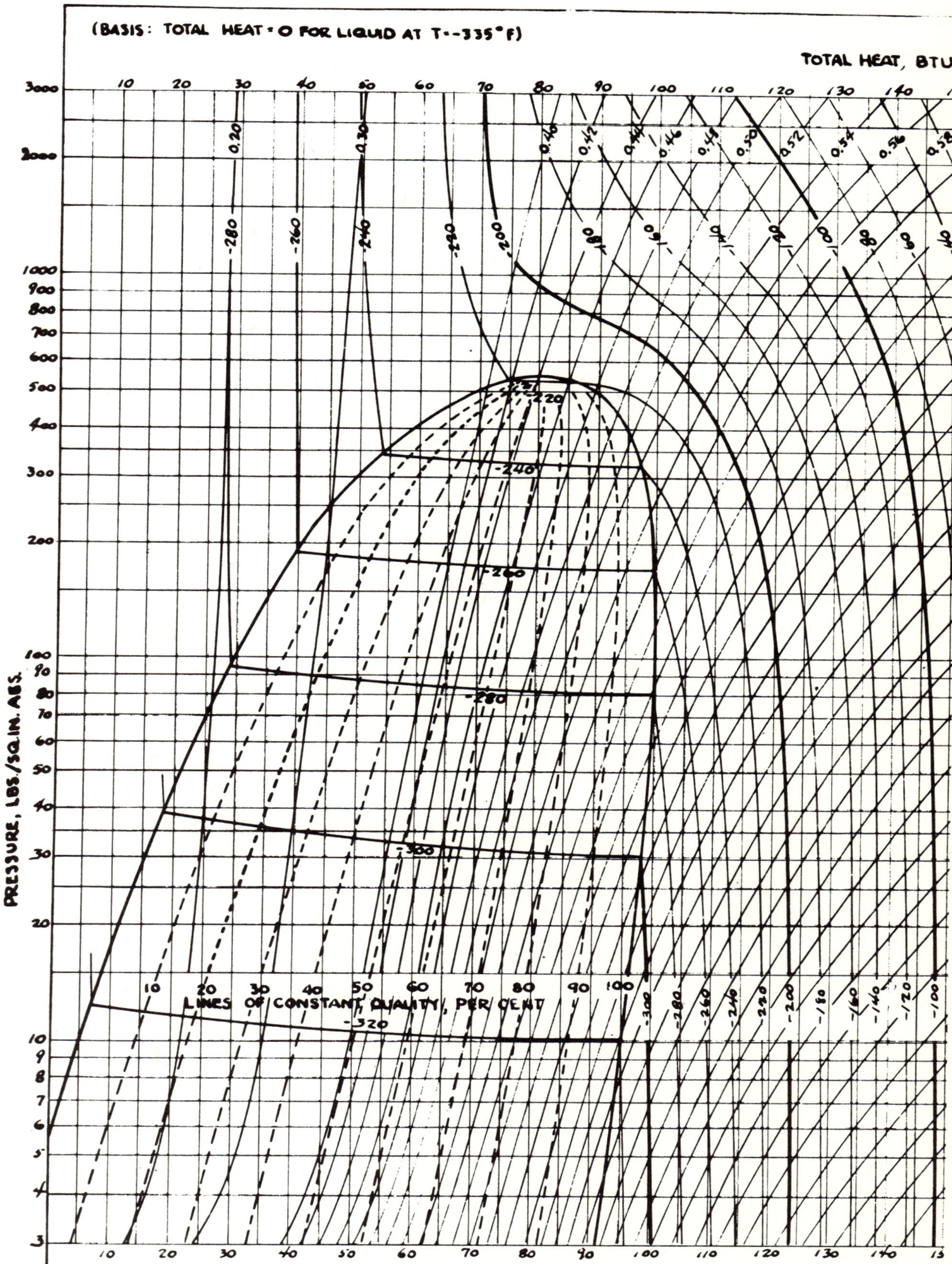

Chart 23. Mollier chart for air.

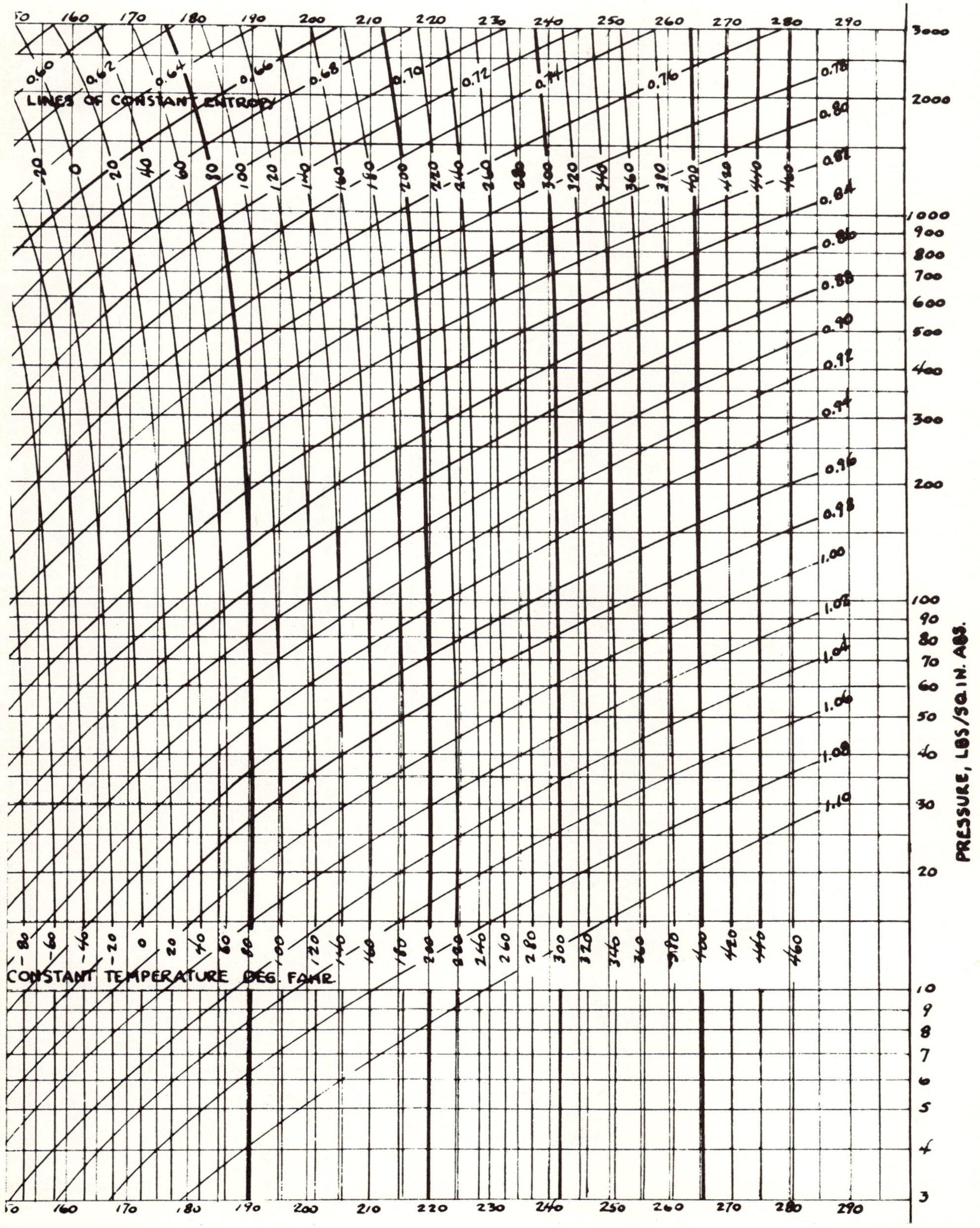

ENTHALPY BTU/LB

Chart 23 (continued)

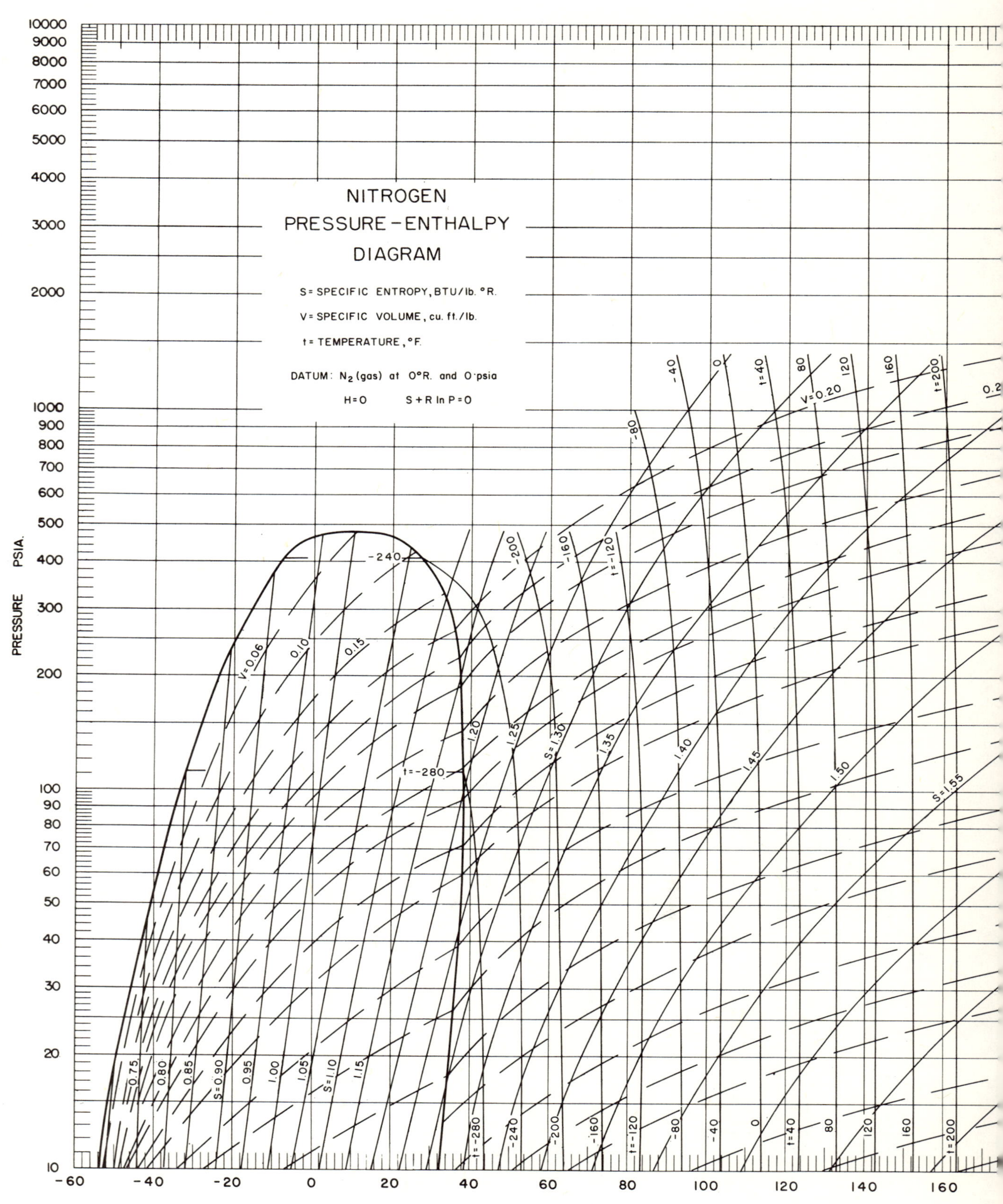

Chart 24. Mollier chart for nitrogen.

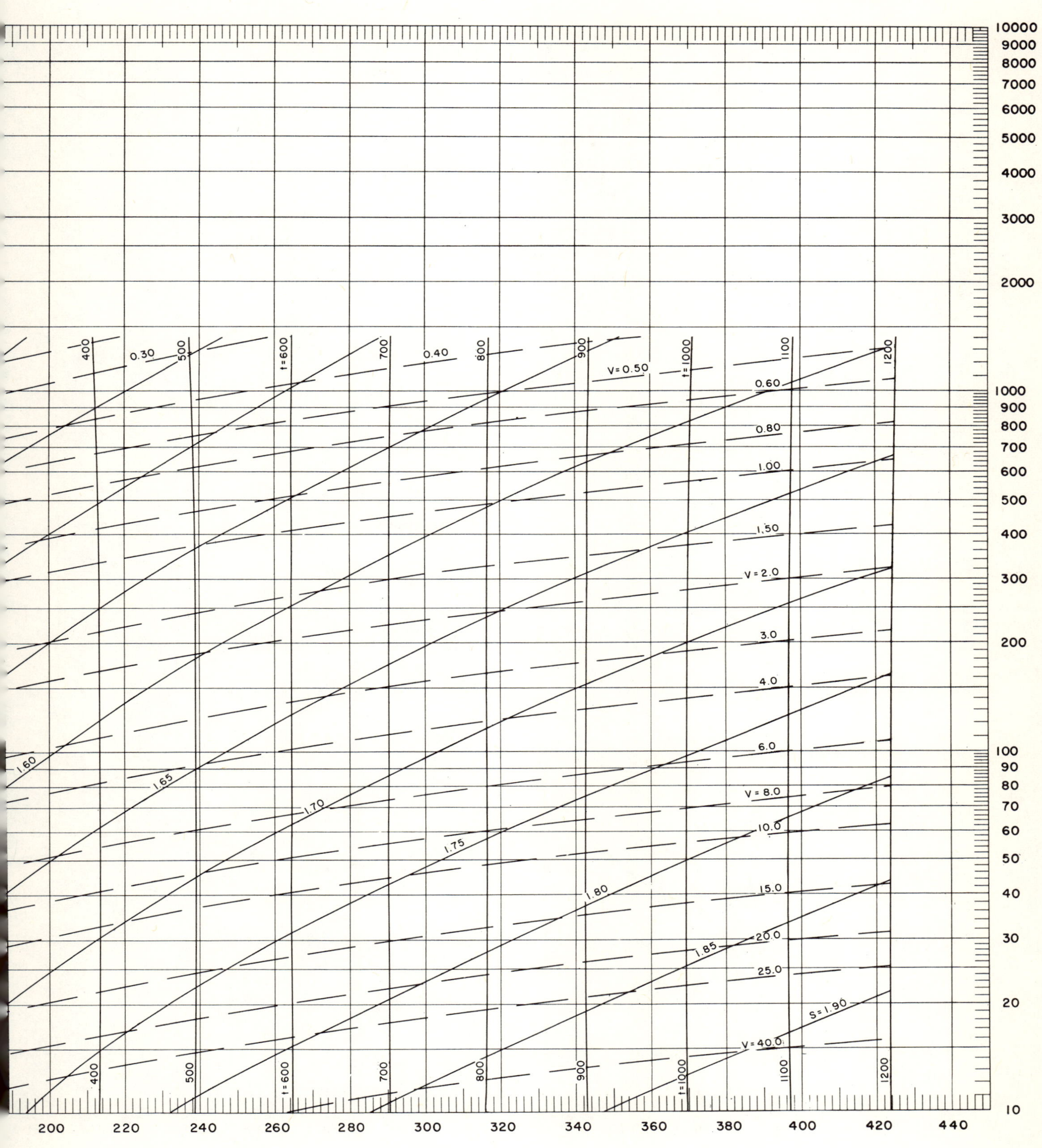

Chart 24 (continued)

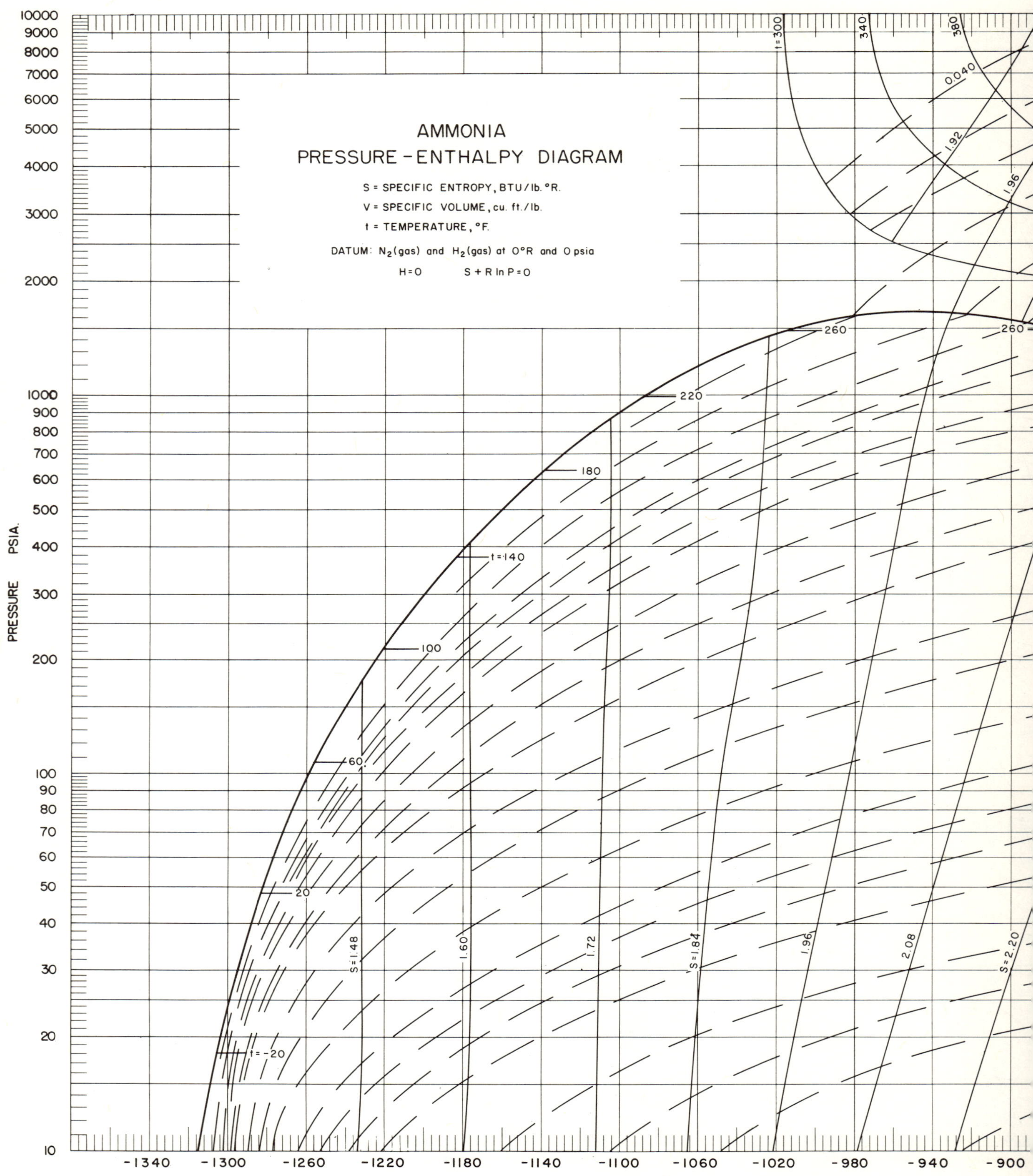

Chart 25. Mollier chart for ammonia.

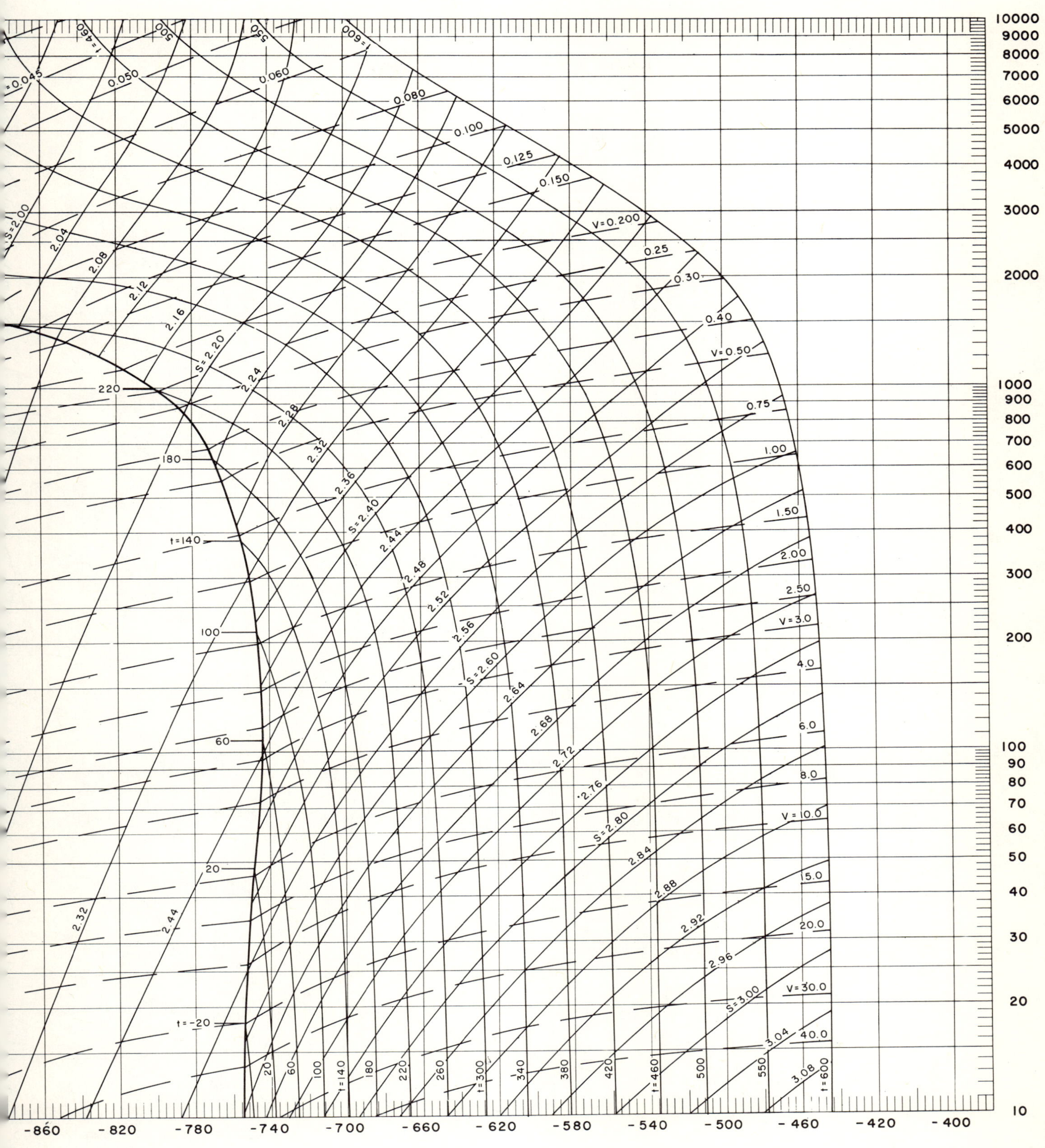

Chart 25 (continued)

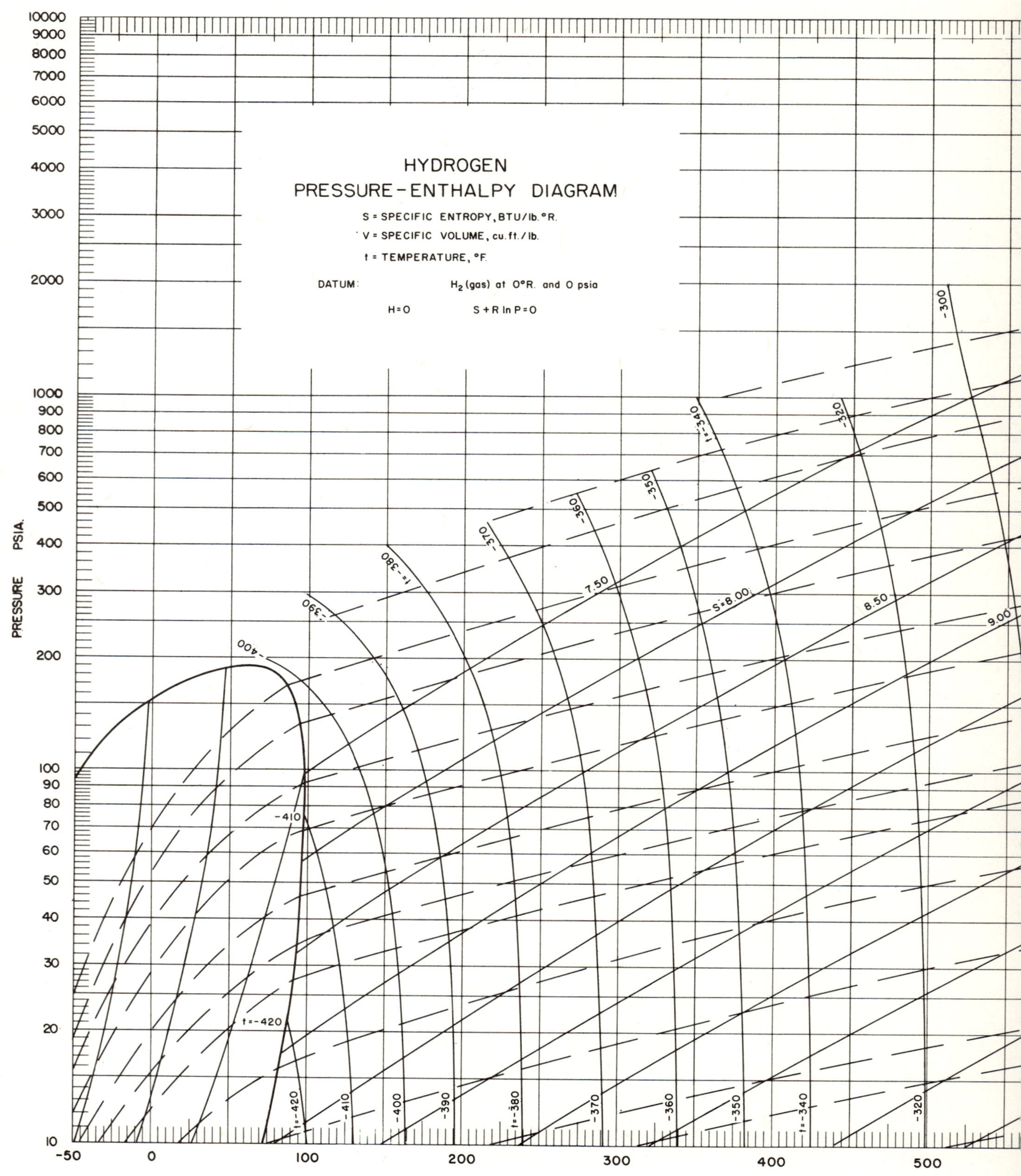

Chart 26. Mollier chart for hydrogen, low range.

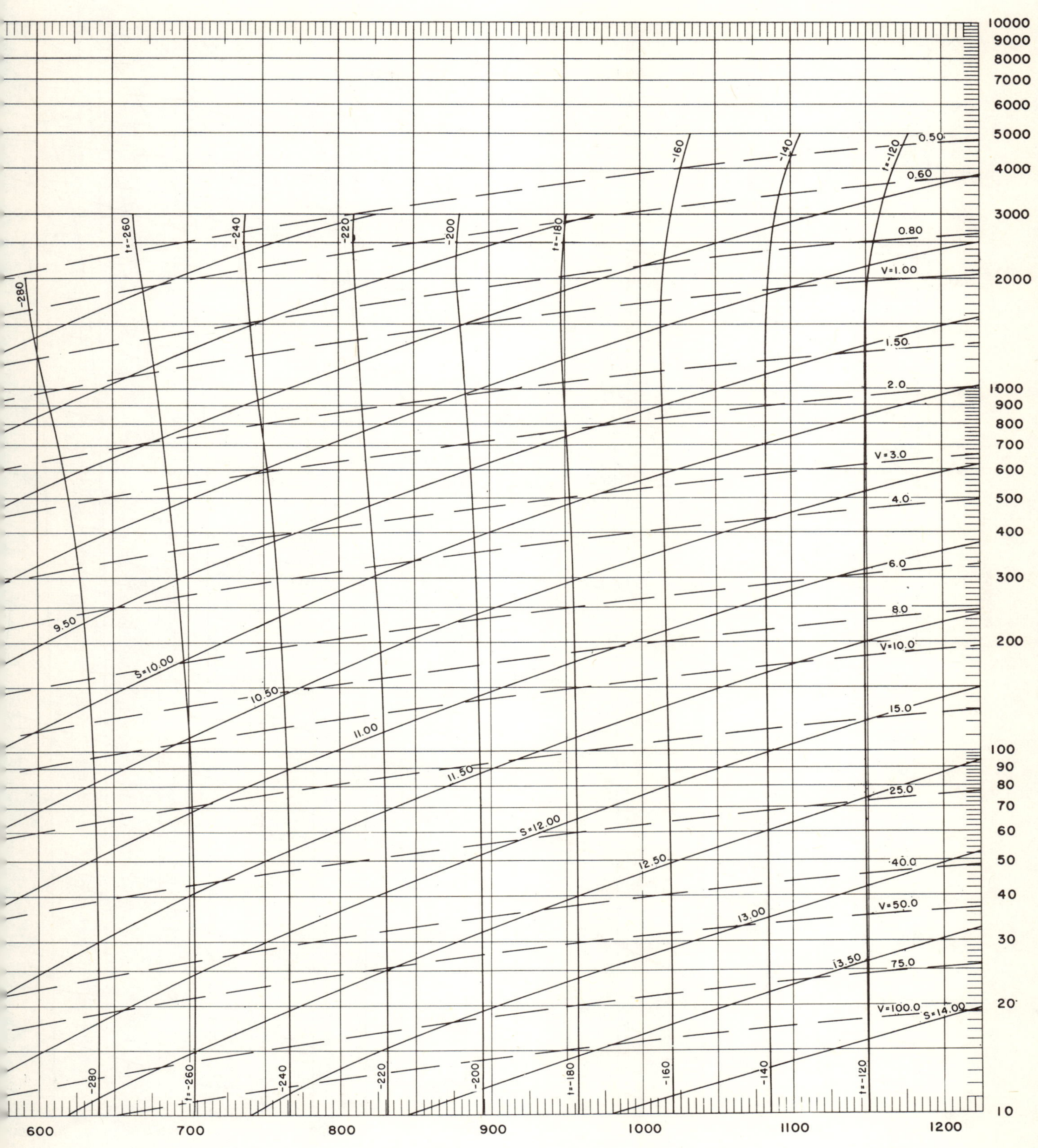

Chart 26 (continued)

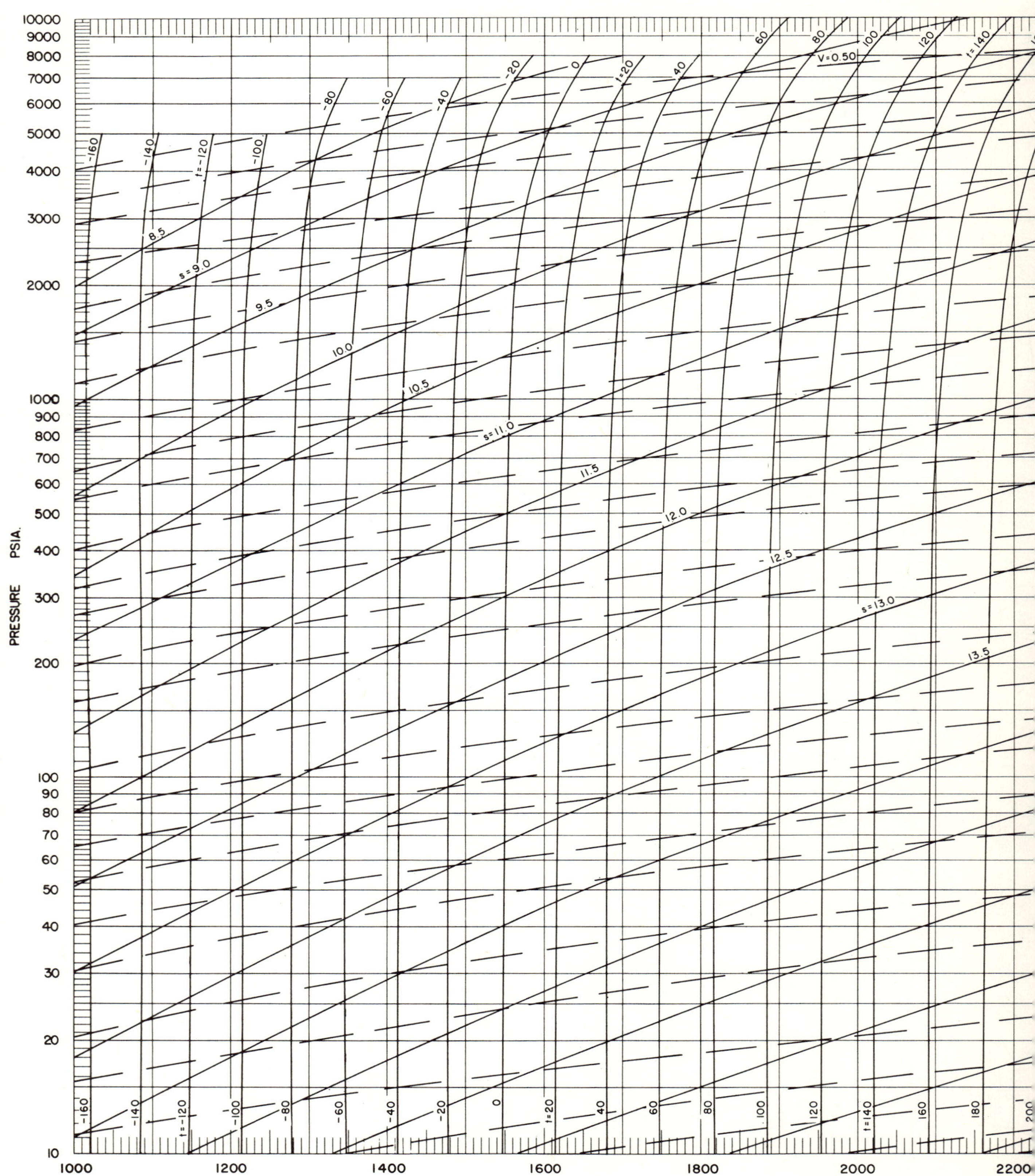

Chart 27. Mollier chart for hydrogen, high range.

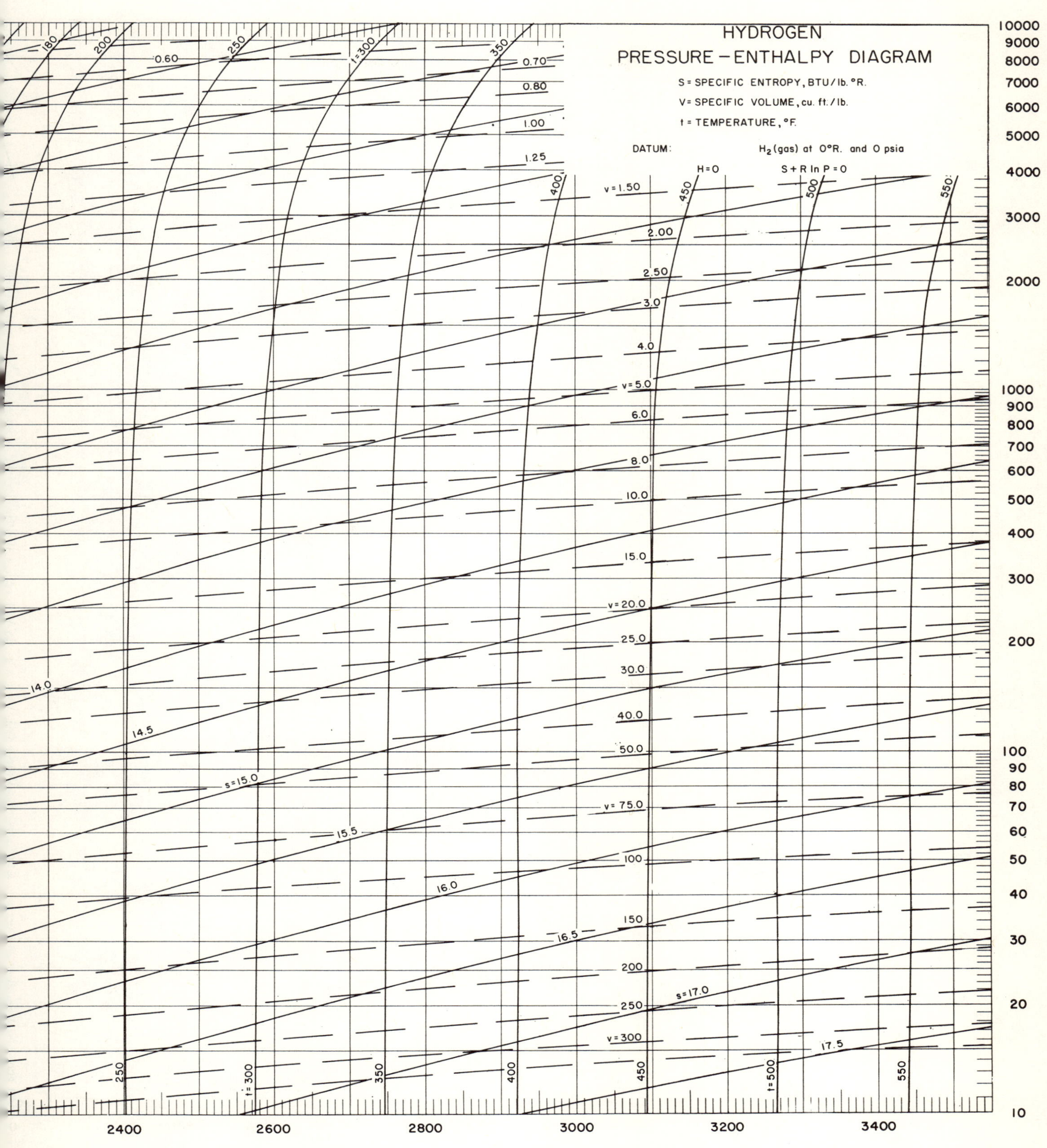

Chart 27 (continued)

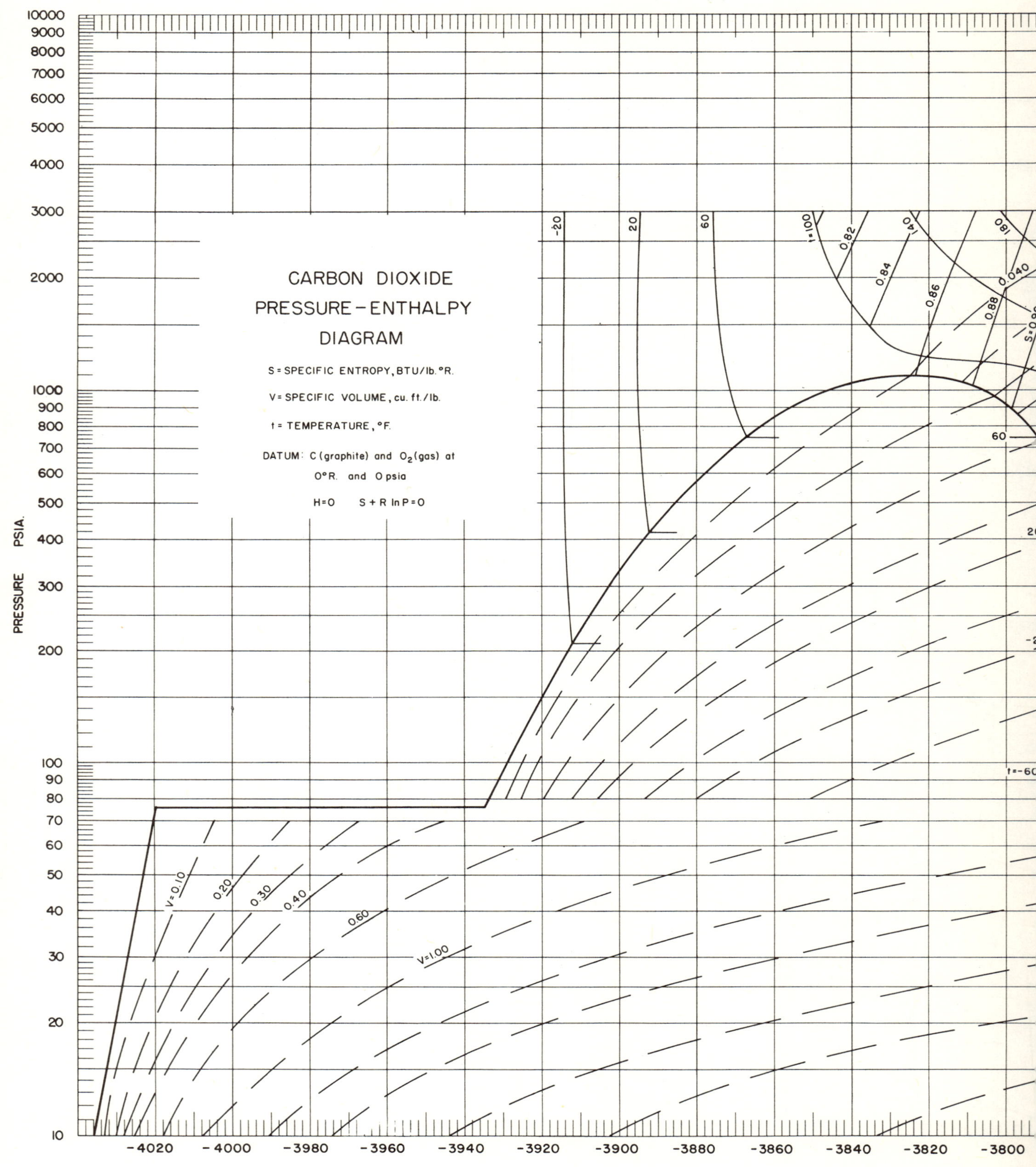

Chart 28. Mollier chart for carbon dioxide.

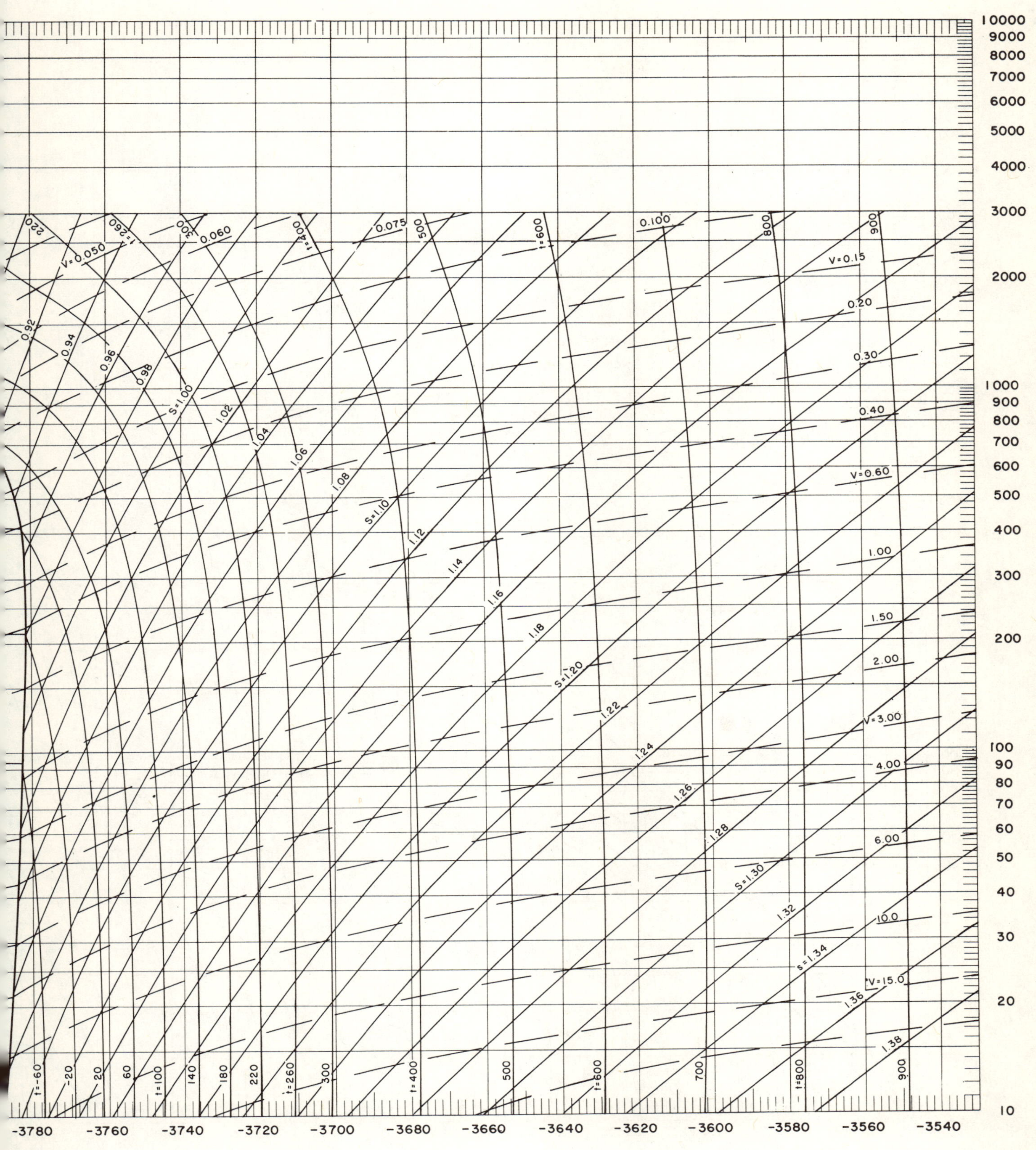

ENTHALPY BTU/LB

Chart 28 (continued)

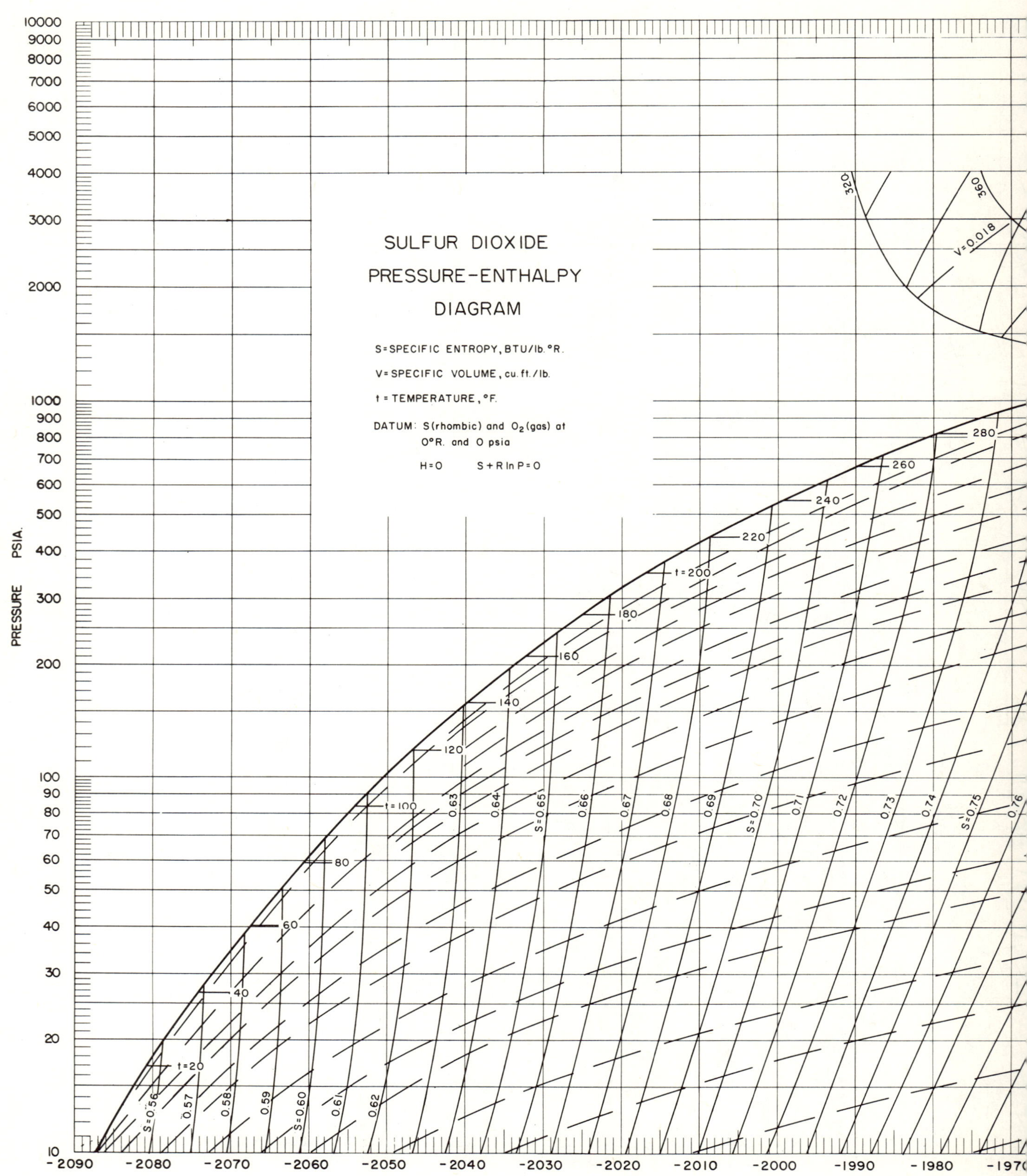

Chart 29. Mollier chart for sulfur dioxide.

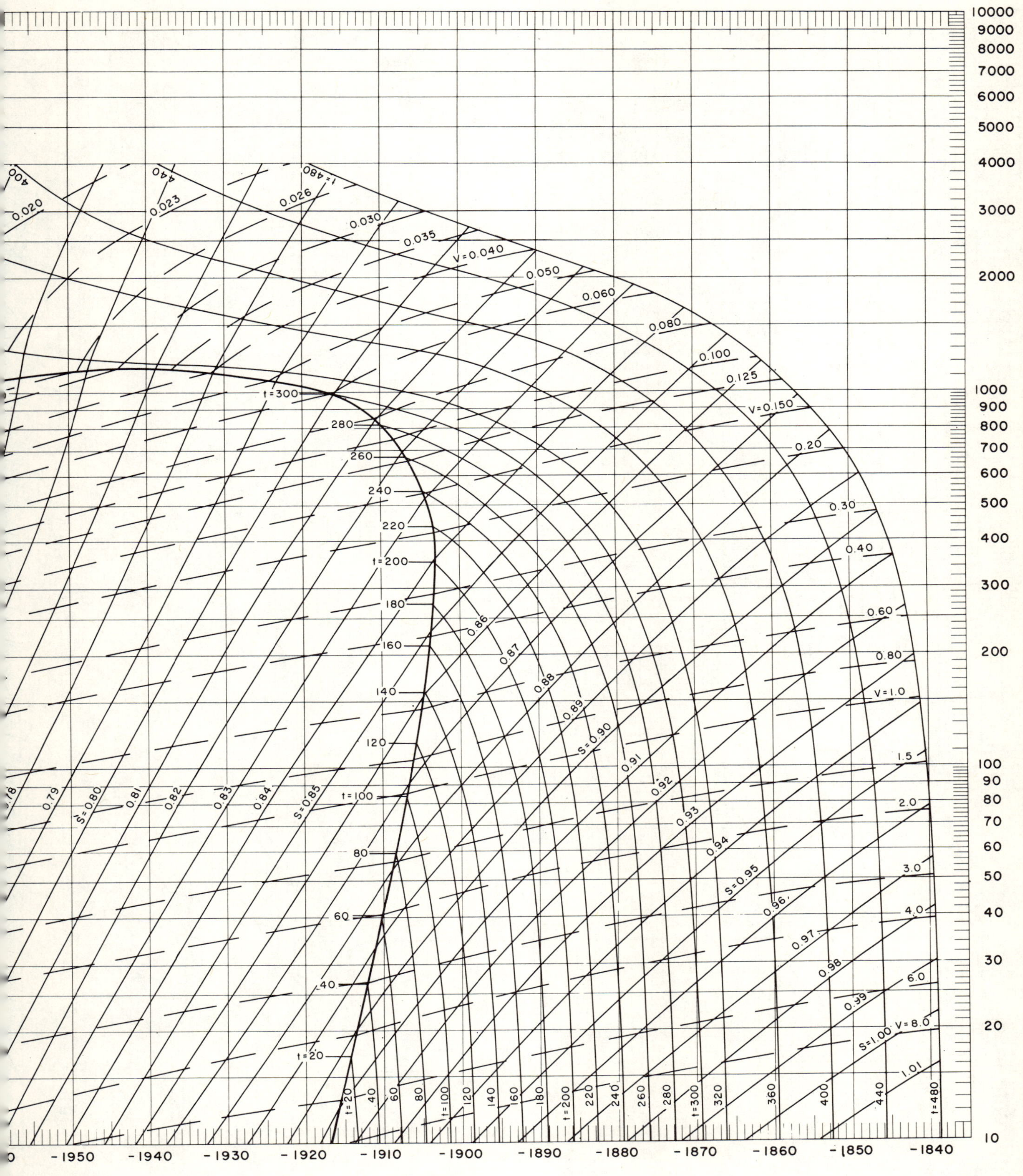

ENTHALPY BTU/LB

Chart 29 (continued)

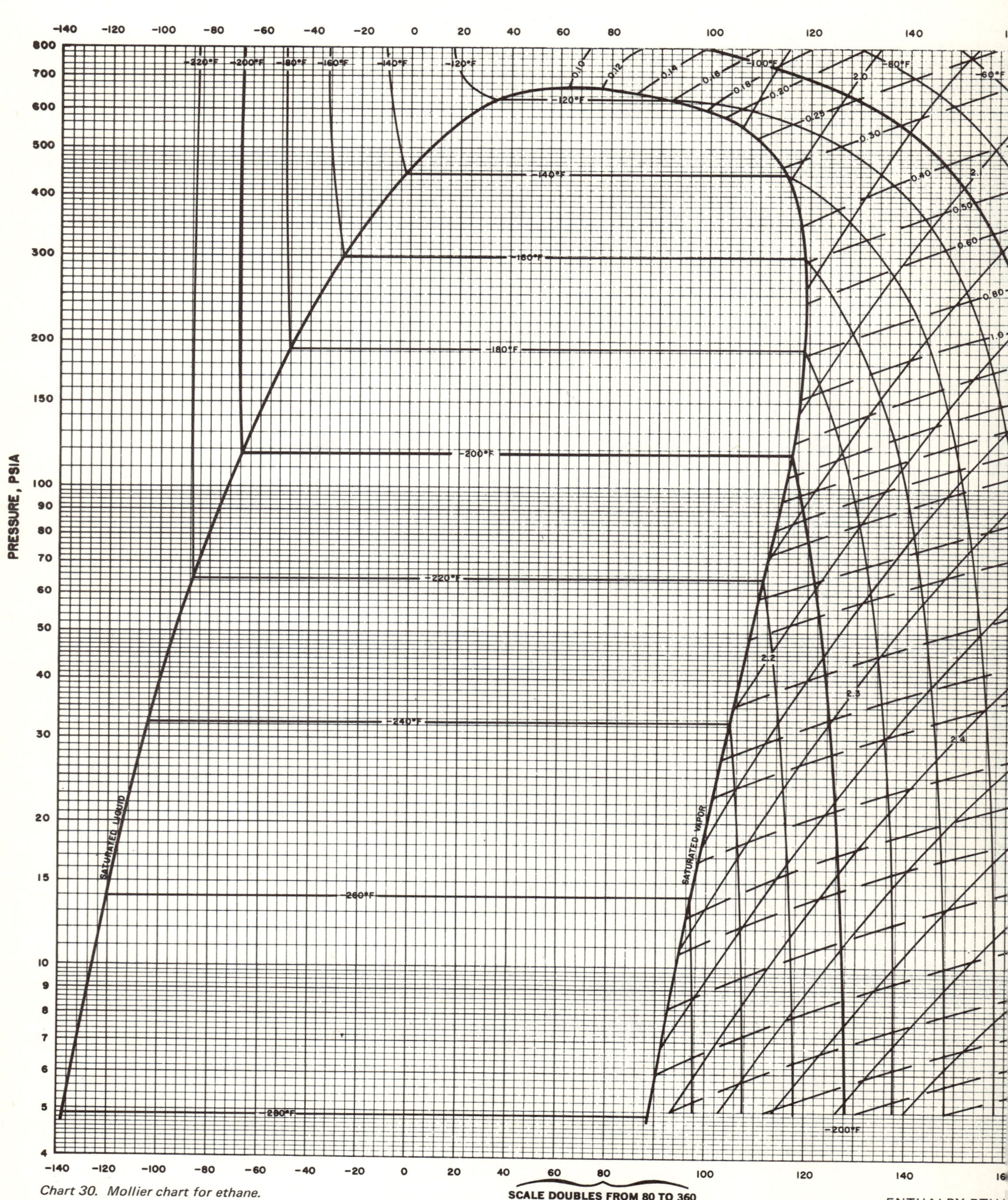

Chart 30. Mollier chart for ethane.

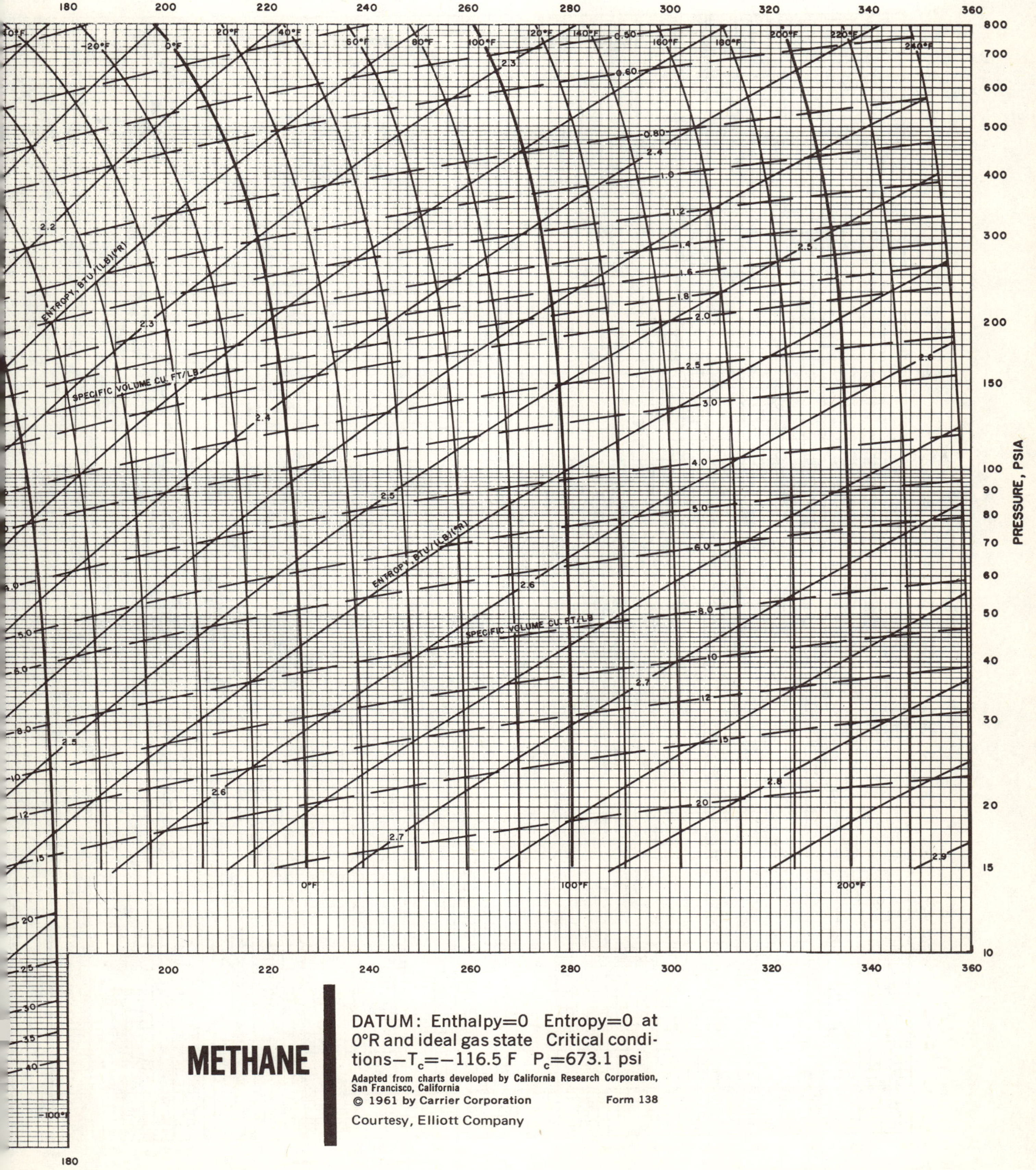

DATUM: Enthalpy=0 Entropy=0 at 0°R and ideal gas state Critical conditions—T_c=−116.5 F P_c=673.1 psi

Adapted from charts developed by California Research Corporation, San Francisco, California

 Form 138

Courtesy, Elliott Company

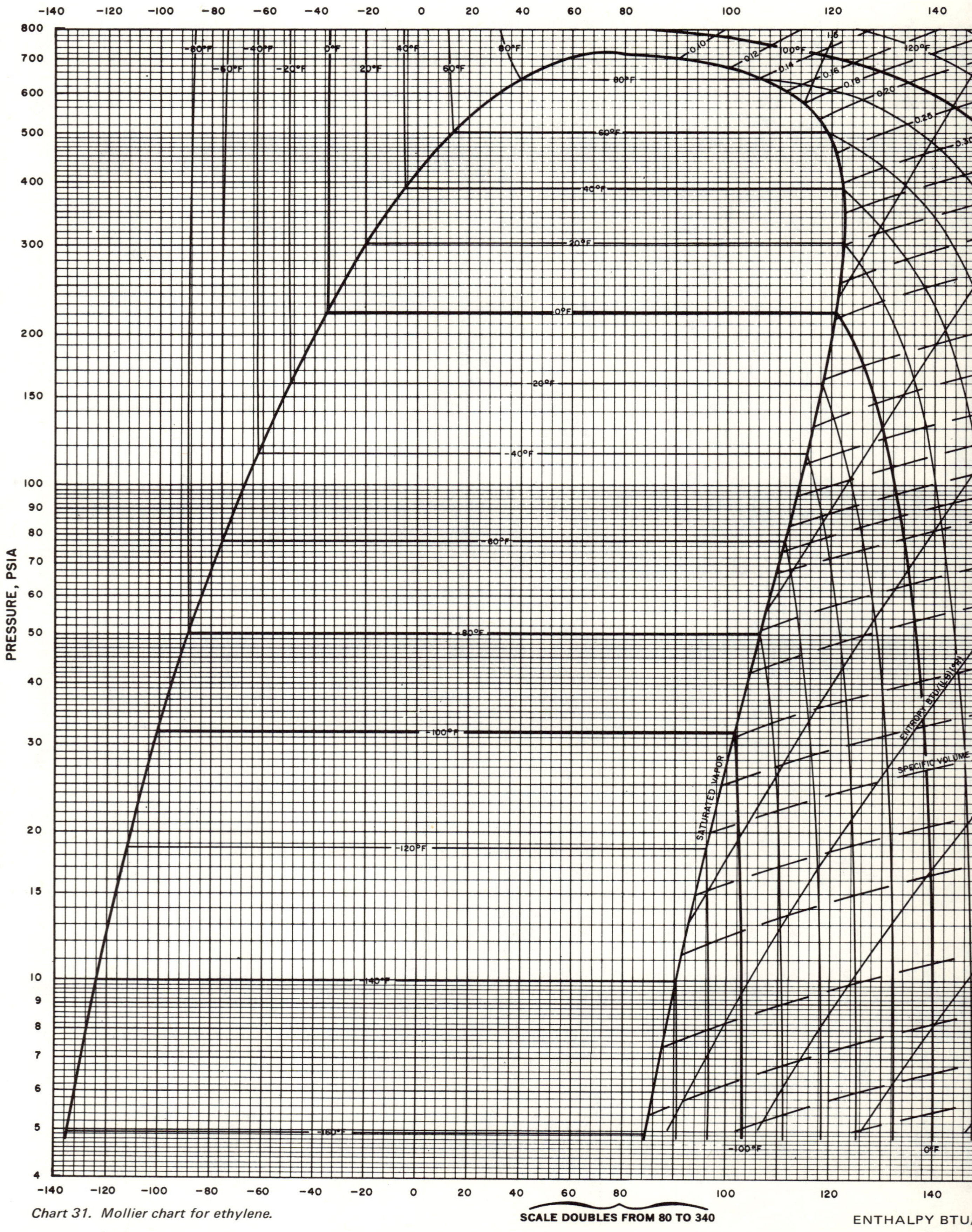

Chart 31. Mollier chart for ethylene.

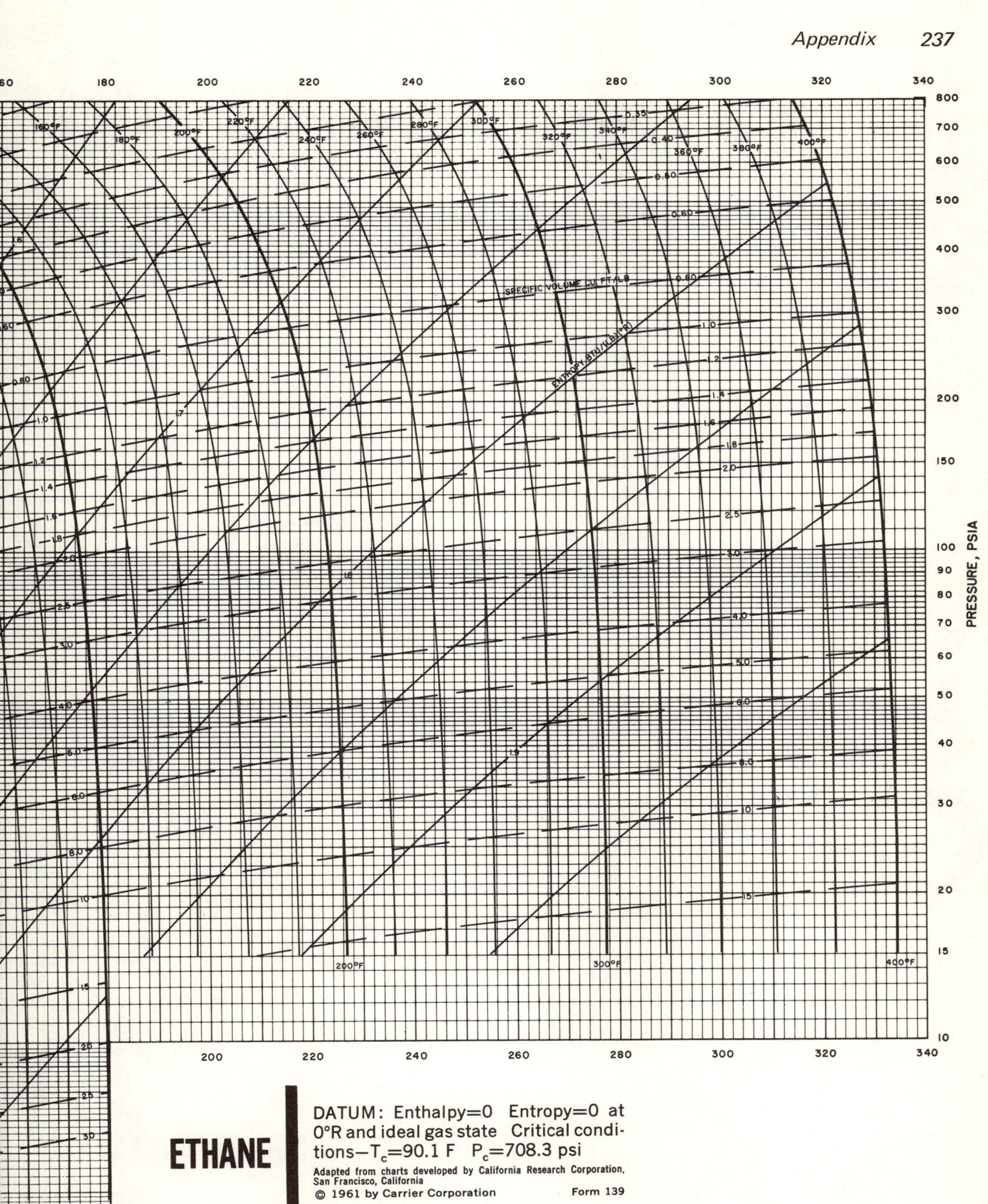

ETHANE

DATUM: Enthalpy=0 Entropy=0 at 0°R and ideal gas state Critical conditions—T_c=90.1 F P_c=708.3 psi

Adapted from charts developed by California Research Corporation, San Francisco, California

 Form 139

Courtesy, Elliott Company

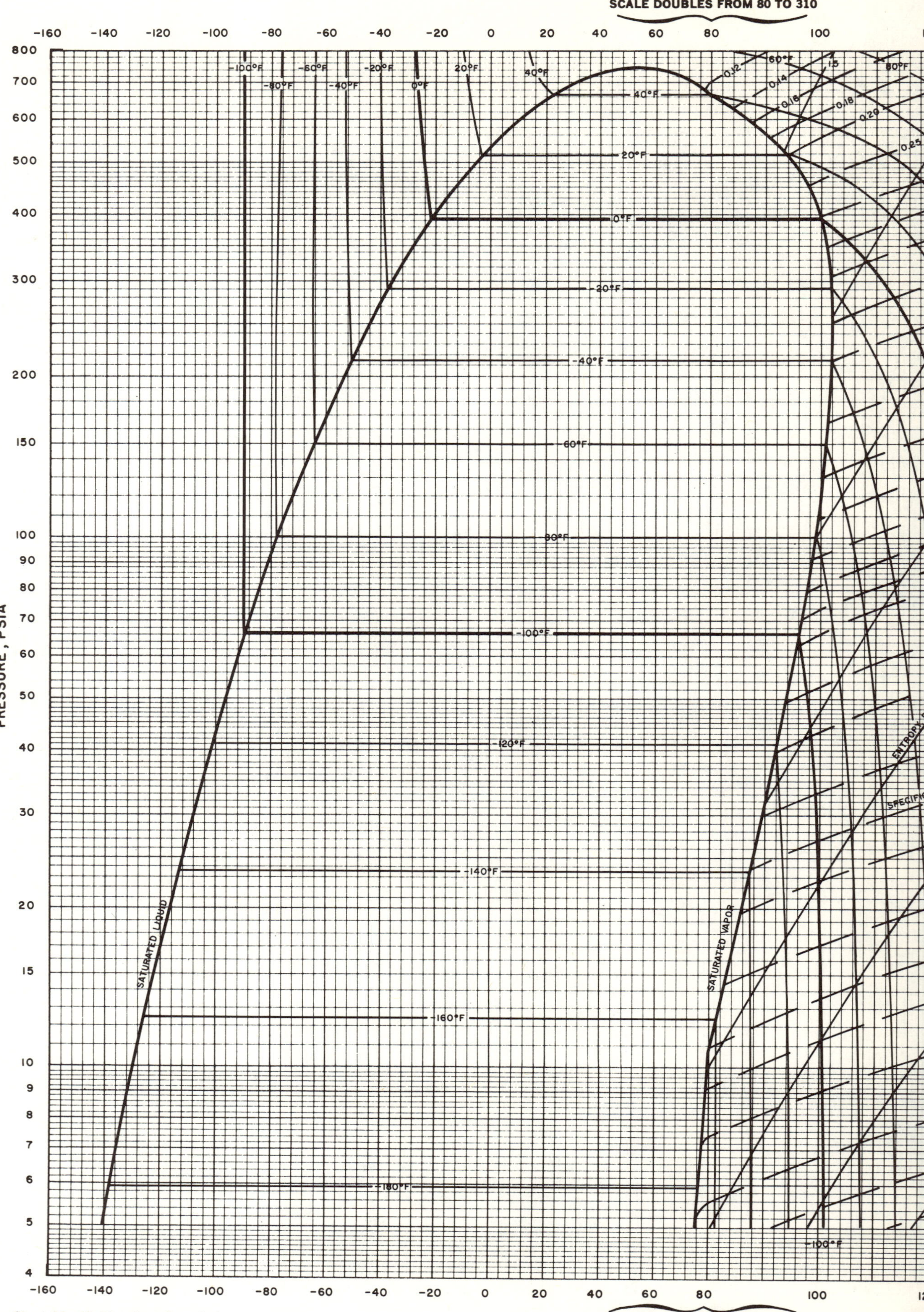

Chart 32. Mollier chart for ethylene.

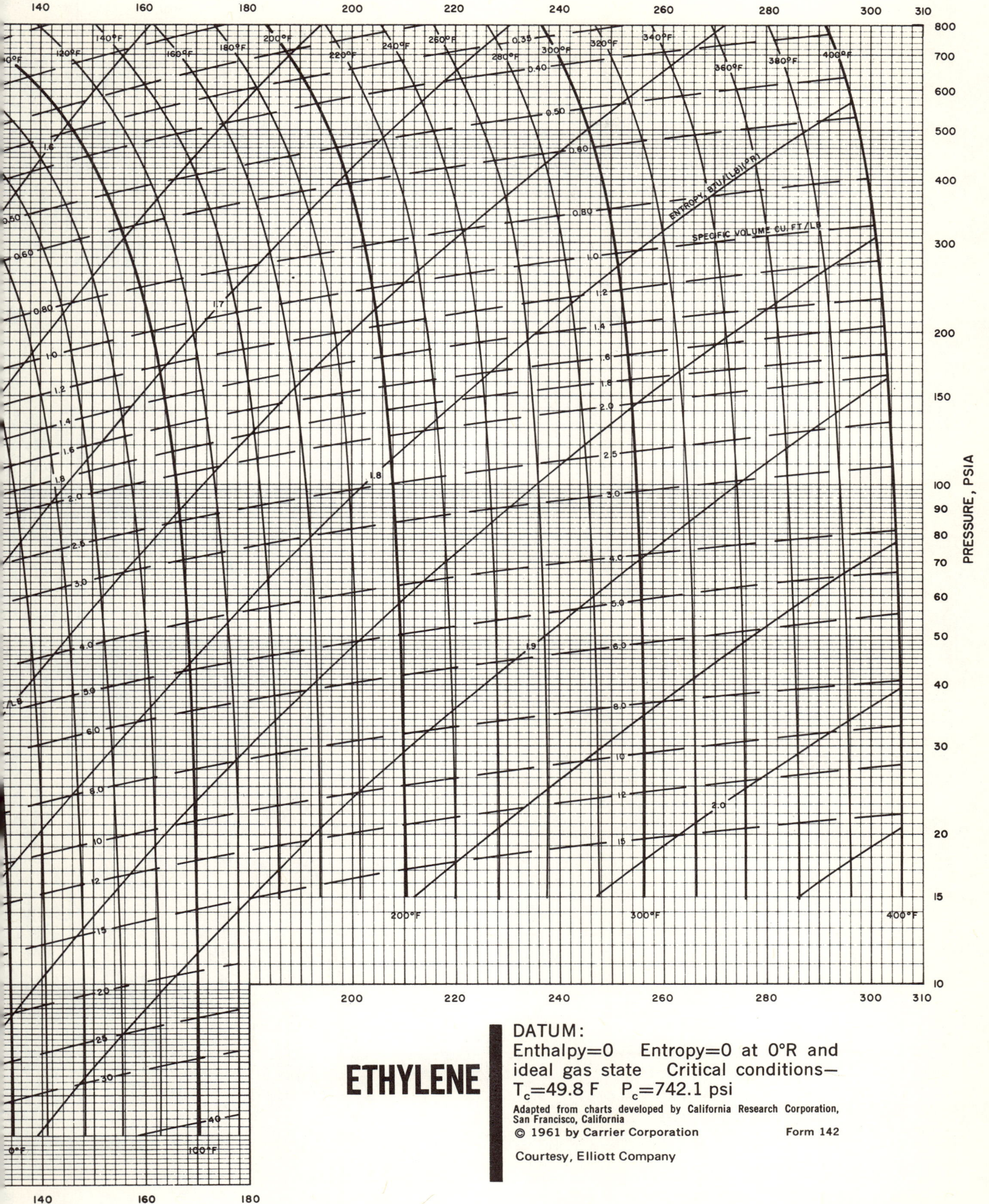

Chart 32 (continued)

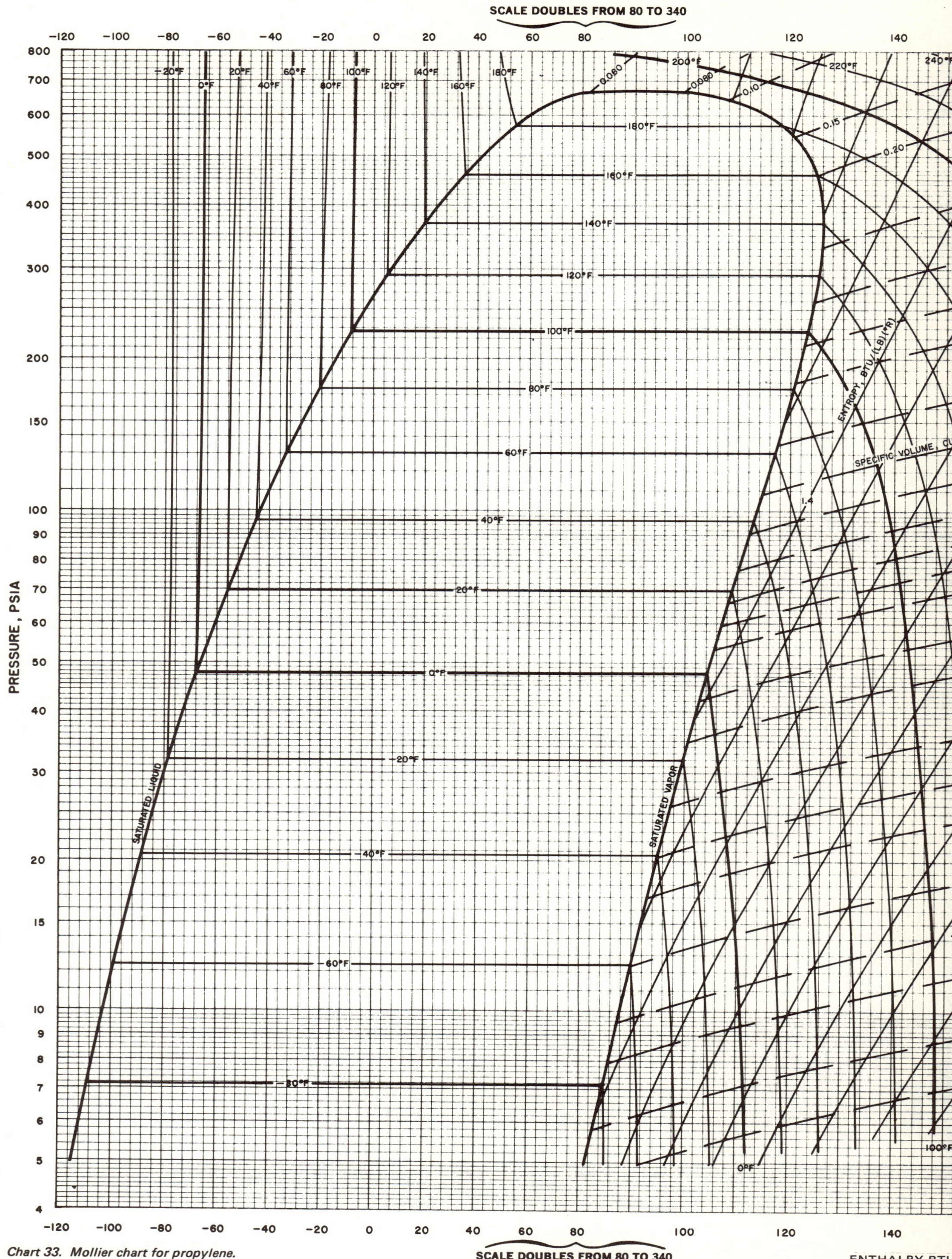

Chart 33. Mollier chart for propylene.

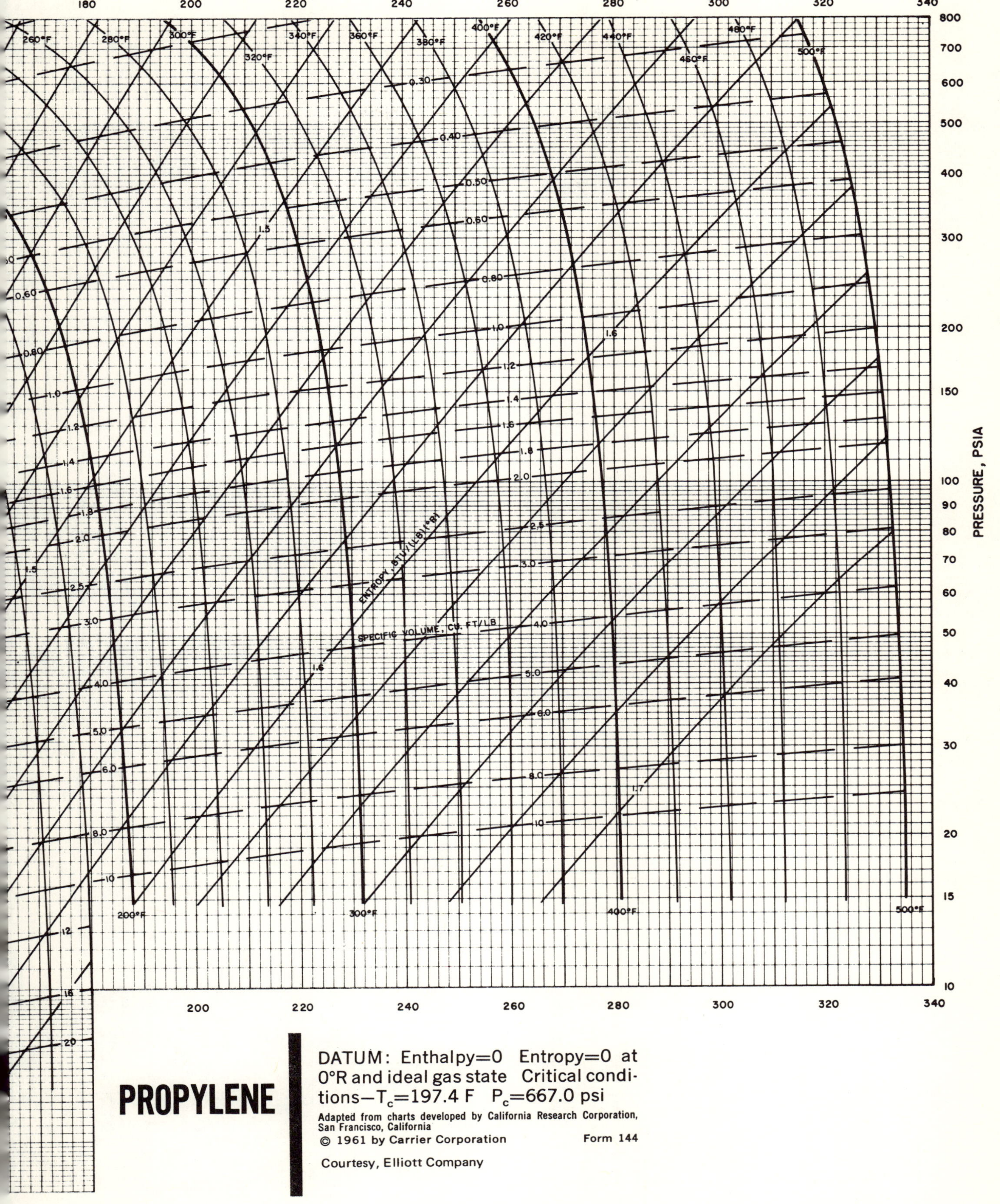

HALPY BTU/LB

Chart 33 (continued)

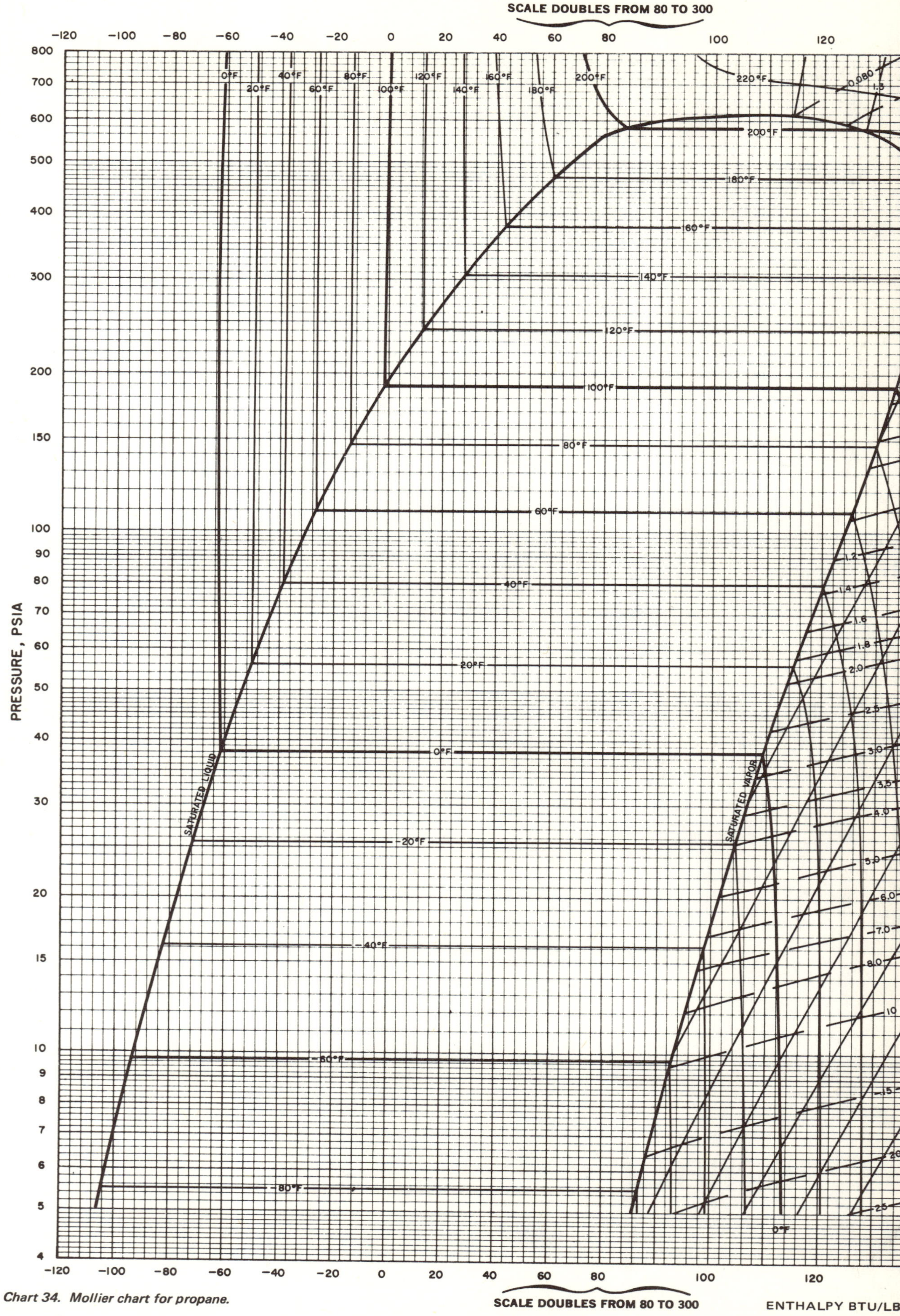

Chart 34. Mollier chart for propane.

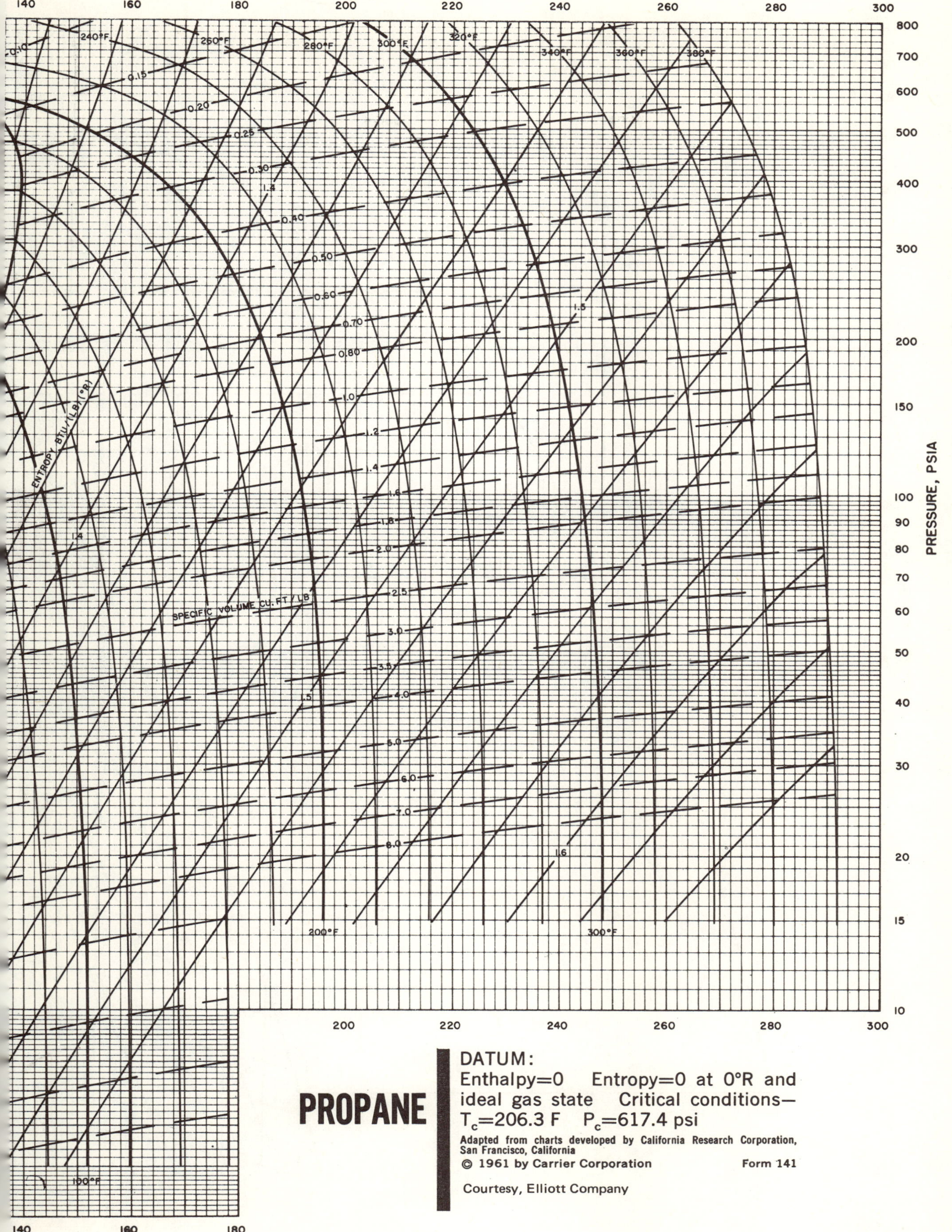

PROPANE

DATUM:
Enthalpy=0 Entropy=0 at 0°R and ideal gas state Critical conditions—
T_c=206.3 F P_c=617.4 psi

Adapted from charts developed by California Research Corporation, San Francisco, California

 Form 141

Courtesy, Elliott Company

ENTHALPY BTU/LB

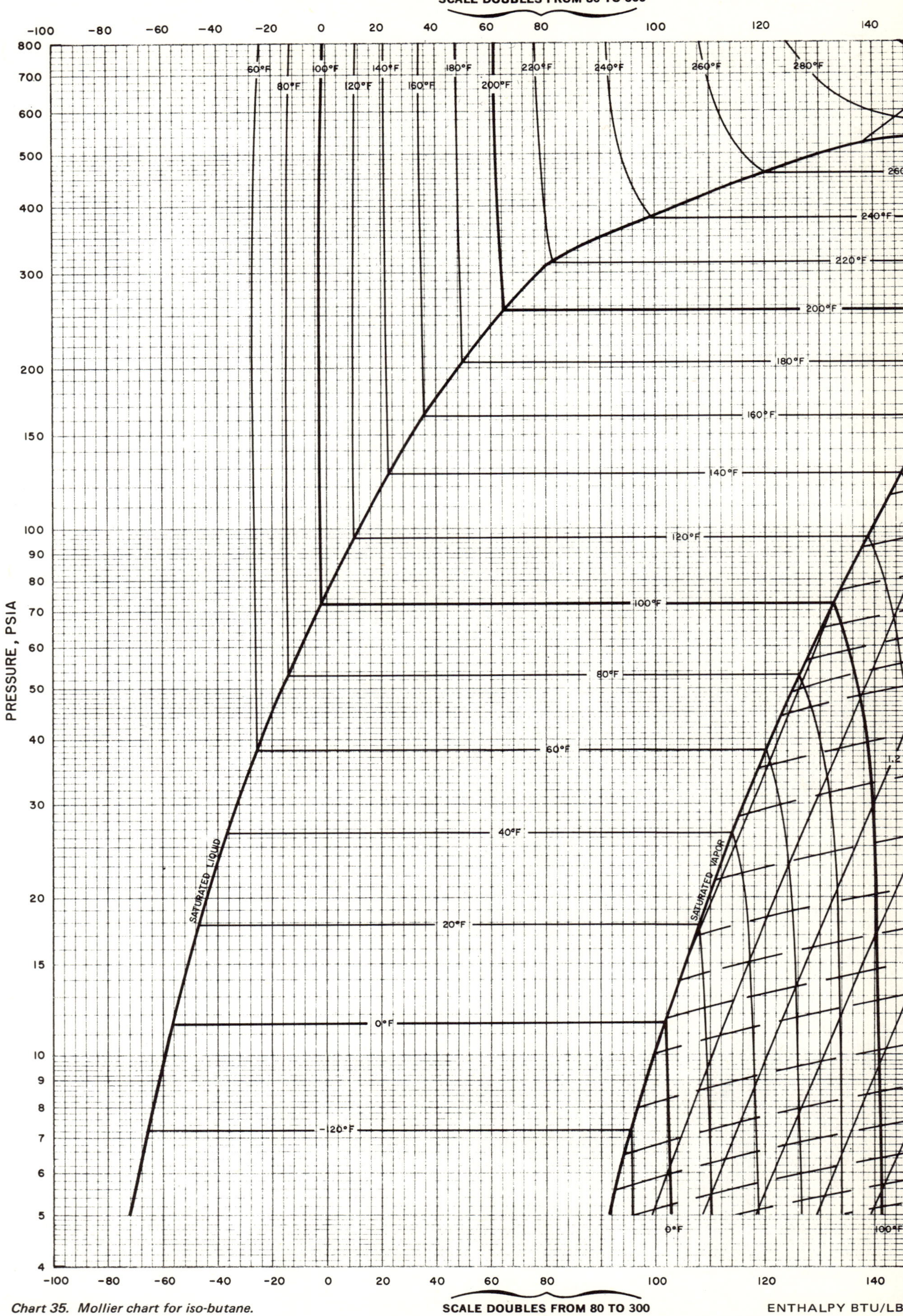

Chart 35. Mollier chart for iso-butane.

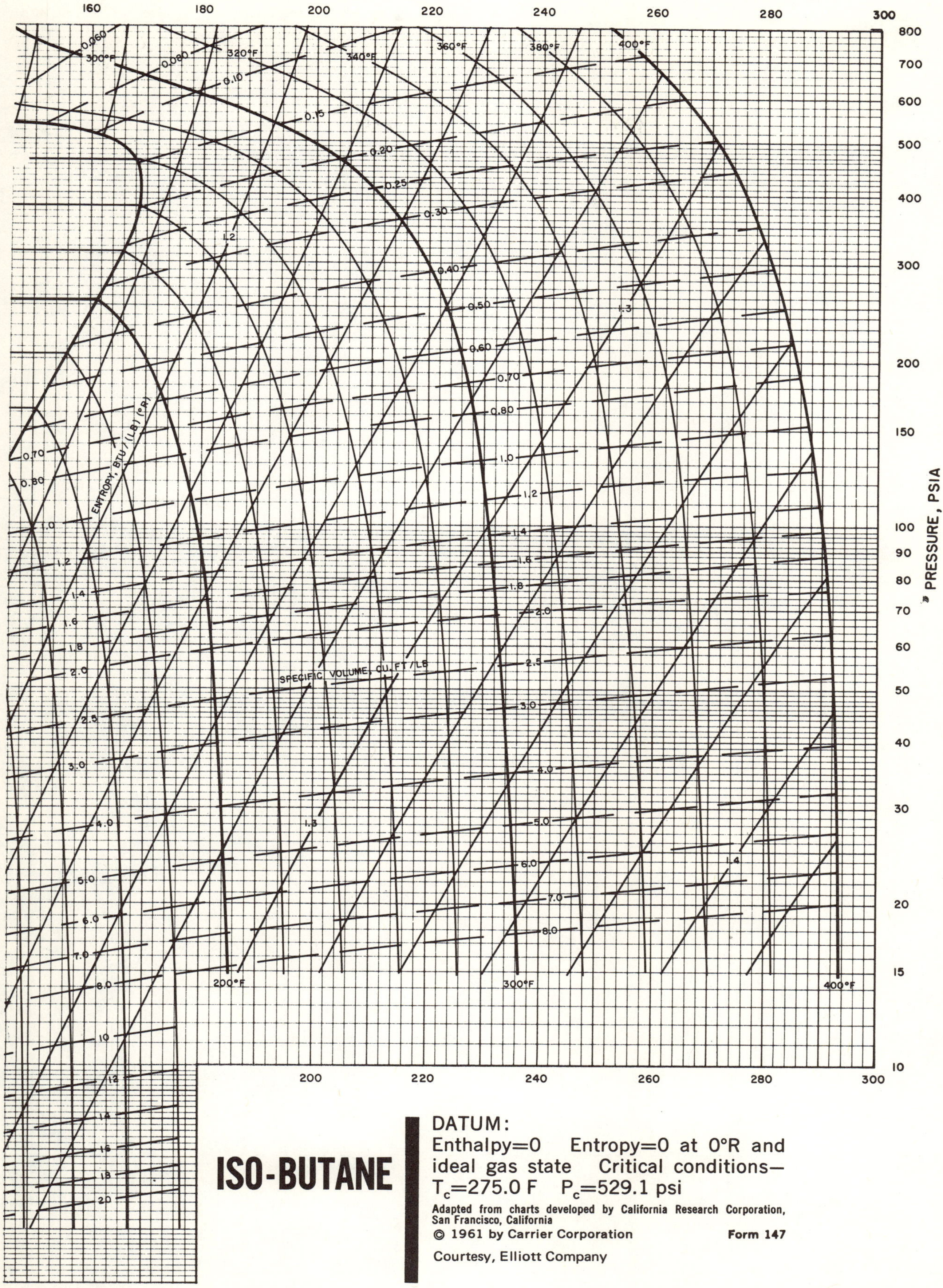

ISO-BUTANE

DATUM:
Enthalpy=0 Entropy=0 at 0°R and ideal gas state Critical conditions—
T_c=275.0 F P_c=529.1 psi

Adapted from charts developed by California Research Corporation, San Francisco, California

 Form 147

Courtesy, Elliott Company

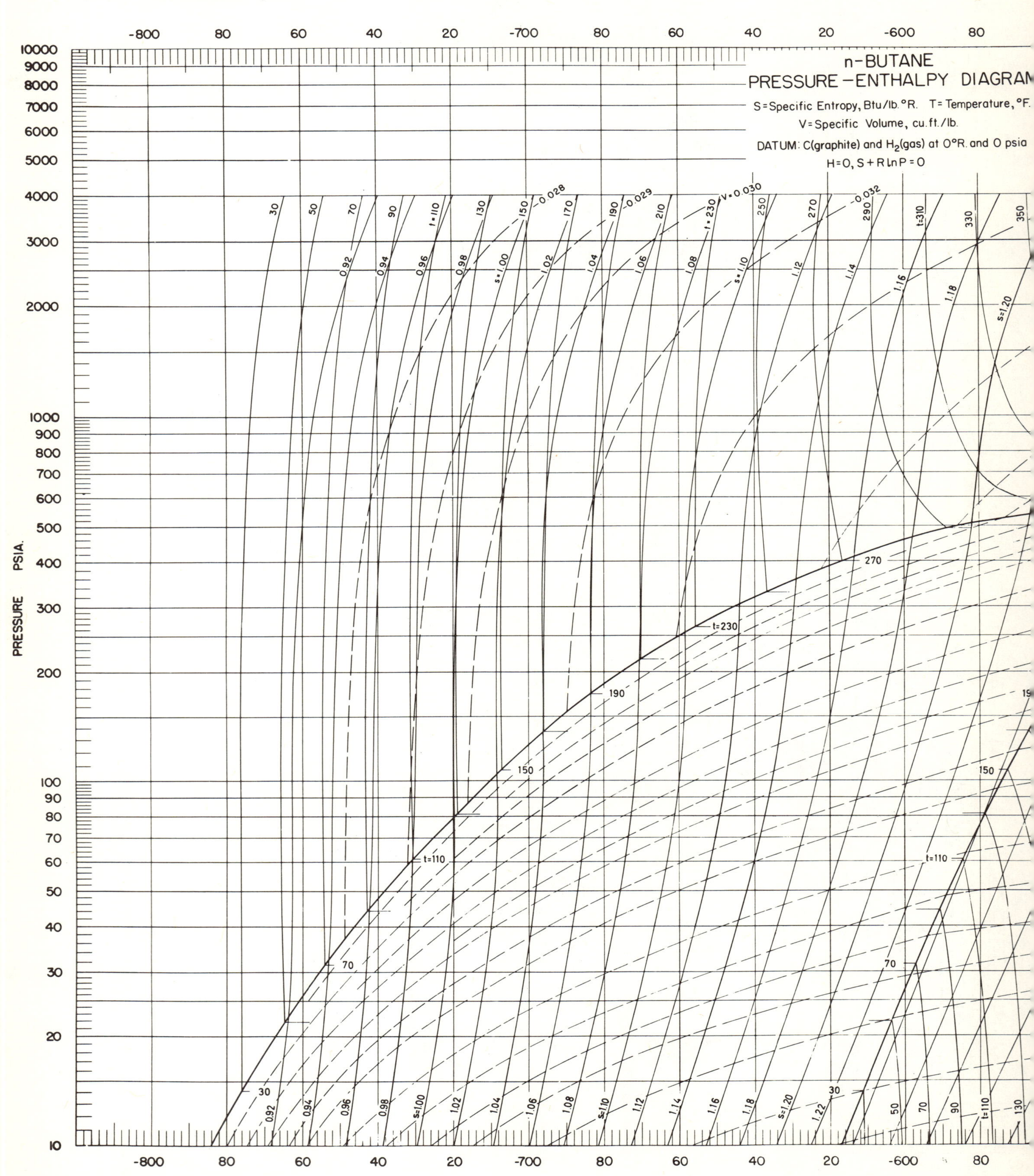

Chart 36. Mollier chart for n-butane.

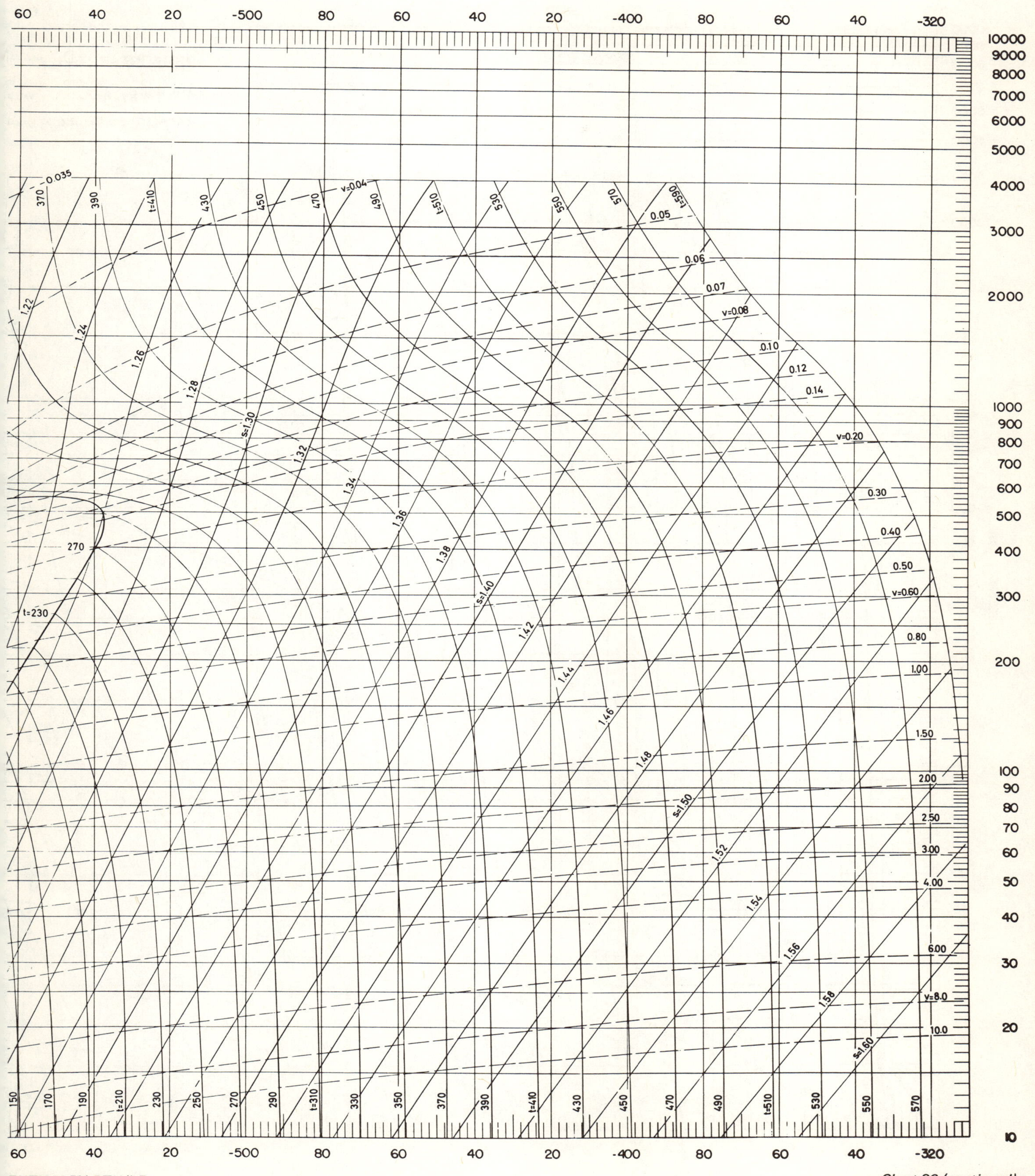

ENTHALPY BTU/LB

Chart 36 (continued)

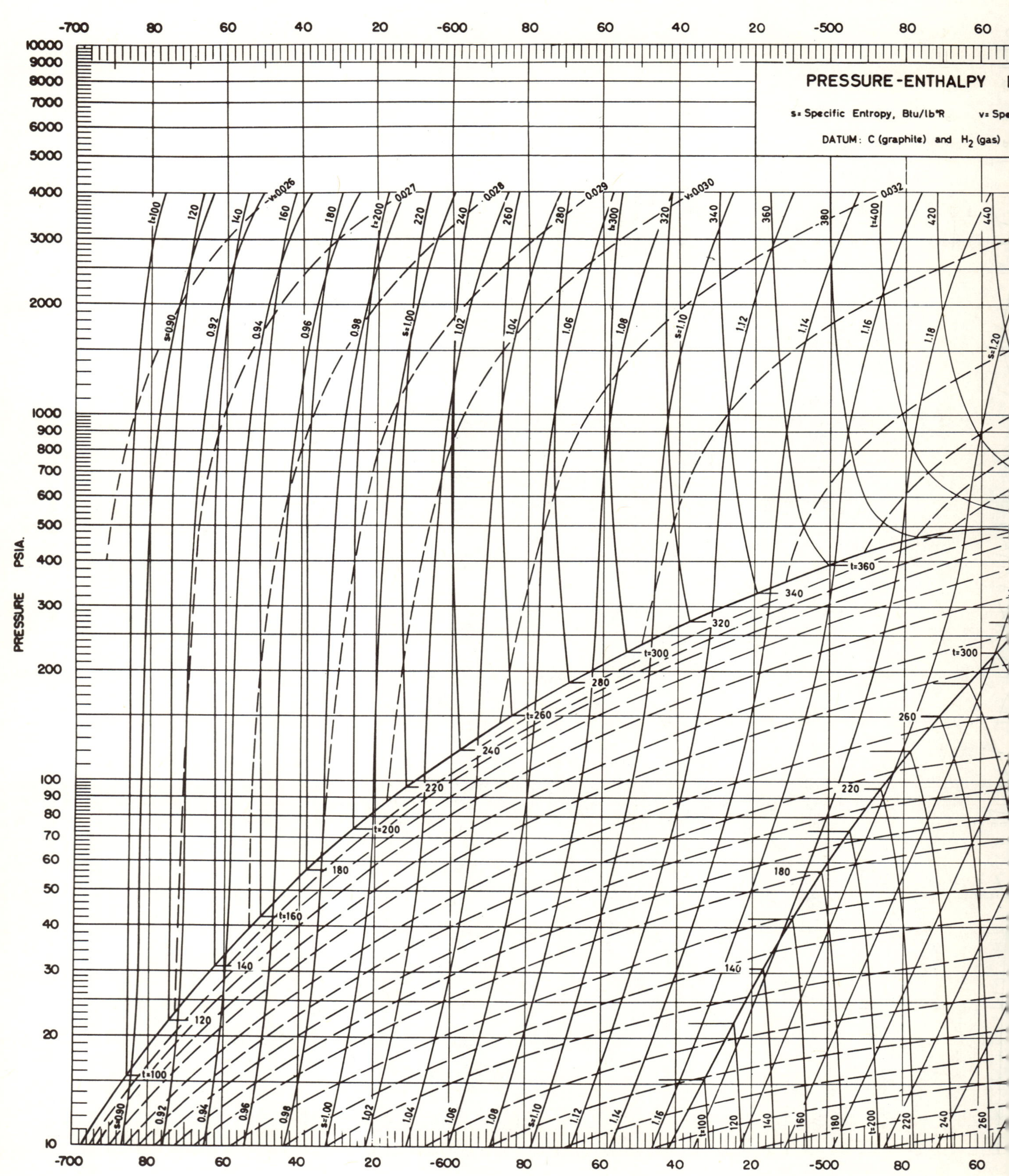

Chart 37. Mollier chart for n-pentane.

ENTHALPY BTU/LB

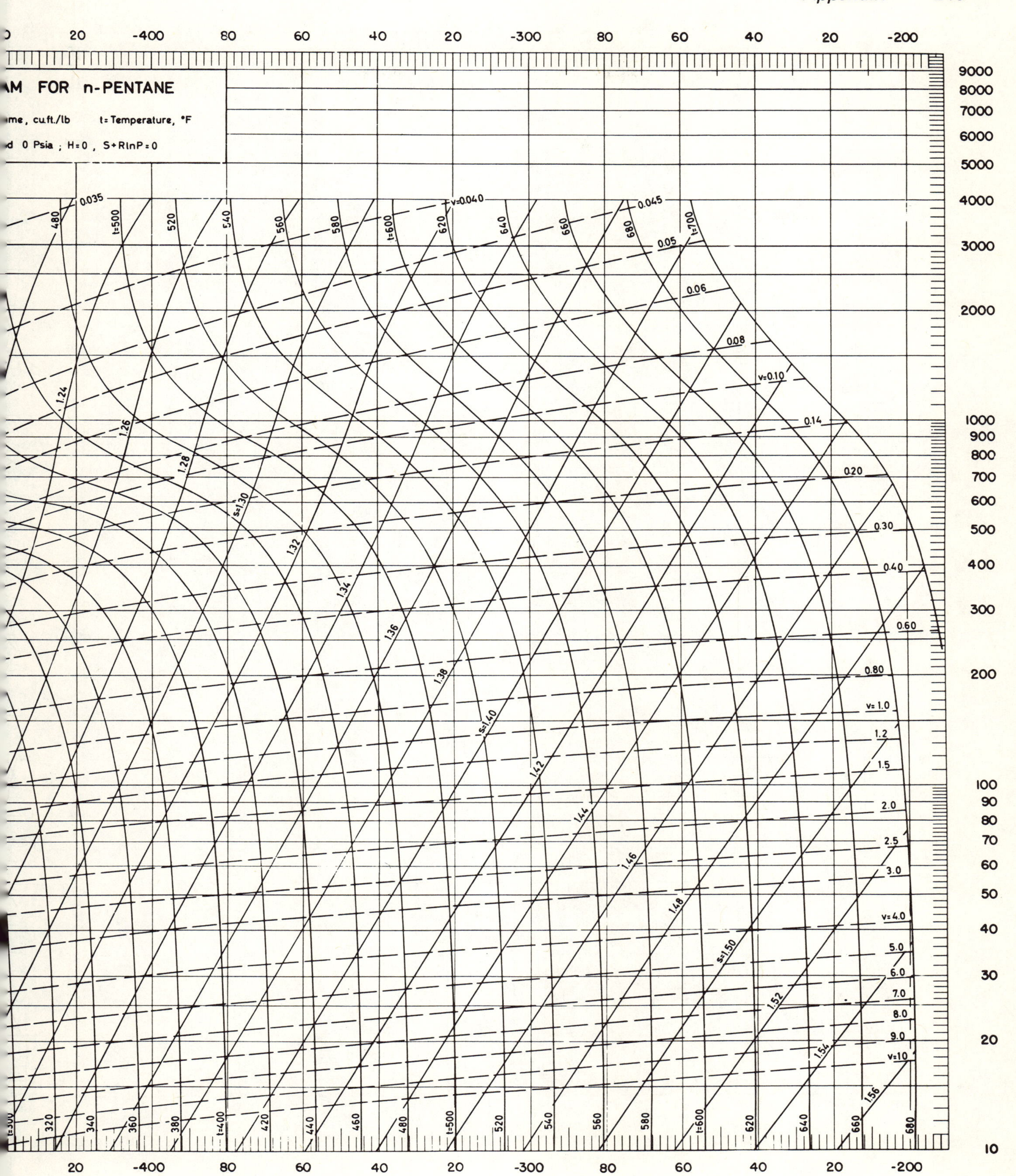

ENTHALPY BTU/LB

Chart 37 (continued)

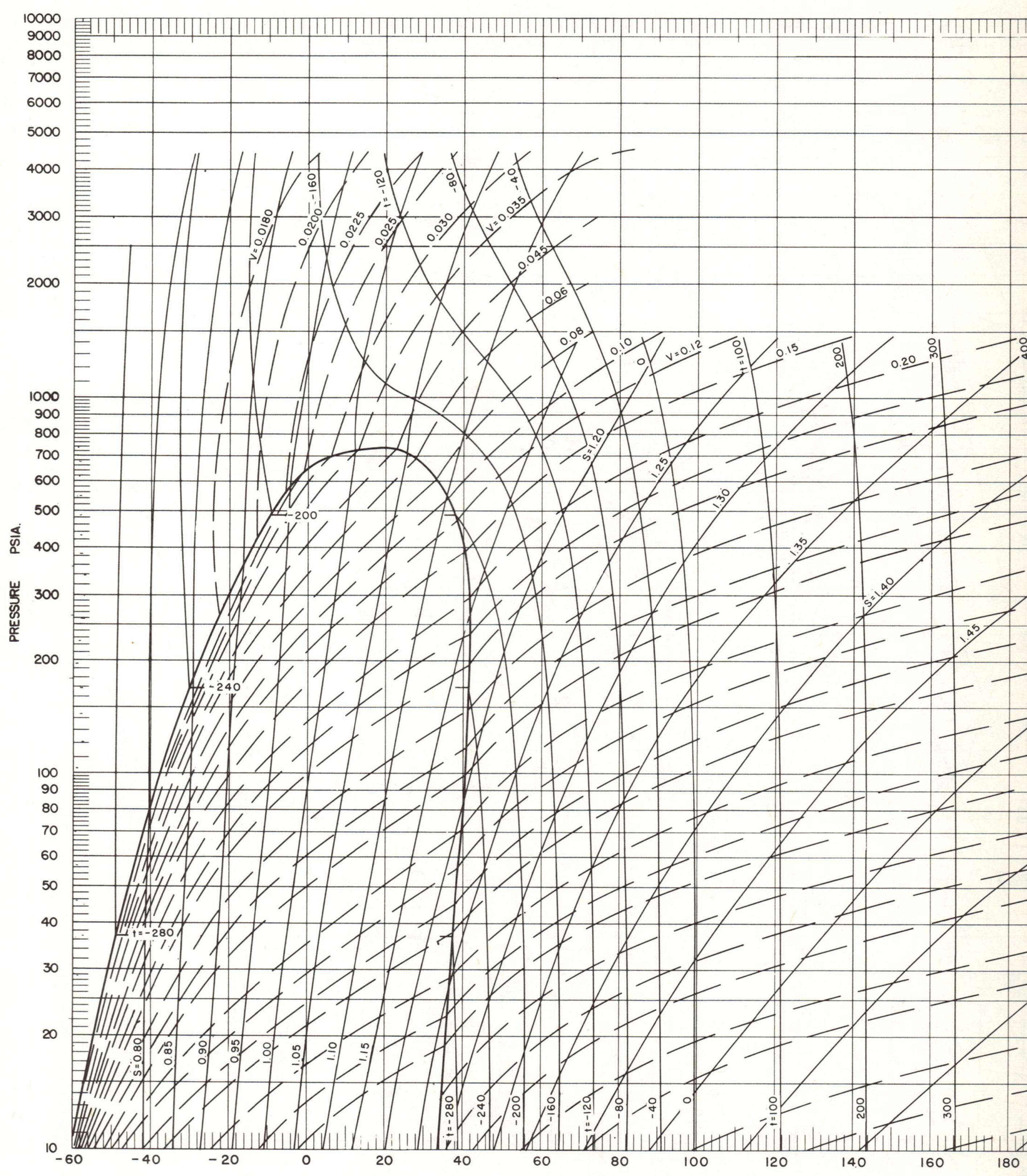

Chart 38. Mollier chart for oxygen.

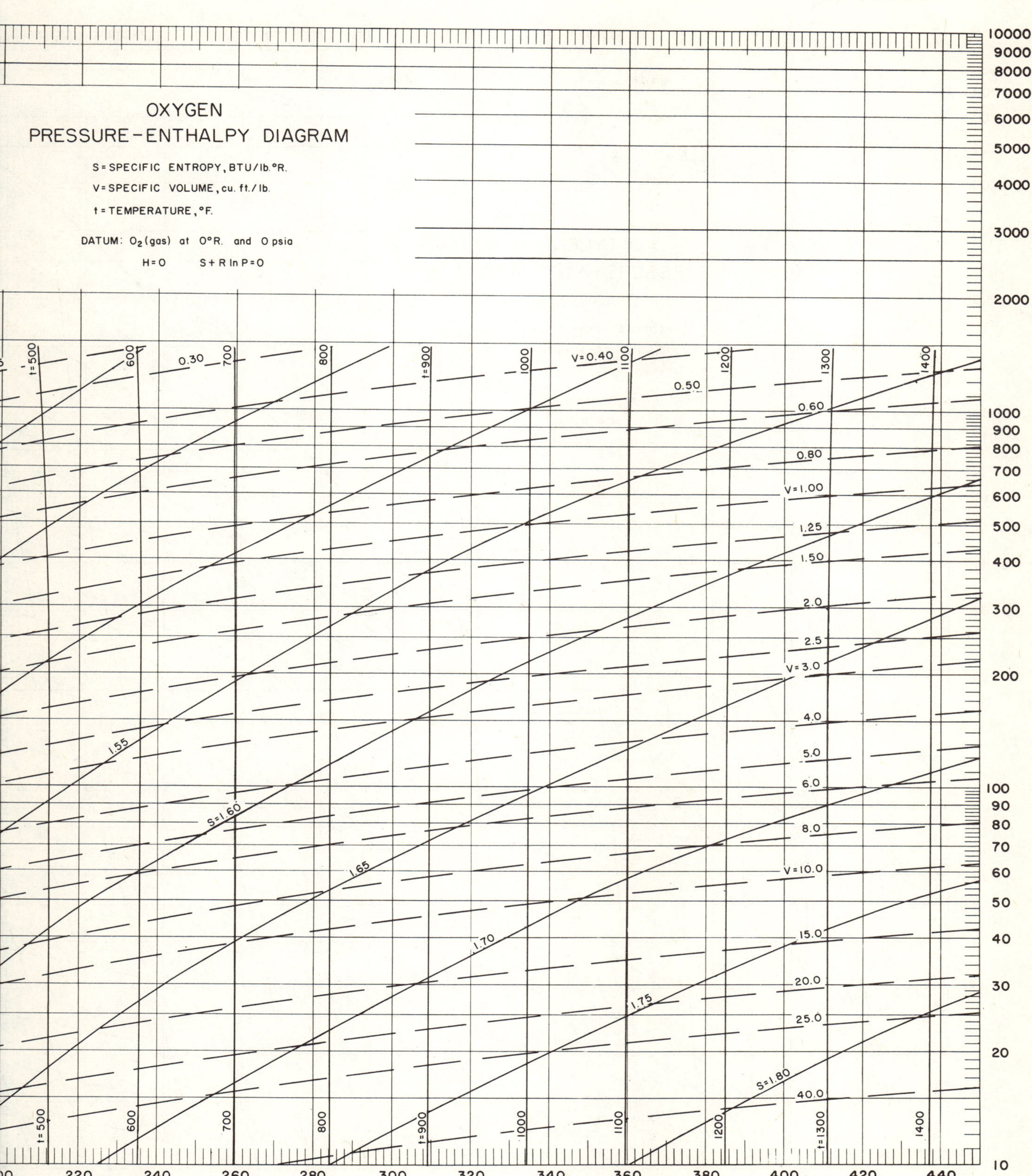

Chart 38 (continued)

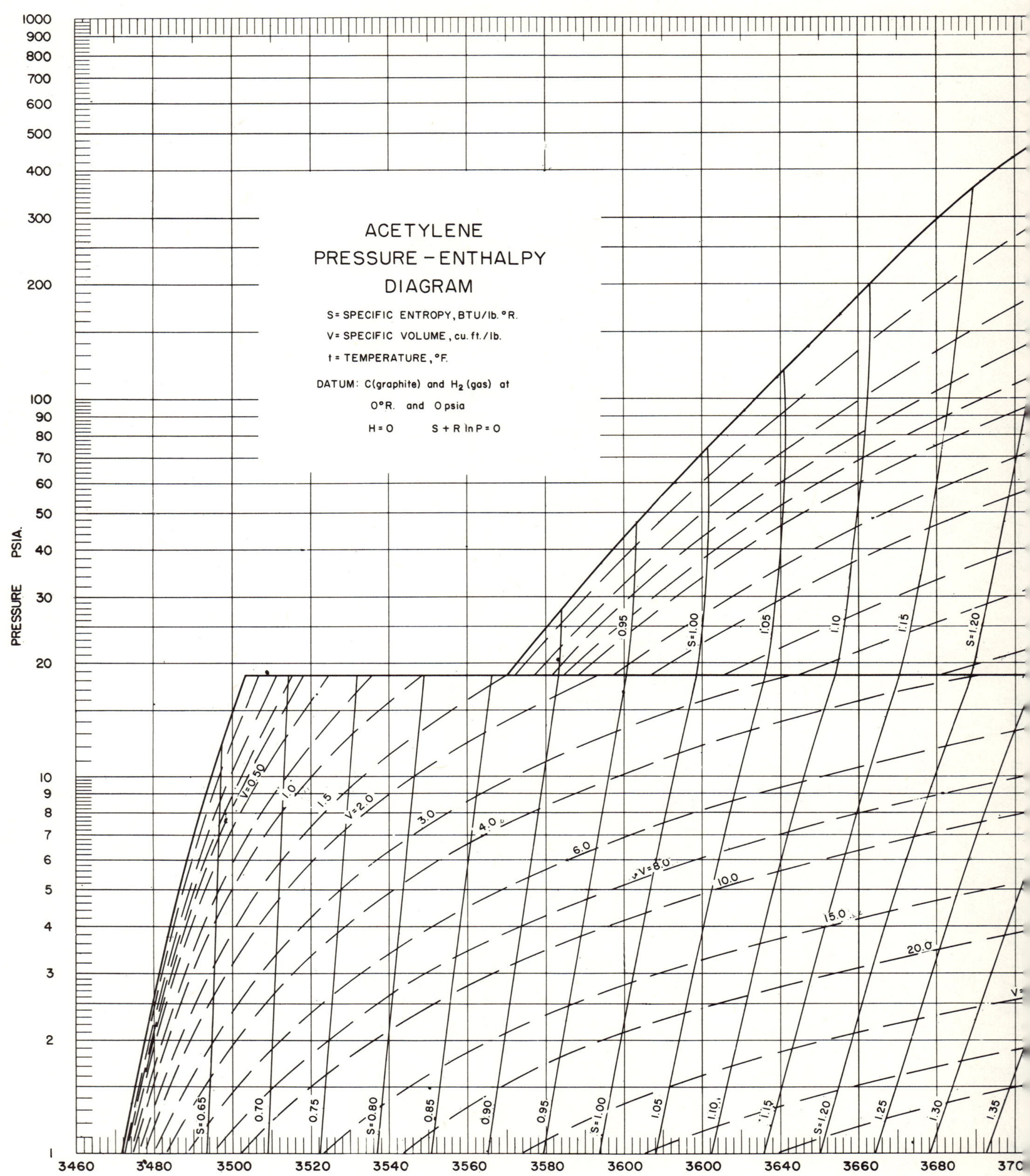

Chart 39. Mollier chart for Acetylene.

ENTHALPY BTU/LB

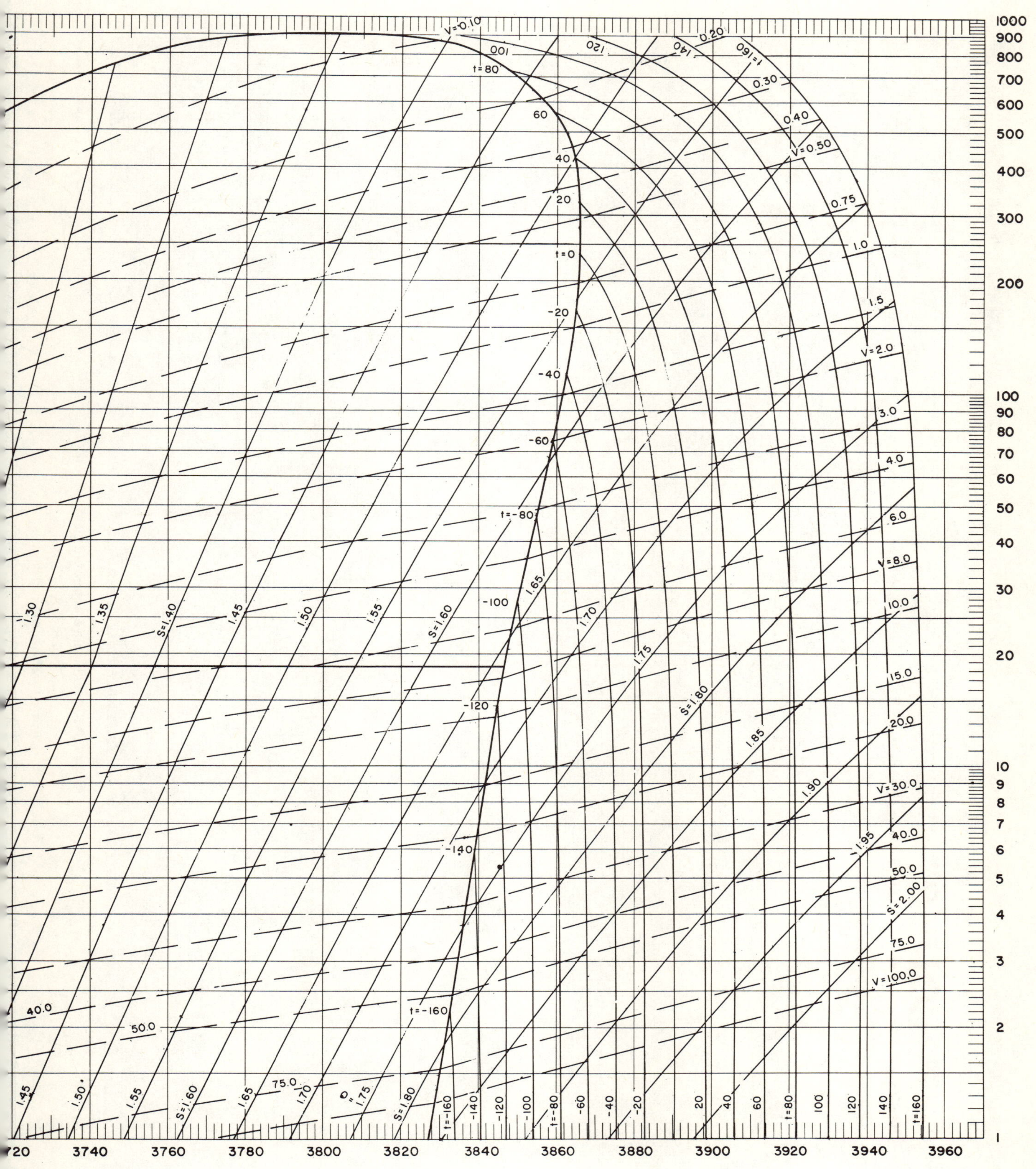

Chart 39 (continued)

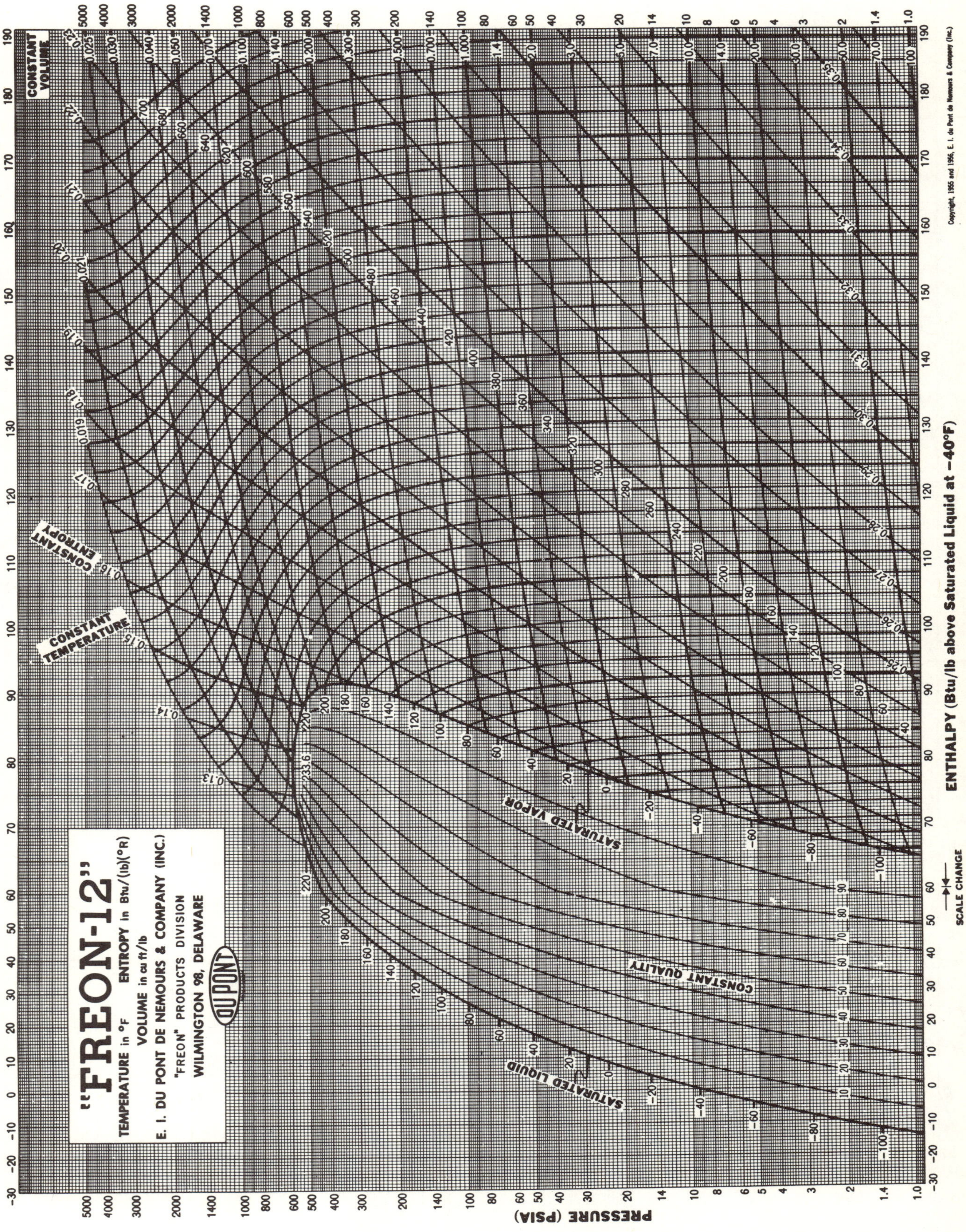

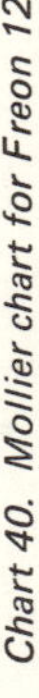

Chart 40. Mollier chart for Freon 12.

1. Crane Company. 1957. *Flow of Fluids*. TP 410.
2. Scheel, Lyman F. 1961. *Gas and Air Compression Machinery*. New York: McGraw-Hill Book Co., Inc.
3. Edmister, W.C., and McGarry, R.J. 1949. "Gas Compression Design." *Chemical Engineering Program* no. 45.
4. Edmister, W.C. 1962. *Applied Hydrocarbon Thermodynamics*. Houston: Gulf Publishing Company.
5. Wilson, G.M. 1964. "Redlich-Kwon Equation of State." *Advanced Cryogenic Engineering* (9-168).
6. Payne, W.H., and Messer, J.H. 1961. *Study of High Clearance Gas Compression*. AGA-GSTS. Denver.
7. Scheel, Lyman F. 1962. *Fallacy of Compression Efficiency Concept*. AGA-GSTS (April). Houston.
8. Ridgway, R.S. 1945. *Effect of Compressibility*. CNGA (March).
9. Natural Gas Processors Suppliers Association. 1966. *Engineering Data Book*. Tulsa, Oklahoma.
10. Hartwick. W. F. 1960. "Compressor Cylinder Cooling." *Oil and Gas Journal* (May 18).
11. O'Neill, P.P., and Wîckli, H.E. 1961. *Predicting Gas Performance of Centrifugal Compressors from Air Test Data*. ASME Paper 61-PID-6.
12. Baljé, O.E. 1960. *Turbomachinery Design Criteria*. ASME Paper 60-WA-231.
13. American Petroleum Institute, "Centrifugal Compressors for General Refinery Services." *API Standard 617*.
14. Prescott, John. 1924. *Applied Elasticity*. New York: Dover Publications.
15. Neill, Heller and Muller. 1968. *Corrosion Experiences of Refinery Compressors*. Meeting of National Association of Corrosion Engineers (March 22).
16. Marks, Lionel, and Baumeister, T. 1967. *Standard Handbook for Mechanical Engineers*. New York: McGraw-Hill Book Co., Inc.
17. Anderson, R.G. 1967. "High-Speed Rotating Parts." *Machine Design* (October 17).
18. Dodge, L. 1963. "Labyrinth Shaft Seals." *Product Engineering* (March 19).
19. Sanborn, L.B. 1967. "Centrifugal Shaft Seals." *Mechanical Engineering* (January).
20. Decker, O. 1967. *Advances in Dynamic Seal Technology*. ASME Paper 67-DE-50.
21. Hartwick, W.F. 1968. *Efficiency Characteristics of Reciprocating Compressors*. ASME Paper 68-WA/DGP-3.
22. Scheel, Lyman F. 1936. "CNGA Compressor Tests." *California Oil World* (November).
23. Dygert, J.C., and Bjorklund, I.S. 1963. "Power Recovery from Fluid Bed Processes." API (May).
24. American Society of Mechanical Engineers. *Theoretical Steam Rate Table*. Bulletin 4707.
25. Scheel, Lyman F. 1968. *Independent Solution for Piston Gas Compression*. ASME Paper 68-FE-46.
26. Service Bureau Corporation. 1969. "CALL/360." *Basic Information and Reference Manual.*
27. Nelson, W.L. "Nelson Cost Indexes." *Oil and Gas Journal* (regular feature).

References

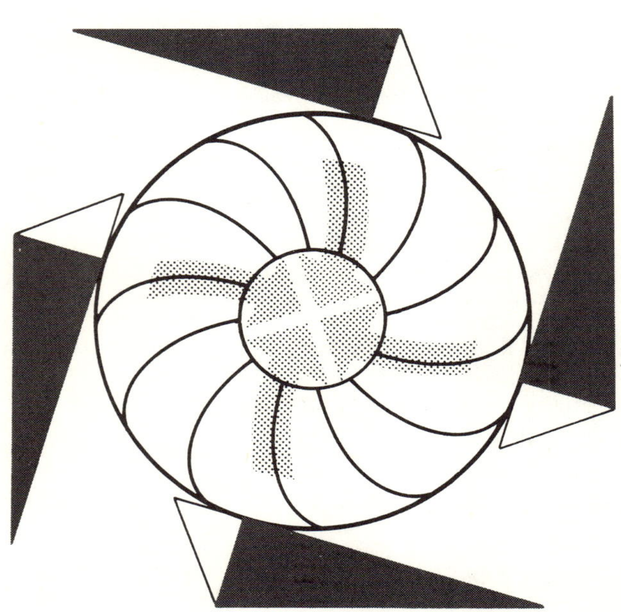

Index